Henry S. Hall, Samuel R. Knight

Elementary Trigonometry

Henry S. Hall, Samuel R. Knight

Elementary Trigonometry

ISBN/EAN: 9783337279516

Printed in Europe, USA, Canada, Australia, Japan

Cover: Foto ©berggeist007 / pixelio.de

More available books at **www.hansebooks.com**

ELEMENTARY TRIGONOMETRY

BY

H. S. HALL, M.A.,

FORMERLY SCHOLAR OF CHRIST'S COLLEGE, CAMBRIDGE;
MASTER OF THE MILITARY SIDE, CLIFTON COLLEGE.

AND

S. R. KNIGHT, B.A., M.B., CH.B.,

FORMERLY SCHOLAR OF TRINITY COLLEGE, CAMBRIDGE;
LATE ASSISTANT MASTER AT MARLBOROUGH COLLEGE.

London:
MACMILLAN AND CO.
AND NEW YORK.
1893

PREFACE.

THE following pages will be found to comprise all the parts of Elementary Trigonometry which can conveniently be treated without the use of infinite series and imaginary quantities.

The chapters have been subdivided into short sections, and the examples to illustrate each section have been very carefully selected and arranged, the earlier ones being easy enough for any reader to whom the subject is new, while the later ones, and the Miscellaneous Examples scattered throughout the book, will furnish sufficient practice for those who intend to pursue the subject further as part of a mathematical education.

No substantial progress in Trigonometry can be made until the fundamental properties of the Trigonometrical Ratios have been thoroughly mastered. To attain this object very considerable practice in easy Identities and Equations is necessary. We have therefore given special prominence to examples of this kind in the early pages; with the same end in view we have postponed the subject of Radian or Circular Measure to a later stage than is usual, believing that it is in every way more satisfactory to dwell on the properties of the trigonometrical ratios, and to exemplify their use in easy problems, than to bewilder a beginner with an angular system the use of which he cannot appreciate, and which at this stage furnishes nothing but practice in easy Arithmetic.

The subject of Logarithms and their application has been treated very fully, and illustrated by a selection of carefully graduated Examples. It is hoped that the examples worked out in this section may serve as useful models for the student, and may do something to cure that inaccuracy in logarithmic work which is so often due to clumsy arrangement.

In the experience of most teachers it is found extremely difficult to get boys to handle problems in Heights and Distances with any degree of confidence and skill. Accordingly we have devoted much thought to the exposition of this part of the subject, and by careful classification of the Examples we have endeavoured to make Chapters VI. and XVII. as easy and attractive as possible.

Very little advance can be expected in Trigonometry until the principal formulæ can be quoted readily, but whether it is advisable for learners to have lists of formulæ compiled for them, so as to be easily accessible at all times, is a matter upon which teachers hold different views. In our own opinion it is distinctly mischievous to furnish such lists; it encourages indolent habits, and fosters a spurious confidence which leads to disaster when the student has to rely solely upon his own knowledge.

In the general arrangement and succession of the different parts of the subject we have been mainly guided by our own long experience in the class room; but as the manuscript and proof-sheets have been read by several skilled teachers, and have been frequently tested by pupils in all stages of proficiency, the hope is entertained that our treatment is such as to enable beginners to take an intelligent interest in the subject from the first, and to acquire a sound elementary knowledge of practical Trigonometry before they encounter the more theoretical difficulties. At the same time, as each chapter is, as far as possible, complete in itself, it will be easy for teachers to adopt a different order of treatment if they prefer it; the full Table of Contents will facilitate the selection of a suitable course of reading, besides furnishing a useful aid to students who are rapidly revising the subject.

We are indebted to several friends for valuable criticism and advice; in particular, we have to thank Mr T. D. Davies of Clifton College for many useful hints, and for some ingenious examples and solutions in Chapters XXIV. and XXV.

H. S. HALL.
S. R. KNIGHT.

November, 1893.

CONTENTS.

Chapter I. MEASUREMENT OF ANGLES.

Chapter II. TRIGONOMETRICAL RATIOS.

Chapter III. RELATIONS BETWEEN THE RATIOS.

Chapter IV. TRIGONOMETRICAL RATIOS OF CERTAIN ANGLES.

Chapter V. SOLUTION OF RIGHT-ANGLED TRIANGLES.

Chapter VI. EASY PROBLEMS.

Chapter VII. RADIAN OR CIRCULAR MEASURE.

Chapter VIII. RATIOS OF ANGLES OF ANY MAGNITUDE.

Chapter IX. VARIATIONS OF THE FUNCTIONS.

Chapter X. CIRCULAR FUNCTIONS OF ALLIED ANGLES.

Chapter XI. FUNCTIONS OF COMPOUND ANGLES.

Chapter XII. TRANSFORMATION OF PRODUCTS AND SUMS.

Chapter XIII. RELATIONS BETWEEN THE SIDES AND ANGLES OF A TRIANGLE.

Chapter XIV. LOGARITHMS.

Chapter XV. THE USE OF LOGARITHMIC TABLES.

Chapter XVI. SOLUTION OF TRIANGLES WITH LOGARITHMS.

Chapter XVII. HEIGHTS AND DISTANCES.

Chapter XVIII. PROPERTIES OF TRIANGLES AND POLYGONS.

Chapter XIX. GENERAL VALUES AND INVERSE FUNCTIONS.

ELEMENTARY TRIGONOMETRY.

CHAPTER I.

MEASUREMENT OF ANGLES.

1. THE word Trigonometry in its primary sense signifies he measurement of triangles. From an early date the science lso included the establishment of the relations which subsist etween the sides, angles, and area of a triangle; but now it has much wider scope and embraces all manner of geometrical and lgebraical investigations carried on through the medium of ertain quantities called trigonometrical ratios, which will be efined in Chap. II. In every branch of Higher Mathematics, hether Pure or Applied, a knowledge of Trigonometry is of the reatest value.

2. **Definition of Angle.** Suppose that the straight line OP the figure is capable of revolving ab it the point O, and ippose that in this way it has assed successively from the posi-on OA to the positions occupied y OB, OC, OD, ..., then the angle etween OA and any position such s OC is measured by the *amount f revolution* which the line OP as undergone in passing from its iitial position OA into its final osition OC.

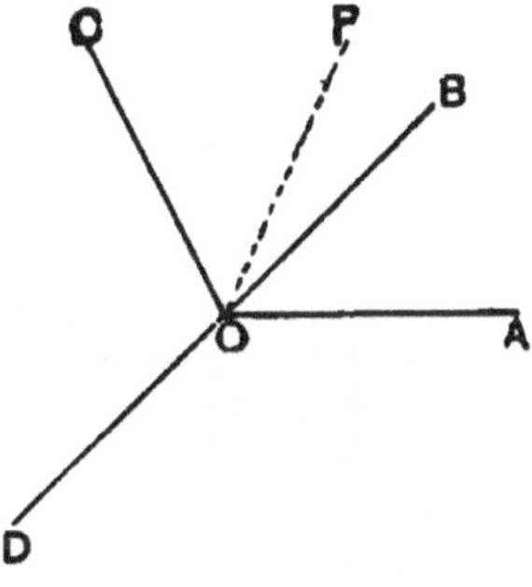

Moreover the line OP may iake any number of complete re-olutions through the original posi-ion OA before taking up its final osition.

It will thus be seen that in Trigonometry angles are not restricted as in Euclid, but may be of any magnitude.

The point O is called the *origin*, and OA the *initial line;* the revolving line OP is known as the *generating line* or the *radius vector.*

3. Measurement of Angles. We must first select some *fixed* unit. The natural unit would be a right angle, but as in practice this is inconveniently large, two systems of measurement have been established, in each of which the unit is a certain fraction of a right angle.

4. Sexagesimal Measure. A right angle is divided into 90 equal parts called *degrees*, a degree into 60 equal parts called *minutes*, a minute into 60 equal parts called *seconds.* An angle is measured by stating the number of degrees, minutes, and seconds which it contains.

For shortness, each of these three divisions, degrees, minutes, seconds, is denoted by a symbol; thus the angle which contains 53 degrees 37 minutes 2·53 seconds is expressed symbolically in the form $53^{\circ}\ 37'\ 2{\cdot}53''$.

5. Centesimal Measure. A right angle is divided into 100 equal parts called *grades*, a grade into 100 equal parts called *minutes*, a minute into 100 equal parts called *seconds.* In this system the angle which contains 53 grades 37 minutes 2·53 seconds is expressed symbolically in the form $53^{g}\ 37^{`}\ 2{\cdot}53^{``}$.

It will be noticed that different accents are used to denote sexagesimal and centesimal minutes and seconds; for though they have the same names, a centesimal minute and second are not the same as a sexagesimal minute and second. Thus a right angle contains 90×60 sexagesimal minutes, whereas it contains 100×100 centesimal minutes.

Sexagesimal Measure is sometimes called the English System, and Centesimal Measure the French System.

6. In *numerical* calculations the sexagesimal measure is always used. The centesimal method was proposed at the time of the French Revolution as part of a general system of decimal measurement, but has never been adopted even in France, as it would have made necessary the alteration of Geographical, Nautical, Astronomical, and other tables prepared according to the sexagesimal method. Beyond giving a few examples in transformation from one system to the other which afford exercise in easy Arithmetic, we shall after this rarely allude to centesimal measure.

In *theoretical* work it is convenient to use another method of measurement, where the unit is the angle subtended at the centre of a circle by an arc whose length is equal to the radius. This system is known as **Circular** or **Radian Measure**, and will be fully explained in Chapter VII.

An angle is usually represented by a single letter, different letters $A, B, C, \ldots, \alpha, \beta, \gamma, \ldots, \theta, \phi, \psi, \ldots$, being used to distinguish different angles. For angles estimated in sexagesimal or centesimal measure these letters are used indifferently, but we shall always denote angles in circular measure by letters taken from the Greek alphabet.

7. *If the number of degrees and grades contained in an angle be D and G respectively, to prove that* $\frac{D}{9} = \frac{G}{10}$.

In sexagesimal measure, the given angle when expressed as the fraction of a right angle is denoted by $\frac{D}{90}$. In centesimal measure, the same fraction is denoted by $\frac{G}{100}$;

$$\therefore \frac{D}{90} = \frac{G}{100}; \text{ that is, } \frac{D}{9} = \frac{G}{10}.$$

8. To pass from one system to the other it is advisable first to express the given angle in terms of a right angle.

In centesimal measure any number of grades, minutes, and seconds may be immediately expressed as the decimal of a right angle. Thus

$$23 \text{ grades} = \tfrac{23}{100} \text{ of a right angle} = \cdot 23 \text{ of a right angle};$$

15 minutes $= \frac{15}{100}$ of a grade $= \cdot 15$ of a grade $= \cdot 0015$ of a right angle;

$$\therefore 23^g \, 15' = \cdot 2315 \text{ of a right angle.}$$

Similarly, $15^g \, 7' \, 53 \cdot 4'' = \cdot 1507534$ of a right angle.

Conversely, any decimal of a right angle can be at once expressed in grades, minutes, and seconds. Thus

$$\begin{aligned} \cdot 2173025 \text{ of a right angle} &= 21 \cdot 73025^g \\ &= 21^g \, 73 \cdot 025' \\ &= 21^g \, 73' \, 2 \cdot 5''. \end{aligned}$$

In practice the intermediate steps are omitted.

Example 1. Reduce $2^g 13' 4{\cdot}5''$ to sexagesimal measure.

This angle $= {\cdot}0213045$ of a right angle
$= 1° 55' 2{\cdot}658''$.

·0213045	of a right angle
90	
1·917405	degrees
60	
55·0443	minutes
60	
2·658	seconds.

Obs. In the Answers we shall express the angles to the nearest tenth of a second, so that the above result would be written $1° 55' 2{\cdot}7''$.

Example 2. Reduce $12° 13' 14{\cdot}3''$ to centesimal measure.

This angle $= {\cdot}13578487\ldots$ of a right angle
$= 13^g 57' 84{\cdot}9''$.

60) 14·3 seconds
60) 13·238333...minutes
90) 12·2206388...degrees
·13578487...of a right angle.

EXAMPLES. I.

Express as the decimal of a right angle

1. $67° 30'$. 2. $11° 15'$. 3. $37^g 50'$.

4. $2° 10' 12''$. 5. $8° 0' 36''$. 6. $2^g 4' 4{\cdot}5''$.

Reduce to centesimal measure

7. $69° 13' 30''$. 8. $19° 0' 45''$. 9. $50° 37' 5{\cdot}7''$.

10. $43° 52' 38{\cdot}1''$. 11. $11° 0' 38{\cdot}4''$. 12. $142° 15' 45''$.

13. $12' 9''$. 14. $3' 26{\cdot}3''$.

Reduce to sexagesimal measure

15. $56^g 87' 50''$. 16. $39^g 6' 25''$. 17. $40^g 1' 25{\cdot}4''$.

18. $1^g 2' 3''$. 19. $3^g 2' 5''$. 20. $8^g 10' 6{\cdot}5''$.

21. $6' 25''$. 22. $37' 5''$.

23. The sum of two angles is 80^g and their difference is $18°$; find the angles in degrees.

24. The number of degrees in a certain angle added to the number of grades in the angle is 152: what is the angle?

25. If the same angle contains in English measure x minutes, and in French measure y minutes, prove that $50x = 27y$.

26. If s and t respectively denote the numbers of sexagesimal and centesimal seconds in any angle, prove that

$$250s = 81t.$$

CHAPTER II.

TRIGONOMETRICAL RATIOS.

9. DEFINITION. **Ratio** is the relation which one quantity bears to another of the *same* kind, the comparison being made by considering what multiple, part or parts, one quantity is of the other.

To find what multiple or part A is of B we divide A by B; hence the ratio of A to B may be measured by the fraction $\frac{A}{B}$.

In order to compare two quantities they must be expressed in terms of the same unit. Thus the ratio of 2 yards to 27 inches is measured by the fraction $\frac{2 \times 3 \times 12}{27}$ or $\frac{8}{3}$.

OBS. **Since a ratio expresses the *number* of times that one quantity contains another, *every ratio is a numerical quantity*.**

10. DEFINITION. If the ratio of any two quantities can be expressed exactly by the ratio of two integers the quantities are said to be **commensurable**; otherwise, they are said to be **incommensurable.** For instance, the quantities $8\frac{4}{5}$ and $5\frac{1}{3}$ are commensurable, while the quantities $\sqrt{2}$ and 3 are incommensurable. But by finding the numerical value of $\sqrt{2}$ we may express the value of the ratio $\sqrt{2} : 3$ by the ratio of two commensurable quantities to any required degree of approximation. Thus to 5 decimal places $\sqrt{2} = 1\cdot41421$, and therefore to the same degree of approximation

$$\sqrt{2} : 3 = 1\cdot41421 : 3 = 141421 : 300000.$$

Similarly, for the ratio of any two incommensurable quantities.

Trigonometrical Ratios.

11. Let PAQ be any acute angle; in AP one of the boundary lines take a point B and draw BC perpendicular to AQ. Thus a right-angled triangle BAC is formed.

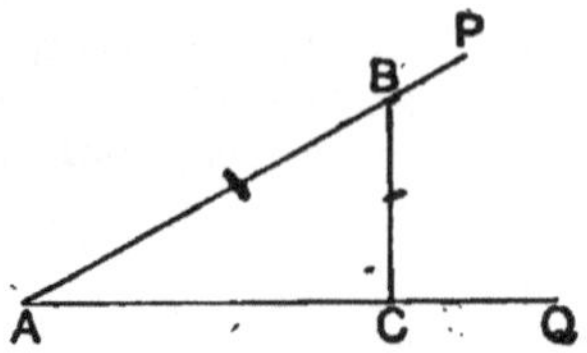

With reference to the angle A the following definitions are employed.

The ratio $\frac{BC}{AB}$ or $\frac{\textit{opposite side}}{\textit{hypotenuse}}$ is called the **sine of A.**

The ratio $\frac{AC}{AB}$ or $\frac{\textit{adjacent side}}{\textit{hypotenuse}}$ is called the **cosine of A.**

The ratio $\frac{BC}{AC}$ or $\frac{\textit{opposite side}}{\textit{adjacent side}}$ is called the **tangent of A.**

The ratio $\frac{AC}{BC}$ or $\frac{\textit{adjacent side}}{\textit{opposite side}}$ is called the **cotangent of A.**

The ratio $\frac{AB}{AC}$ or $\frac{\textit{hypotenuse}}{\textit{adjacent side}}$ is called the **secant of A.**

The ratio $\frac{AB}{BC}$ or $\frac{\textit{hypotenuse}}{\textit{opposite side}}$ is called the **cosecant of A.**

These six ratios are known as the **trigonometrical ratios.** It will be shewn later that as long as the anglo remains the same the trigonometrical ratios remain the same. [Art. 19.]

12. Instead of writing in full the words *sine, cosine, tangent, cotangent, secant, cosecant*, abbreviations are adopted. Thus the above definitions may be more conveniently expressed and arranged as follows:

$$\sin A = \frac{BC}{AB}, \qquad \operatorname{cosec} A = \frac{AB}{BC},$$

$$\cos A = \frac{AC}{AB}, \qquad \sec A = \frac{AB}{AC},$$

$$\tan A = \frac{BC}{AC}, \qquad \cot A = \frac{AC}{BC}.$$

In addition to these six ratios, two others, the *versed sine* and *coversed sine* are sometimes used; they are written vers A and covers A and are thus defined:

$$\text{vers } A = 1 - \cos A, \quad \text{covers } A = 1 - \sin A.$$

13. In Chapter VIII. the definitions of the trigonometrical ratios will be extended to the case of angles of any magnitude, but for the present we confine our attention to the consideration of acute angles.

14. Although the verbal form of the definitions of the trigonometrical ratios given in Art. 11 may be helpful to the student at first, he will gain no freedom in their use until he is able to write down from the figure any ratio at sight.

In the adjoining figure, PQR is a right-angled triangle in which $PQ = 13$, $PR = 5$, $QR = 12$.

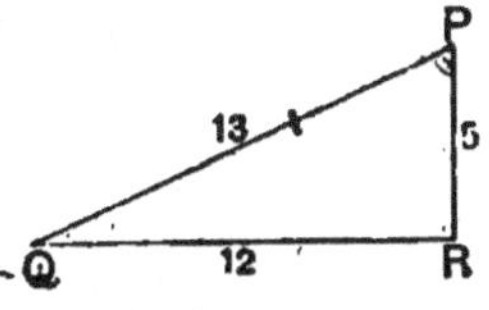

Since PQ is the greatest side, R is the right angle. The trigonometrical ratios of the angles P and Q may be written down at once; for example,

$$\sin Q = \frac{PR}{PQ} = \frac{5}{13}, \qquad \cos Q = \frac{QR}{PQ} = \frac{12}{13},$$

$$\tan P = \frac{QR}{PR} = \frac{12}{5}, \qquad \sec P = \frac{PQ}{QR} = \frac{13}{12}.$$

15. It is important to observe that *the trigonometrical ratios of an angle are numerical quantities.* Each one of them represents the *ratio of one length to another*, and they must themselves never be regarded as lengths.

16. In every right-angled triangle the hypotenuse is the greatest side; hence from the definitions of Art. 11 it will be seen that those ratios which have the hypotenuse in the *denominator* can never be greater than unity, while those which have the hypotenuse in the *numerator* can never be less than unity. Those ratios which do not involve the hypotenuse are not thus restricted in value, for either of the two sides which subtend the acute angles may be the greater. Hence

the sine and cosine of an angle can never be greater than 1;

the cosecant and secant of an angle can never be less than 1;

the tangent and cotangent may have any numerical value.

17. Let ABC be a right-angled triangle having the right angle at A; then by Euc. I. 47,

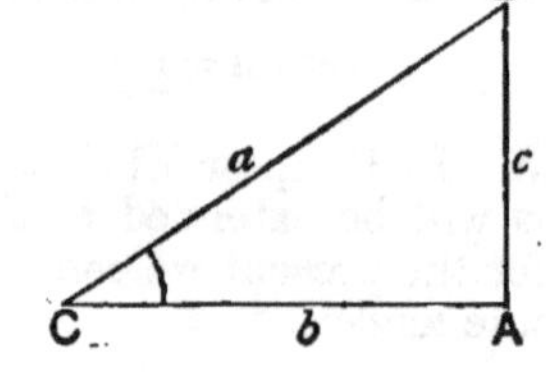

the sq. on BC

=sum of sqq. on AC and AB,

or, more briefly,

$$BC^2 = AC^2 + AB^2.$$

When we use this latter mode of expression it is understood that the sides AB, AC, BC are expressed in terms of some common unit, and the above statement may be regarded as a *numerical relation* connecting the numbers of units of length in the three sides of a right-angled triangle.

It is usual to denote the numbers of units of length in the sides opposite the angles A, B, C by the letters a, b, c respectively. Thus in the above figure we have $a^2 = b^2 + c^2$, so that if the lengths of two sides of a right-angled triangle are known, this equation will give the length of the third side.

Example 1. ABC is a right-angled triangle of which C is the right angle; if $a=3$, $b=4$, find c, and also $\sin A$ and $\cot B$.

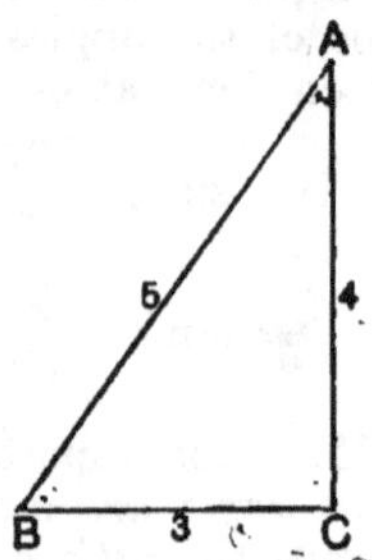

Here $c^2 = a^2 + b^2 = (3)^2 + (4)^2 = 9 + 16 = 25$;

$$\therefore c = 5.$$

Also $$\sin A = \frac{BC}{AB} = \frac{3}{5};$$

$$\cot B = \frac{BC}{AC} = \frac{3}{4}.$$

Example 2. A ladder 17 ft. long is placed with its foot at a distance of 8 ft. from the wall of a house and just reaches a window-sill. Find the height of the window-sill, and the sine and tangent of the angle which the ladder makes with the wall.

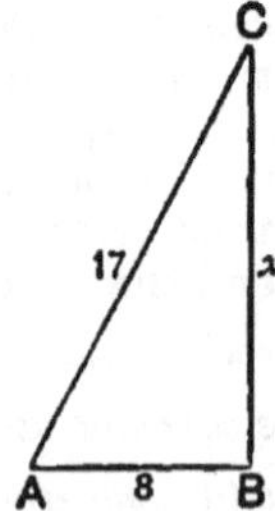

Let AC be the ladder, and BC the wall.

Let x be the number of feet in BC;

then $x^2 = (17)^2 - (8)^2 = (17+8)(17-8) = 25 \times 9$;

$$\therefore x = 5 \times 3 = 15.$$

Also $$\sin C = \frac{AB}{AC} = \frac{8}{17};$$

$$\tan C = \frac{AB}{BC} = \frac{8}{15}.$$

18. The following important proposition depends upon the property of similar triangles proved in Euc. VI. 4. The student who has not read the sixth Book of Euclid should not fail to notice the result arrived at, even if he is unable at this stage to understand the proof.

19. *To prove that the trigonometrical ratios remain unaltered so long as the angle remains the same.*

Let AOP be any acute angle. In OP take any points B and

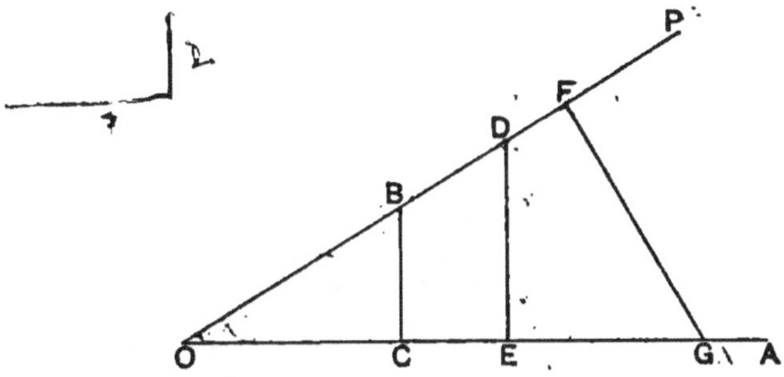

D, and draw BC and DE perpendicular to OA. Also take any point F in OP and draw FG at right angles to OP.

From the triangle BOC, $\quad \sin POA = \dfrac{BC}{OB}$;

from the triangle DOE, $\quad \sin POA = \dfrac{DE}{OD}$;

from the triangle FOG, $\quad \sin POA = \dfrac{FG}{OG}$.

But the triangles BOC, DOE, FOG are equiangular;

$$\therefore \quad \frac{BC}{OB} = \frac{DE}{OD} = \frac{FG}{OG}. \qquad \text{[Euc. VI. 4.]}$$

Thus the sine of the angle POA is the same whether it is obtained from the triangle BOC, or from the triangle DOE, or from the triangle FOG.

A similar proof holds for each of the other trigonometrical ratios. These ratios are therefore independent of the length of the revolving line and depend only on the magnitude of the angle.

20. If A denote any acute angle, we have proved that all the trigonometrical ratios of A depend only on the magnitude of the angle A and not upon the lengths of the lines which bound the angle. It may easily be seen that a change made in the value of A will produce a consequent change in the values of all the trigonometrical ratios of A. This point will be discussed more fully in Chap. IX.

DEFINITION. Any expression which involves a variable quantity x, and whose value is dependent on that of x is called a **function of x.**

Hence the trigonometrical *ratios* may also be defined as trigonometrical *functions*; for the present we shall chiefly employ the term *ratio*, but in a later part of the subject the idea of ratio is gradually lost and the term *function* becomes more appropriate.

21. The use of the principle proved in Art. 19 is well shewn in the following example, where the trigonometrical ratios are employed as a connecting link between the lines and angles.

Example. ABC is a right-angled triangle of which A is the right angle. BD is drawn perpendicular to BC and meets CA produced in D: if $AB=12$, $AC=16$, $BC=20$, find BD and CD.

From the right-angled triangle CBD,

$$\frac{BD}{BC}=\tan C;$$

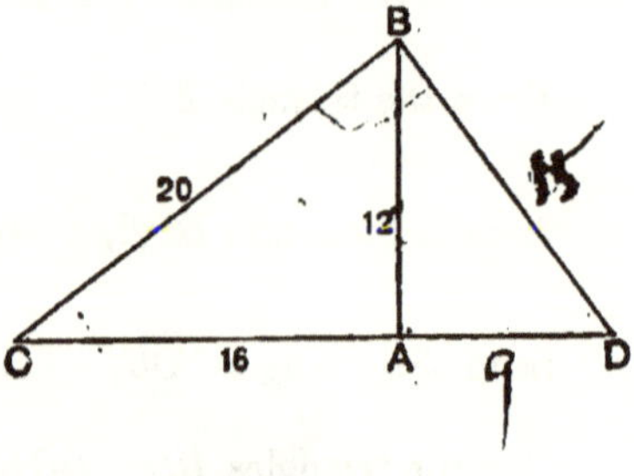

from the right-angled triangle ABC,

$$\frac{AB}{AC}=\tan C;$$

$$\therefore \frac{BD}{BC}=\frac{AB}{AC};$$

$$\therefore \frac{BD}{20}=\frac{12}{16}; \text{ whence } BD=15.$$

Again,
$$\frac{CD}{CB}=\sec C=\frac{BC}{CA};$$

$$\therefore \frac{CD}{20}=\frac{20}{16}; \text{ whence } CD=25.$$

The same results can be obtained by the help of Euc. VI. 8.

EXAMPLES. II.

1. The sides AB, BC, CA of a right-angled triangle are 17, 15, 8 respectively; write down the values of $\sin A$, $\sec A$, $\tan B$, $\sec B$.

2. The sides PQ, QR, RP of a right-angled triangle are 13, 5, 12 respectively: write down the values of $\cot P$, $\operatorname{cosec} Q$, $\cos Q$, $\cos P$.

3. ABC is a triangle in which A is a right angle; if $b=15$, $c=20$, find a, $\sin C$, $\cos B$, $\cot C$, $\sec C$.

4. ABC is a triangle in which B is a right angle; if $a=24$, $b=25$, find c, $\sin C$, $\tan A$, $\operatorname{cosec} A$.

5. The sides ED, EF, DF of a right-angled triangle are 35, 37, 12 respectively: write down the values of $\sec E$, $\sec F$, $\cot E$, $\sin F$.

6. The hypotenuse of a right-angled triangle is 15 inches, and one of the sides is 9 inches: find the third side and the sine, cosine and tangent of the angle opposite to it.

7. Find the hypotenuse AB of a right-angled triangle in which $AC=7$, $BC=24$. Write down the sine and cosine of A, and shew that the sum of their squares is equal to 1.

8. A ladder 41 ft. long is placed with its foot at a distance of 9 ft. from the wall of a house and just reaches a window-sill. Find the height of the window-sill, and the sine and cotangent of the angle which the ladder makes with the ground.

9. A ladder is 29 ft. long; how far must its foot be placed from a wall so that the ladder may just reach the top of the wall which is 21 ft. from the ground? Write down all the trigonometrical ratios of the angle between the ladder and the wall.

10. $ABCD$ is a square; C is joined to E, the middle point of AD: find all the trigonometrical ratios of the angle ECD.

11. $ABCD$ is a quadrilateral in which the diagonal AC is at right angles to each of the sides AB, CD: if $AB=15$, $AC=36$, $AD=85$, find $\sin ABC$, $\sec ACB$, $\cos CDA$, $\operatorname{cosec} DAC$.

12. $PQRS$ is a quadrilateral in which the angle PSR is a right angle. If the diagonal PR is at right angles to RQ, and $RP=20$, $RQ=21$, $RS=16$, find $\sin PRS$, $\tan RPS$, $\cos RPQ$, $\operatorname{cosec} PQR$.

CHAPTER III.

RELATIONS BETWEEN THE TRIGONOMETRICAL RATIOS.

22. Reciprocal relations between certain ratios.

(1) Let ABC be a triangle, right-angled at C;

then $$\sin A = \frac{BC}{AB} = \frac{a}{c},$$

and $$\operatorname{cosec} A = \frac{AB}{BC} = \frac{c}{a};$$

$$\therefore \quad \sin A \times \operatorname{cosec} A = \frac{a}{c} \times \frac{c}{a} = 1.$$

Thus $\sin A$ and $\operatorname{cosec} A$ are reciprocals;

$$\therefore \quad \sin A = \frac{1}{\operatorname{cosec} A},$$

and $$\operatorname{cosec} A = \frac{1}{\sin A}.$$

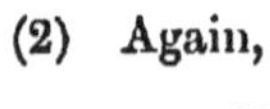

(2) Again,

$$\cos A = \frac{AC}{AB} = \frac{b}{c}, \quad \text{and} \quad \sec A = \frac{AB}{AC} = \frac{c}{b};$$

$$\therefore \quad \cos A \times \sec A = \frac{b}{c} \times \frac{c}{b} = 1;$$

$$\therefore \quad \cos A = \frac{1}{\sec A}, \quad \text{and} \quad \sec A = \frac{1}{\cos A}.$$

(3) Also

$$\tan A = \frac{BC}{AC} = \frac{a}{b}, \quad \text{and} \quad \cot A = \frac{AC}{BC} = \frac{b}{a};$$

$$\therefore \quad \tan A \times \cot A = \frac{a}{b} \times \frac{b}{a} = 1;$$

$$\therefore \quad \tan A = \frac{1}{\cot A}, \quad \text{and} \quad \cot A = \frac{1}{\tan A}.$$

23. *To express* tan A *and* cot A *in terms of* sin A *and* cos A.

From the adjoining figure we have

$$\tan A = \frac{BC}{AC} = \frac{a}{b} = \frac{a}{c} \div \frac{b}{c}$$

$$= \sin A \div \cos A\,;$$

$$\therefore\ \tan A = \frac{\sin A}{\cos A}.$$

Again,

$$\cot A = \frac{AC}{BC} = \frac{b}{a} = \frac{b}{c} \div \frac{a}{c}$$

$$= \cos A \div \sin A\,;$$

$$\therefore\ \cot A = \frac{\cos A}{\sin A}\,;$$

which is also evident from the reciprocal relation $\cot A = \dfrac{1}{\tan A}$.

Example. Prove that cosec A tan A = sec A.

$$\operatorname{cosec} A \tan A = \frac{1}{\sin A} \times \frac{\sin A}{\cos A} = \frac{1}{\cos A}$$

$$= \sec A.$$

24. We frequently meet with expressions which involve the square and other powers of the trigonometrical ratios, such as $(\sin A)^2$, $(\tan A)^3$, ... It is usual to write these in the shorter forms $\sin^2 A$, $\tan^3 A$, ...

Thus

$$\tan^2 A = (\tan A)^2 = \left(\frac{\sin A}{\cos A}\right)^2$$

$$= \frac{(\sin A)^2}{(\cos A)^2} = \frac{\sin^2 A}{\cos^2 A}.$$

Example. Shew that $\sin^2 A \sec A \cot^2 A = \cos A$.

$$\sin^2 A \sec A \cot^2 A = \sin^2 A \times \frac{1}{\cos A} \times \left(\frac{\cos A}{\sin A}\right)^2$$

$$= \sin^2 A \times \frac{1}{\cos A} \times \frac{\cos^2 A}{\sin^2 A}$$

$$= \cos A,$$

by cancelling factors common to numerator and denominator.

25. *To prove that* $\sin^2 A + \cos^2 A = 1$.

Let BAC be any acute angle; draw BC perpendicular to

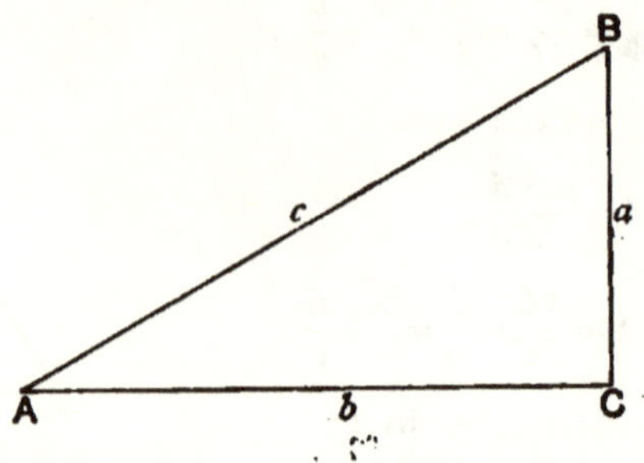

AC, and denote the sides of the right-angled triangle ABC by a, b, c.

By definition, $$\sin A = \frac{BC}{AB} = \frac{a}{c};$$

and $$\cos A = \frac{AC}{AB} = \frac{b}{c};$$

$$\therefore \sin^2 A + \cos^2 A = \frac{a^2}{c^2} + \frac{b^2}{c^2} = \frac{a^2+b^2}{c^2}$$

$$= \frac{c^2}{c^2}$$

$$= 1.$$

COR. $\sin^2 A = 1 - \cos^2 A$, $\sin A = \sqrt{1-\cos^2 A}$;
$\cos^2 A = 1 - \sin^2 A$, $\cos A = \sqrt{1-\sin^2 A}$.

Example 1. Prove that $\cos^4 A - \sin^4 A = \cos^2 A - \sin^2 A$.

$$\cos^4 A - \sin^4 A = (\cos^2 A + \sin^2 A)(\cos^2 A - \sin^2 A)$$
$$= \cos^2 A - \sin^2 A,$$

since the first factor is equal to 1.

Example 2. Prove that $\cot \alpha \sqrt{1-\cos^2\alpha} = \cos \alpha$.

$$\cot \alpha \sqrt{1-\cos^2\alpha} = \cot \alpha \times \sin \alpha$$
$$= \frac{\cos \alpha}{\sin \alpha} \times \sin \alpha = \cos \alpha.$$

26. *To prove that* $\sec^2 A = 1 + \tan^2 A$.

With the figure of the previous article, we have

$$\sec A = \frac{AB}{AC} = \frac{c}{b};$$

$$\therefore \quad \sec^2 A = \frac{c^2}{b^2} = \frac{b^2 + a^2}{b^2}$$

$$= 1 + \frac{a^2}{b^2}$$

$$= 1 + \tan^2 A.$$

Cor. $\sec^2 A - \tan^2 A = 1$, $\quad \sec A = \sqrt{1 + \tan^2 A}$,

$\tan^2 A = \sec^2 A - 1$, $\quad \tan A = \sqrt{\sec^2 A - 1}$.

Example. Prove that $\cos A \sqrt{\sec^2 A - 1} = \sin A$.

$$\cos A \sqrt{\sec^2 A - 1} = \cos A \times \tan A$$

$$= \cos A \times \frac{\sin A}{\cos A}$$

$$= \sin A.$$

27. *To prove that* $\operatorname{cosec}^2 A = 1 + \cot^2 A$.

With the figure of Art. 25, we have

$$\operatorname{cosec} A = \frac{AB}{BC} = \frac{c}{a};$$

$$\therefore \quad \operatorname{cosec}^2 A = \frac{c^2}{a^2} = \frac{a^2 + b^2}{a^2}$$

$$= 1 + \frac{b^2}{a^2}$$

$$= 1 + \cot^2 A.$$

Cor. $\operatorname{cosec}^2 A - \cot^2 A = 1$, $\quad \operatorname{cosec} A = \sqrt{1 + \cot^2 A}$,

$\cot^2 A = \operatorname{cosec}^2 A - 1$, $\quad \cot A = \sqrt{\operatorname{cosec}^2 A - 1}$.

Example. Prove that $\cot^4 \alpha - 1 = \operatorname{cosec}^4 \alpha - 2 \operatorname{cosec}^2 \alpha$.

$$\cot^4 \alpha - 1 = (\cot^2 \alpha + 1)(\cot^2 \alpha - 1)$$

$$= \operatorname{cosec}^2 \alpha \, (\operatorname{cosec}^2 \alpha - 1 - 1)$$

$$= \operatorname{cosec}^2 \alpha \, (\operatorname{cosec}^2 \alpha - 2)$$

$$= \operatorname{cosec}^4 \alpha - 2 \operatorname{cosec}^2 \alpha.$$

28. The formulæ proved in the last three articles are n independent, for they are merely different ways of expressing i trigonometrical symbols the property of a right-angled triangl proved in Euc. I. 47.

29. It will be useful here to collect the formulæ proved i this chapter.

I. $\operatorname{cosec} A \times \sin A = 1, \quad \operatorname{cosec} A = \dfrac{1}{\sin A}, \quad \sin A = \dfrac{1}{\operatorname{cosec} A};$

$\sec A \times \cos A = 1, \quad \sec A = \dfrac{1}{\cos A}, \quad \cos A = \dfrac{1}{\sec A};$

$\cot A \times \tan A = 1, \quad \cot A = \dfrac{1}{\tan A}, \quad \tan A = \dfrac{1}{\cot A}.$

II. $\tan A = \dfrac{\sin A}{\cos A}, \quad \cot A = \dfrac{\cos A}{\sin A}.$

III. $\sin^2 A + \cos^2 A = 1,$

$\sec^2 A = 1 + \tan^2 A,$

$\operatorname{cosec}^2 A = 1 + \cot^2 A.$

Easy Identities.

30. We shall now exemplify the use of these fundamental formulæ in proving *identities*. An identity asserts that two expressions are always equal, and the proof of this equality is called "proving the identity." Some easy illustrations have already been given in this chapter. The general method of procedure is to choose one of the expressions given (usually the more complicated of the two) and to shew by successive transformations that it can be made to assume the form of the other.

Example 1. Prove that $\sin^2 A \cot^2 A + \cos^2 A \tan^2 A = 1$.

Here it will be found convenient to express all the trigonometrical ratios in terms of the sine and cosine.

$$\text{The first side} = \sin^2 A \cdot \frac{\cos^2 A}{\sin^2 A} + \cos^2 A \cdot \frac{\sin^2 A}{\cos^2 A}$$

$$= \cos^2 A + \sin^2 A$$

$$= 1.$$

Example 2. Prove that $\sec^4\theta - \sec^2\theta = \tan^2\theta + \tan^4\theta$.

The form of this identity at once suggests that we should use the secant-tangent formula of Art. 26; hence

$$\begin{aligned}\text{the first side} &= \sec^2\theta\,(\sec^2\theta - 1)\\ &= (1+\tan^2\theta)\tan^2\theta\\ &= \tan^2\theta + \tan^4\theta.\end{aligned}$$

EXAMPLES. III. a.

Prove the following identities :

1. $\sin A \cot A = \cos A$.
2. $\cos A \tan A = \sin A$.
3. $\cot A \sec A = \operatorname{cosec} A$.
4. $\sin A \sec A = \tan A$.
5. $\cos A \operatorname{cosec} A = \cot A$.
6. $\cot A \sec A \sin A = 1$.
7. $(1-\cos^2 A)\operatorname{cosec}^2 A = 1$.
8. $(1-\sin^2 A)\sec^2 A = 1$.
9. $\cot^2\theta\,(1-\cos^2\theta) = \cos^2\theta$.
10. $(1-\cos^2\theta)\sec^2\theta = \tan^2\theta$.
11. $\tan a\sqrt{1-\sin^2 a} = \sin a$.
12. $\operatorname{cosec} a\sqrt{1-\sin^2 a} = \cot a$.
13. $(1+\tan^2 A)\cos^2 A = 1$.
14. $(\sec^2 A - 1)\cot^2 A = 1$.
15. $(1-\cos^2\theta)(1+\tan^2\theta) = \tan^2\theta$.
16. $\cos a \operatorname{cosec} a\sqrt{\sec^2 a - 1} = 1$.
17. $\sin^2 A\,(1+\cot^2 A) = 1$.
18. $(\operatorname{cosec}^2 A - 1)\tan^2 A = 1$.
19. $(1-\cos^2 A)(1+\cot^2 A) = 1$.
20. $\sin a \sec a\sqrt{\operatorname{cosec}^2 a - 1} = 1$.
21. $\cos a\sqrt{\cot^2 a + 1} = \sqrt{\operatorname{cosec}^2 a - 1}$.
22. $\sin^2\theta\cot^2\theta + \sin^2\theta = 1$.
23. $(1+\tan^2\theta)(1-\sin^2\theta) = 1$.
24. $\sin^2\theta\sec^2\theta = \sec^2\theta - 1$.
25. $\operatorname{cosec}^2\theta\tan^2\theta - 1 = \tan^2\theta$.

Prove the following identities:

26. $\dfrac{1}{\sec^2 A}+\dfrac{1}{\operatorname{cosec}^2 A}=1.$

27. $\dfrac{1}{\cos^2 A}-\dfrac{1}{\cot^2 A}=1.$

28. $\dfrac{\sin A}{\operatorname{cosec} A}+\dfrac{\cos A}{\sec A}=1.$

29. $\dfrac{\sec A}{\cos A}-\dfrac{\tan A}{\cot A}=1.$

30. $\sin^4 a-\cos^4 a=2\sin^2 a-1=1-2\cos^2 a.$

31. $\sec^4 a-1=2\tan^2 a+\tan^4 a.$

32. $\operatorname{cosec}^4 a-1=2\cot^2 a+\cot^4 a.$

33. $(\tan a \operatorname{cosec} a)^2-(\sin a \sec a)^2=1.$

34. $(\sec\theta\cot\theta)^2-(\cos\theta\operatorname{cosec}\theta)^2=1.$

35. $\tan^2\theta-\cot^2\theta=\sec^2\theta-\operatorname{cosec}^2\theta.$

31. The foregoing examples have required little more than a direct application of the fundamental formulæ; we shall now give some identities offering a greater variety of treatment.

Example 1. Prove that $\sec^2 A+\operatorname{cosec}^2 A=\sec^2 A\operatorname{cosec}^2 A.$

The first side $=\dfrac{1}{\cos^2 A}+\dfrac{1}{\sin^2 A}=\dfrac{\sin^2 A+\cos^2 A}{\cos^2 A\sin^2 A}$

$$=\frac{1}{\cos^2 A\sin^2 A}=\sec^2 A\operatorname{cosec}^2 A.$$

Occasionally it is found convenient to prove the equality of the two expressions by reducing each to the same form.

Example 2. Prove that

$$\sin^3 A\tan A+\cos^3 A\cot A+2\sin A\cos A=\tan A+\cot A.$$

The first side $=\sin^2 A\cdot\dfrac{\sin A}{\cos A}+\cos^2 A\dfrac{\cos A}{\sin A}+2\sin A\cos A$

$$=\frac{\sin^4 A+\cos^4 A+2\sin^2 A\cos^2 A}{\sin A\cos A}$$

$$=\frac{(\sin^2 A+\cos^2 A)^2}{\sin A\cos A}=\frac{1}{\sin A\cos A}.$$

The second side $=\dfrac{\sin A}{\cos A}+\dfrac{\cos A}{\sin A}=\dfrac{\sin^2 A+\cos^2 A}{\cos A\sin A}$

$$=\frac{1}{\sin A\cos A}.$$

Thus each side of the identity $=\dfrac{1}{\sin A\cos A}.$

Example 3. Prove that $\dfrac{\tan\alpha-\cot\beta}{\tan\beta-\cot\alpha}=\tan\alpha\cot\beta$.

$$\text{The first side}=\frac{\tan\alpha-\cot\beta}{\dfrac{1}{\cot\beta}-\dfrac{1}{\tan\alpha}}=\frac{\tan\alpha-\cot\beta}{\dfrac{\tan\alpha-\cot\beta}{\tan\alpha\cot\beta}}$$

$$=\frac{\tan\alpha-\cot\beta}{1}\times\frac{\tan\alpha\cot\beta}{\tan\alpha-\cot\beta}$$

$$=\tan\alpha\cot\beta.$$

The transformations in the successive steps are usually suggested by the form into which we wish to bring the result. For instance, in this last example we might have proved the identity by substituting for the tangent and cotangent in terms of the sine and cosine. This however is not the best method, for the form in which the right-hand side is given suggests that we should retain $\tan\alpha$ and $\cot\beta$ unchanged throughout the work.

EXAMPLES. III. b.

Prove the following identities:

1. $\dfrac{\sin\alpha\cot^2\alpha}{\cos\alpha}=\dfrac{1}{\tan\alpha}$.

2. $\dfrac{\sec^2\alpha\cot\alpha}{\operatorname{cosec}^2\alpha}=\tan\alpha$.

3. $1-\operatorname{vers}\theta=\sin\theta\cot\theta$.

4. $\operatorname{vers}\theta\sec\theta=\sec\theta-1$.

5. $\sec\theta-\tan\theta\sin\theta=\cos\theta$.

6. $\tan\theta+\cot\theta=\sec\theta\operatorname{cosec}\theta$.

7. $\sqrt{1+\cot^2 A}\,.\,\sqrt{\sec^2 A-1}\,.\,\sqrt{1-\sin^2 A}=1$.

8. $(\cos\theta+\sin\theta)^2+(\cos\theta-\sin\theta)^2=2$.

9. $(1+\tan\theta)^2+(1-\tan\theta)^2=2\sec^2\theta$.

10. $(\cot\theta-1)^2+(\cot\theta+1)^2=2\operatorname{cosec}^2\theta$.

11. $\sin^2 A\,(1+\cot^2 A)+\cos^2 A\,(1+\tan^2 A)=2$.

12. $\cos^2 A\,(\sec^2 A-\tan^2 A)+\sin^2 A\,(\operatorname{cosec}^2 A-\cot^2 A)=1$.

13. $\cot^2\alpha+\cot^4\alpha=\operatorname{cosec}^4\alpha-\operatorname{cosec}^2\alpha$.

14. $\dfrac{\tan^2\alpha}{1+\tan^2\alpha}\cdot\dfrac{1+\cot^2\alpha}{\cot^2\alpha}=\sin^2\alpha\sec^2\alpha$.

15. $\dfrac{1}{1-\sin\alpha}+\dfrac{1}{1+\sin\alpha}=2\sec^2\alpha$.

Prove the following identities:

16. $\dfrac{\tan a}{\sec a-1}+\dfrac{\tan a}{\sec a+1}=2\operatorname{cosec} a.$

17. $\dfrac{1}{1+\sin^2 a}+\dfrac{1}{1+\operatorname{cosec}^2 a}=1.$

18. $(\sec\theta+\operatorname{cosec}\theta)(\sin\theta+\cos\theta)=\sec\theta\operatorname{cosec}\theta+2.$

19. $(\cos\theta-\sin\theta)(\operatorname{cosec}\theta-\sec\theta)=\sec\theta\operatorname{cosec}\theta-2.$

20. $(1+\cot\theta+\operatorname{cosec}\theta)(1+\cot\theta-\operatorname{cosec}\theta)=2\cot\theta.$

21. $(\sec\theta+\tan\theta-1)(\sec\theta-\tan\theta+1)=2\tan\theta.$

22. $(\sin A+\operatorname{cosec} A)^2+(\cos A+\sec A)^2=\tan^2 A+\cot^2 A+7.$

23. $(\sec^2 A+\tan^2 A)(\operatorname{cosec}^2 A+\cot^2 A)=1+2\sec^2 A\operatorname{cosec}^2 A.$

24. $(1-\sin A+\cos A)^2=2(1-\sin A)(1+\cos A).$

25. $\sin A(1+\tan A)+\cos A(1+\cot A)=\sec A+\operatorname{cosec} A.$

26. $\cos\theta(\tan\theta+2)(2\tan\theta+1)=2\sec\theta+5\sin\theta.$

27. $(\tan\theta+\sec\theta)^2=\dfrac{1+\sin\theta}{1-\sin\theta}.$

28. $\dfrac{2\sin\theta\cos\theta-\cos\theta}{1-\sin\theta+\sin^2\theta-\cos^2\theta}=\cot\theta.$

29. $\cot^2\theta\,.\,\dfrac{\sec\theta-1}{1+\sin\theta}+\sec^2\theta\,.\,\dfrac{\sin\theta-1}{1+\sec\theta}=0.$

[*The following examples contain functions of two angles; in each case the two angles are quite independent of each other.*]

30. $\tan^2 a+\sec^2\beta=\sec^2 a+\tan^2\beta.$

31. $\dfrac{\tan a+\cot\beta}{\cot a+\tan\beta}=\dfrac{\tan a}{\tan\beta}.$

32. $\dfrac{\tan a-\cot\beta}{\cot a-\tan\beta}=-\dfrac{\cot\beta}{\cot a}.$

33. $\cot a\tan\beta(\tan a+\cot\beta)=\cot a+\tan\beta.$

34. $\sin^2 a\cos^2\beta-\cos^2 a\sin^2\beta=\sin^2 a-\sin^2\beta.$

35. $\sec^2 a\tan^2\beta-\tan^2 a\sec^2\beta=\tan^2\beta-\tan^2 a.$

36. $(\sin a\cos\beta+\cos a\sin\beta)^2+(\cos a\cos\beta-\sin a\sin\beta)^2=1.$

32. By means of the relations collected together in Art. 29, all the trigonometrical ratios can be expressed in terms of any one.

Example 1. Express all the trigonometrical ratios of A in terms of $\tan A$.

We have $\cot A = \dfrac{1}{\tan A}$;

$$\sec A = \sqrt{1+\tan^2 A};$$

$$\cos A = \frac{1}{\sec A} = \frac{1}{\sqrt{1+\tan^2 A}};$$

$$\sin A = \frac{\sin A}{\cos A}\cos A = \tan A \cos A = \frac{\tan A}{\sqrt{1+\tan^2 A}};$$

$$\operatorname{cosec} A = \frac{1}{\sin A} = \frac{\sqrt{1+\tan^2 A}}{\tan A}.$$

Obs. In writing down the ratios we choose the simplest and most natural order. For instance, $\cot A$ is obtained at once by the *reciprocal relation* connecting the tangent and cotangent: $\sec A$ comes immediately from the tangent-secant formula; the remaining three ratios now readily follow.

Example 2. Given $\cos A = \dfrac{5}{13}$, find $\operatorname{cosec} A$ and $\cot A$.

$$\operatorname{cosec} A = \frac{1}{\sin A} = \frac{1}{\sqrt{1-\cos^2 A}}$$

$$= \frac{1}{\sqrt{1-\left(\frac{5}{13}\right)^2}} = \frac{1}{\sqrt{1-\frac{25}{169}}} = \frac{1}{\sqrt{\frac{144}{169}}} = \frac{1}{\frac{12}{13}} = \frac{13}{12}.$$

$$\cot A = \frac{\cos A}{\sin A} = \cos A \times \operatorname{cosec} A$$

$$= \frac{5}{13} \times \frac{13}{12} = \frac{5}{12}.$$

33. *It is always possible to describe a right-angled triangle when two sides are given:* for the third side can be found by Euc. I. 47, and the construction can then be effected by Euc. I. 22. We can thus readily obtain all the trigonometrical ratios when one is given, or express all in terms of any one.

Example 1. Given $\cos A = \frac{5}{13}$, find $\operatorname{cosec} A$ and $\cot A$.

Take a right-angled triangle PQR, of which Q is the right angle, having the hypotenuse $PR = 13$ units, and $PQ = 5$ units.

Let $QR = x$ units; then

$$x^2 = (13)^2 - (5)^2 = (13+5)(13-5)$$
$$= 18 \times 8 = 9 \times 2 \times 8;$$
$$\therefore\ x = 3 \times 4 = 12.$$

Now $$\cos RPQ = \frac{PQ}{PR} = \frac{5}{13},$$

so that $$\angle RPQ = A.$$

Hence $$\operatorname{cosec} A = \frac{PR}{QR} = \frac{13}{12},$$

and $$\cot A = \frac{PQ}{QR} = \frac{5}{12}.$$ [Compare Art. 32, Ex. 2.]

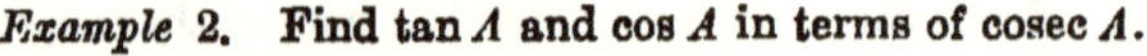

Example 2. Find $\tan A$ and $\cos A$ in terms of $\operatorname{cosec} A$.

Take a triangle PQR right-angled at Q, and having $\angle RPQ = A$. For shortness, denote $\operatorname{cosec} A$ by c.

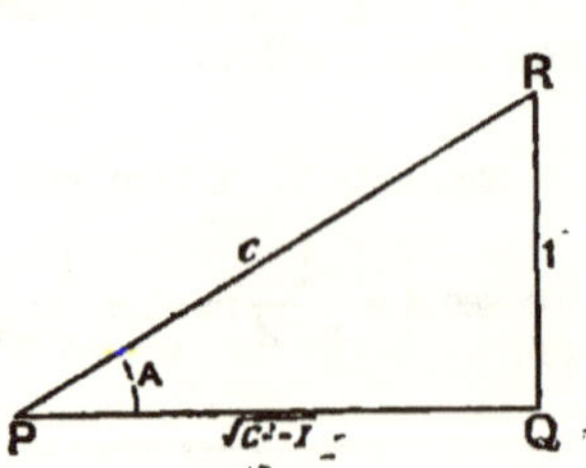

Then $$\operatorname{cosec} A = c = \frac{c}{1};$$

but $$\operatorname{cosec} A = \frac{PR}{QR};$$

$$\therefore \frac{PR}{QR} = \frac{c}{1}.$$

Let QR be taken as the unit of measurement;

then $QR = 1$, and therefore $PR = c$.

Let PQ contain x units; then

$$x^2 = c^2 - 1, \text{ so that } x = \sqrt{c^2 - 1}.$$

Hence $$\tan A = \frac{QR}{PQ} = \frac{1}{\sqrt{c^2-1}} = \frac{1}{\sqrt{\operatorname{cosec}^2 A - 1}},$$

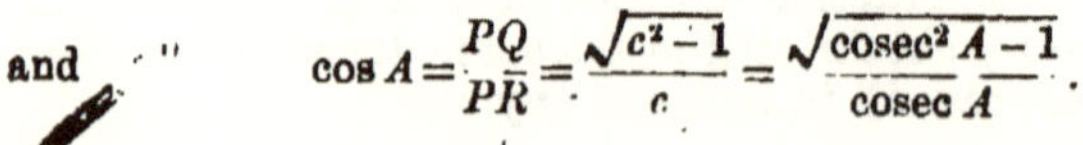

and $$\cos A = \frac{PQ}{PR} = \frac{\sqrt{c^2-1}}{c} = \frac{\sqrt{\operatorname{cosec}^2 A - 1}}{\operatorname{cosec} A}.$$

EXAMPLES. III. c.

1. Given $\sin A = \frac{1}{2}$, find $\sec A$ and $\cot A$.

2. Given $\tan A = \frac{4}{3}$, find $\sin A$ and $\cos A$.

3. Find $\cot\theta$ and $\sin\theta$ when $\sec\theta = 4$.

4. If $\tan a = \frac{1}{2}$, find $\sec a$ and $\operatorname{cosec} a$.

5. Find the sine and cotangent of an angle whose secant is 7.

6. If $25 \sin A = 7$, find $\tan A$ and $\sec A$.

7. Express $\sin A$ and $\tan A$ in terms of $\cos A$.

8. Express $\operatorname{cosec} a$ and $\cos a$ in terms of $\cot a$.

9. Find $\sin\theta$ and $\cot\theta$ in terms of $\sec\theta$.

10. Express all the trigonometrical ratios of A in terms of $\sin A$.

11. Given $\sin A - \cos A = 0$, find $\operatorname{cosec} A$.

12. If $\sin A = \frac{m}{n}$, prove that $\sqrt{n^2 - m^2} \cdot \tan A = m$.

13. If $p \cot\theta = \sqrt{q^2 - p^2}$, find $\sin\theta$.

14. When $\sec A = \frac{m^2+1}{2m}$, find $\tan A$ and $\sin A$.

15. Given $\tan A = \frac{2pq}{p^2 - q^2}$, find $\cos A$ and $\operatorname{cosec} A$.

16. If $\sec a = \frac{13}{5}$, find the value of $\frac{2 \sin a - 3 \cos a}{4 \sin a - 9 \cos a}$.

17. If $\cot\theta = \frac{p}{q}$, find the value of $\frac{p \cos\theta - q \sin\theta}{p \cos\theta + q \sin\theta}$.

CHAPTER IV.

TRIGONOMETRICAL RATIOS OF CERTAIN ANGLES.

34. Trigonometrical Ratios of 45°.

Let BAC be a right-angled isosceles triangle, with the right angle at C; so that $B = A = 45°$.

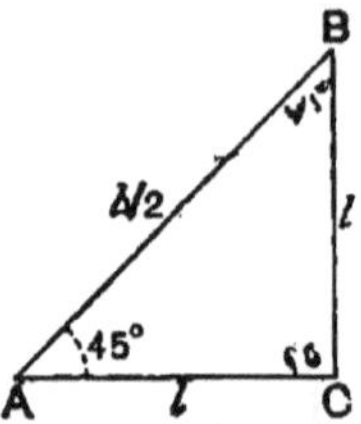

Let each of the equal sides contain l units,

then $$AC = BC = l.$$

Also $$AB^2 = l^2 + l^2 = 2l^2;$$

$$\therefore \quad AB = l\sqrt{2}.$$

$$\therefore \quad \sin 45° = \frac{BC}{AB} = \frac{l}{l\sqrt{2}} = \frac{1}{\sqrt{2}};$$

$$\cos 45° = \frac{AC}{AB} = \frac{l}{l\sqrt{2}} = \frac{1}{\sqrt{2}};$$

$$\tan 45° = \frac{BC}{AC} = \frac{l}{l} = 1.$$

The other three ratios are the reciprocals of these; thus

$$\operatorname{cosec} 45° = \sqrt{2}, \quad \sec 45° = \sqrt{2}, \quad \cot 45° = 1;$$

or they may be read off from the figure.

35. Trigonometrical Ratios of 60° and 30°.

Let ABC be an equilateral triangle; thus each of its angles is 60°.

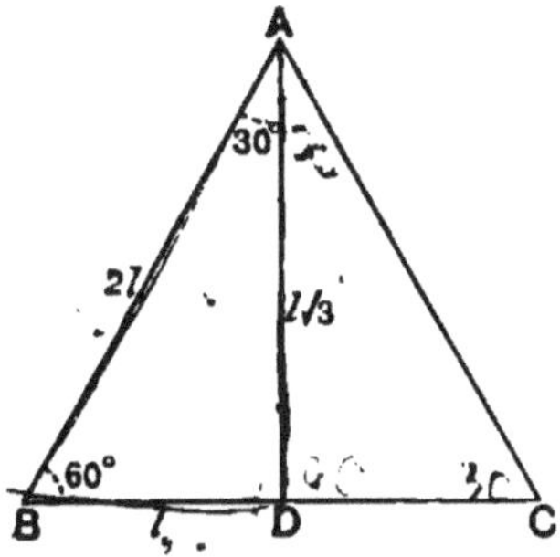

Bisect $\angle BAC$ by AD meeting BC at D; then $\angle BAD = 30°$.

By Euc. I. 4, the triangles ABD, ACD are equal in all respects; therefore $BD = CD$, and the angles at D are right angles.

In the right-angled triangle ADB, let $BD = l$; then

$$AB = BC = 2l;$$

$$\therefore \quad AD^2 = 4l^2 - l^2 = 3l^2;$$

$$\therefore \quad AD = l\sqrt{3}.$$

$$\therefore \quad \sin 60° = \frac{AD}{AB} = \frac{l\sqrt{3}}{2l} = \frac{\sqrt{3}}{2};$$

$$\cos 60° = \frac{BD}{AB} = \frac{l}{2l} = \frac{1}{2};$$

$$\tan 60° = \frac{AD}{BD} = \frac{l\sqrt{3}}{l} = \sqrt{3}.$$

Again,

$$\sin 30° = \frac{BD}{AB} = \frac{l}{2l} = \frac{1}{2};$$

$$\cos 30° = \frac{AD}{AB} = \frac{l\sqrt{3}}{2l} = \frac{\sqrt{3}}{2};$$

$$\tan 30° = \frac{BD}{AD} = \frac{l}{l\sqrt{3}} = \frac{1}{\sqrt{3}}.$$

The other ratios may be read off from the figure.

36. The trigonometrical ratios of 45°, 60°, 30° occur very frequently; it is therefore important that the student should be able to quote readily their numerical values. The exercise which follows will furnish useful practice.

At first it will probably be found safer to make use of the accompanying diagrams than to trust to the memory.

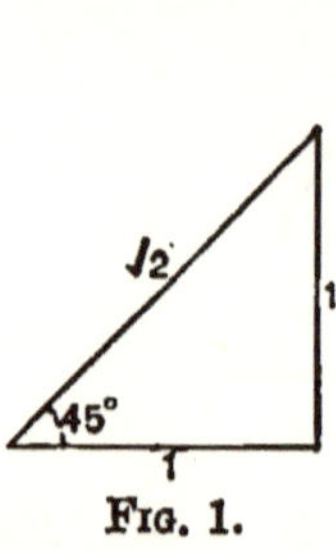

Fig. 1.

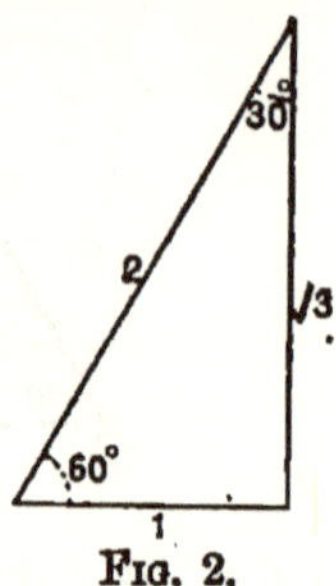

Fig. 2.

The trigonometrical ratios of 45° can be read off from Fig. 1; those of 60° and 30° from Fig. 2.

Example 1. Find the values of $\sec^3 45°$ and $\sin 60° \cot 30° \tan 45°$.

$$\sec^3 45° = (\sec 45°)^3 = (\sqrt{2})^3 = \sqrt{2} \times \sqrt{2} \times \sqrt{2} = 2\sqrt{2}.$$

$$\sin 60° \cot 30° \tan 45° = \frac{\sqrt{3}}{2} \times \sqrt{3} \times 1 = \frac{3}{2}.$$

Example 2. Find the value of

$$2 \cot 45° + \cos^3 60° - 2 \sin^4 60° + \tfrac{3}{4} \tan^2 30°.$$

The value $= (2 \times 1) + \left(\frac{1}{2}\right)^3 - 2\left(\frac{\sqrt{3}}{2}\right)^4 + \frac{3}{4}\left(\frac{1}{\sqrt{3}}\right)^2$

$= 2 + \frac{1}{8} - 2\left(\frac{3}{4}\right)^2 + \frac{3}{4}\left(\frac{1}{3}\right)$

$= 2 + \frac{1}{8} - \frac{9}{8} + \frac{1}{4} = 1\frac{1}{4}.$

EXAMPLES. IV. a.

Find the numerical value of

1. $\tan^2 60° + 2 \tan^2 45°$.
2. $\tan^3 45° + 4 \cos^3 60°$.
3. $2 \operatorname{cosec}^2 45° - 3 \sec^2 30°$.
4. $\cot 60° \tan 30° + \sec^2 45°$.

5. $2 \sin 30° \cos 30° \cot 60°$.

6. $\tan^2 45° \sin 60° \tan 30° \tan^2 60°$.

7. $\tan^2 60° + 4 \cos^2 45° + 3 \sec^2 30°$.

8. $\frac{1}{2} \operatorname{cosec}^2 60° + \sec^2 45° - 2 \cot^2 60°$.

9. $\tan^2 30° + 2 \sin 60° + \tan 45° - \tan 60° + \cos^2 30°$.

10. $\cot^2 45° + \cos 60° - \sin^2 60° - \frac{3}{4} \cot^2 60°$.

11. $3 \tan^2 30° + \frac{4}{3} \cos^2 30° - \frac{1}{2} \sec^2 45° - \frac{1}{3} \sin^2 60°$.

12. $\cos 60° - \tan^2 45° + \frac{3}{4} \tan^2 30° + \cos^2 30° - \sin 30°$.

13. $\frac{1}{3} \sin^2 60° - \frac{1}{2} \sec 60° \tan^2 30° + \frac{4}{3} \sin^2 45° \tan^2 60°$.

14. If $\tan^2 45° - \cos^2 60° = x \sin 45° \cos 45° \tan 60°$, find x.

15. Find x from the equation

$$x \sin 30° \cos^2 45° = \frac{\cot^2 30° \sec 60° \tan 45°}{\operatorname{cosec}^2 45° \operatorname{cosec} 30°}.$$

37. Definition. The **complement** of an angle is its *defect from* a right angle.

Two angles are said to be **complementary** when their sum is a right angle.

Thus in every right-angled triangle, each acute angle is the complement of the other. For in the figure of the next article, if B is the right angle, the sum of A and C is 90°.

$$\therefore \quad C = 90° - A, \text{ and } A = 90° - C.$$

Trigonometrical Ratios of Complementary Angles.

38. Let ABC be a right-angled triangle, of which B is the right angle; then the angles at A and C are complementary, so that $C = 90° - A$.

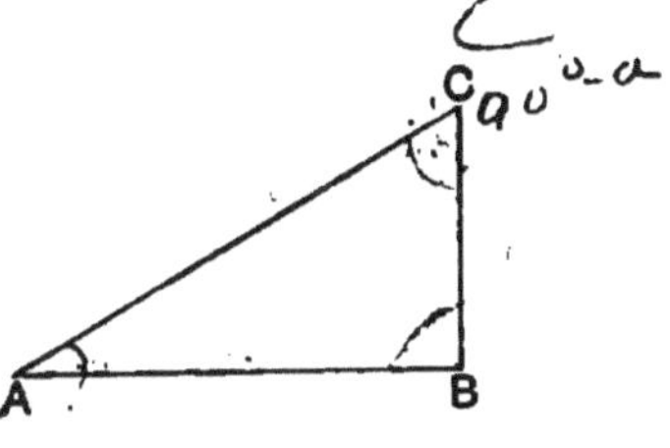

$$\therefore \quad \sin(90° - A) = \sin C = \frac{AB}{AC} = \cos A;$$

$$\text{and } \cos(90° - A) = \cos C = \frac{BC}{AC} = \sin A.$$

Similarly, it may be proved that

$$\left.\begin{aligned} \tan(90° - A) &= \cot A, \\ \cot(90° - A) &= \tan A; \end{aligned}\right\} \text{ and } \left.\begin{aligned} \sec(90° - A) &= \operatorname{cosec} A, \\ \operatorname{cosec}(90° - A) &= \sec A. \end{aligned}\right\}$$

39. If we define the co-sine, co-tangent, co-secant, as the co-functions of the angle, the foregoing results may be embodied in a single statement:

each function of an angle is equal to the corresponding co-function of its complement.

As an illustration of this we may refer to Art. 35, from which it will be seen that

$$\sin 60^\circ = \cos 30^\circ = \frac{\sqrt{3}}{2};$$

$$\sin 30^\circ = \cos 60^\circ = \frac{1}{2};$$

$$\tan 60^\circ = \cot 30^\circ = \sqrt{3}.$$

Example 1. Find a value of A when $\cos 2A = \sin 3A$.

Since $\cos 2A = \sin(90^\circ - 2A)$,

the equation becomes $\sin(90^\circ - 2A) = \sin 3A$;

$$\therefore\ 90^\circ - 2A = 3A;$$

whence $A = 18^\circ$.

Thus *one* value of A which satisfies the equation is $A = 18^\circ$. In a later chapter we shall be able to solve the equation more completely, and shew that there are other values of A which satisfy it.

Example 2. Prove that $\sec A \sec(90^\circ - A) = \tan A + \tan(90^\circ - A)$.

Here it will be found easier to begin with the expression on the right side of the identity.

The second side $= \tan A + \cot A$

$$= \frac{\sin A}{\cos A} + \frac{\cos A}{\sin A} = \frac{\sin^2 A + \cos^2 A}{\cos A \sin A}$$

$$= \frac{1}{\cos A \sin A}$$

$$= \sec A \operatorname{cosec} A = \sec A \sec(90^\circ - A).$$

EXAMPLES. IV. b.

Find the complements of the following angles:

1. $67^\circ\,30'$. 2. $25^\circ\,30''$. 3. $10^\circ\,1'\,3''$.

4. $45^\circ - A$. 5. $45^\circ + B$. 6. $30^\circ - B$.

7. In a triangle C is 50° and A is the complement of 10°; find B.

8. In a triangle A is the complement of 40°; and B is the complement of 20°; find C.

Find a value of A in each of the following equations:

9. $\sin A = \cos 4A$.
10. $\cos 3A = \sin 7A$.
11. $\tan A = \cot 3A$.
12. $\cot A = \tan A$.
13. $\cot A = \tan 2A$.
14. $\sec 5A = \operatorname{cosec} A$.

Prove the following identities:

15. $\sin(90^\circ - A)\cot(90^\circ - A) = \sin A$.

16. $\sin A \tan(90^\circ - A)\sec(90^\circ - A) = \cot A$.

17. $\cos A \tan A \tan(90^\circ - A)\operatorname{cosec}(90^\circ - A) = 1$.

18. $\sin A \cos(90^\circ - A) + \cos A \sin(90^\circ - A) = 1$.

19. $\cos(90^\circ - A)\operatorname{cosec}(90^\circ - A) = \tan A$.

20. $\operatorname{cosec}^2(90^\circ - A) = 1 + \sin^2 A \operatorname{cosec}^2(90^\circ - A)$.

21. $\sin A \cot A \cot(90^\circ - A)\sec(90^\circ - A) = 1$.

22. $\sec(90^\circ - A) - \cot A \cos(90^\circ - A)\tan(90^\circ - A) = \sin A$.

23. $\tan^2 A \sec^2(90^\circ - A) - \sin^2 A \operatorname{cosec}^2(90^\circ - A) = 1$.

24. $\tan(90^\circ - A) + \cot(90^\circ - A) = \operatorname{cosec} A \operatorname{cosec}(90^\circ - A)$.

25. $\dfrac{\sin(90^\circ - A)}{\sec(90^\circ - A)} \cdot \dfrac{\tan(90^\circ - A)}{\cos A} = \cos A$.

26. $\dfrac{\operatorname{cosec}^2 A \tan^2 A}{\cot(90^\circ - A)} \cdot \dfrac{\cot A}{\sec^2 A} = \sec^2(90^\circ - A) - 1$.

27. $\dfrac{\cot(90^\circ - A)}{\operatorname{cosec}^2 A} \cdot \dfrac{\sec A \cot^3 A}{\sin^2(90^\circ - A)} = \sqrt{\tan^2 A + 1}$.

28. $\dfrac{\cos^2(90^\circ - A)}{\operatorname{vers} A} = 1 + \sin(90^\circ - A)$.

29. $\dfrac{\cot^2 A \sin^2(90^\circ - A)}{\cot A + \cos A} = \tan(90^\circ - A) - \cos A$.

30. If $x \sin(90^\circ - A)\cot(90^\circ - A) = \cos(90^\circ - A)$, find x.

31. Find the value of x which will satisfy

$$\sec A \operatorname{cosec}(90^\circ - A) - x \cot(90^\circ - A) = 1.$$

Easy Trigonometrical Equations.

40. As a further exercise in using the formulæ of Art. 29 and the numerical values of the functions of 45°, 60°, 30°, we shall now give some examples in trigonometrical equations.

Example 1. Solve $4\cos A = 3\sec A$.

By expressing the secant in terms of the cosine, we have

$$4\cos A = \frac{3}{\cos A},$$

$$4\cos^2 A = 3,$$

$$\cos A = \pm\frac{\sqrt{3}}{2}.$$

$$\therefore \cos A = \frac{\sqrt{3}}{2} \text{ ....(1)},$$

or

$$\cos A = -\frac{\sqrt{3}}{2} \text{ ....(2)}.$$

Since $\cos 30° = \frac{\sqrt{3}}{2}$, we see from (1) that $A = 30°$.

The student will be able to understand the meaning of the negative result in (2) after he has read Chap. VIII.

Example 2. Solve $3\sec^2\theta = 8\tan\theta - 2$.

Since $\sec^2\theta = 1 + \tan^2\theta$,

we have $3(1 + \tan^2\theta) = 8\tan\theta - 2$,

or $3\tan^2\theta - 8\tan\theta + 5 = 0$.

This is a quadratic equation in which $\tan\theta$ is the unknown quantity, and it may be solved by any of the rules for solving quadratic equations.

Thus $(\tan\theta - 1)(3\tan\theta - 5) = 0$,

therefore *either* $\tan\theta - 1 = 0$ .... (1),

or $3\tan\theta - 5 = 0$ .... (2).

From (1), $\tan\theta = 1$, so that $\theta = 45°$.

From (2), $\tan\theta = \frac{5}{3}$, a result which we cannot interpret at present.

41. When an equation involves more than two functions, it will usually be best to express each function in terms of the sine and cosine.

Example. Solve $3 \tan \theta + \cot \theta = 5 \operatorname{cosec} \theta$.

We have
$$\frac{3 \sin \theta}{\cos \theta} + \frac{\cos \theta}{\sin \theta} = \frac{5}{\sin \theta},$$
$$3 \sin^2 \theta + \cos^2 \theta = 5 \cos \theta,$$
$$3(1 - \cos^2 \theta) + \cos^2 \theta = 5 \cos \theta,$$
$$2 \cos^2 \theta + 5 \cos \theta - 3 = 0,$$
$$(2 \cos \theta - 1)(\cos \theta + 3) = 0;$$

therefore *either* $2 \cos \theta - 1 = 0$ ....(1),

or $\cos \theta + 3 = 0$ ....(2).

From (1), $\cos \theta = \frac{1}{2}$, so that $\theta = 60°$.

From (2), $\cos \theta = -3$, a result which must be rejected as *impossible*, because the numerical value of the cosine of an angle can never be greater than unity. [Art. 16.]

EXAMPLES. IV. c.

Find a solution of each of the following equations:

1. $2 \sin \theta = \operatorname{cosec} \theta$.
2. $\tan \theta = 3 \cot \theta$.
3. $\sec \theta = 4 \cos \theta$.
4. $\sec \theta - \operatorname{cosec} \theta = 0$.
5. $4 \sin \theta = 3 \operatorname{cosec} \theta$.
6. $\operatorname{cosec}^2 \theta = 4$.
7. $\sqrt{2} \cos \theta = \cot \theta$.
8. $\tan \theta = 2 \sin \theta$.
9. $\sec^2 \theta = 2 \tan^2 \theta$.
10. $\operatorname{cosec}^2 \theta = 4 \cot^2 \theta$.
11. $\sec^2 \theta = 3 \tan^2 \theta - 1$.
12. $\sec^2 \theta + \tan^2 \theta = 7$.
13. $\cot^2 \theta + \operatorname{cosec}^2 \theta = 3$.
14. $2(\cos^2 \theta - \sin^2 \theta) = 1$.
15. $2 \cos^2 \theta + 4 \sin^2 \theta = 3$.
16. $6 \cos^2 \theta = 1 + \cos \theta$.
17. $4 \sin \theta = 12 \sin^2 \theta - 1$.
18. $2 \sin^2 \theta = 3 \cos \theta$.
19. $\tan \theta = 4 - 3 \cot \theta$.
20. $\cos^2 \theta - \sin^2 \theta = 2 - 5 \cos \theta$.
21. $\cot \theta + \tan \theta = 2 \sec \theta$.
22. $4 \operatorname{cosec} \theta + 2 \sin \theta = 9$.
23. $\tan \theta - \cot \theta = \operatorname{cosec} \theta$.
24. $2 \cos \theta + 2\sqrt{2} = 3 \sec \theta$.
25. $2 \sin \theta \tan \theta + 1 = \tan \theta + 2 \sin \theta$.
26. $6 \tan \theta - 5\sqrt{3} \sec \theta + 12 \cot \theta = 0$.
27. If $\tan \theta + 3 \cot \theta = 4$, prove that $\tan \theta = 1$ or 3.
28. Find $\cot \theta$ from the equation
$$\operatorname{cosec}^2 \theta + \cot^2 \theta = 3 \cot \theta.$$

MISCELLANEOUS EXAMPLES. A.

1. Express as the decimal of a right angle

(1) $25^g\ 37'\ 6{\cdot}4''$; (2) $63^\circ\ 21'\ 36''$.

2. Shew that

$$\sin A \cos A \tan A + \cos A \sin A \cot A = 1.$$

3. A ladder 29 ft. long just reaches a window at a height of 21 ft. from the ground: find the cosine and cosecant of the angle made by the ladder with the ground.

4. If $\operatorname{cosec} A = \frac{17}{15}$, find $\tan A$ and $\sec A$.

5. Shew that $\operatorname{cosec}^2 A - \cot A \cos A \operatorname{cosec} A - 1 = 0$.

---

6. Reduce to sexagesimal measure

(1) $17^g\ 18'\ 75''$; (2) $\cdot 0003$ of a right angle.

7. ABC is a triangle in which B is a right angle; if $c=9$, $a=40$, find b, $\cot A$, $\sec A$, $\sec C$.

8. Which of the following statements are possible and which impossible?

(1) $4 \sin \theta = 1$; (2) $2 \sec \theta = 1$; (3) $7 \tan \theta = 40$.

9. Prove that $\cos \theta \operatorname{vers} \theta (\sec \theta + 1) = \sin^2 \theta$.

10. Express $\sec \alpha$ and $\operatorname{cosec} \alpha$ in terms of $\cot \alpha$.

---

11. Find the numerical value of

$$3 \tan^2 30^\circ + \frac{1}{4} \sec 60^\circ + 5 \cot^2 45^\circ - \frac{2}{3} \sin^2 60^\circ.$$

12. If $\tan \alpha = \frac{m}{n}$, find $\sin \alpha$ and $\sec \alpha$.

13. If m sexagesimal minutes are equivalent to n centesimal minutes, prove that $m = \cdot 54n$.

14. If $\sin A = \frac{4}{5}$, prove that $\tan A + \sec A = 3$, when A is an acute angle.

15. Shew that

$$\cot(90° - A) \cot A \cos(90° - A) \tan(90° - A) = \cos A.$$

---

16. PQR is a triangle in which P is a right angle; if $PQ = 21$, $PR = 20$, find $\tan Q$ and $\operatorname{cosec} Q$.

17. Shew that $(\tan \alpha - \cot \alpha) \sin \alpha \cos \alpha = 1 - 2\cos^2 \alpha$.

18. Find a value of θ which satisfies the equation

$$\sec 6\theta = \operatorname{cosec} 3\theta.$$

19. Prove that

$$\tan^2 60° - 2\tan^2 45° = \cot^2 30° - 2\sin^2 30° - \frac{3}{4}\operatorname{cosec}^2 45°.$$

20. Solve the equations:

(1) $3\sin\theta = 2\cos^2\theta$; (2) $5\cot\theta - \operatorname{cosec}^2\theta = 3$.

---

21. Prove that $1 + 2\sec^2 A \tan^2 A - \sec^4 A - \tan^4 A = 0$.

22. In the equation

$$8\sin^2\theta - 10\sin\theta + 3 = 0,$$

shew that one value of θ is impossible, and find the other value.

23. In a triangle ABC right-angled at C, prove that

$$\tan A + \tan B = \frac{c^2}{ab}.$$

24. If $\cot A = c$, shew that $c + c^{-1} = \sec A \operatorname{cosec} A$.

25. Shew that $\dfrac{\sqrt{2 \operatorname{vers} A - \operatorname{vers}^2 A}}{1 - \operatorname{vers} A} = \tan A$.

CHAPTER V.

SOLUTION OF RIGHT-ANGLED TRIANGLES.

42. EVERY triangle has six *parts*, namely, three sides and three angles. In Trigonometry it is usual to denote the three angles by the capital letters A, B, C, and the lengths of the sides respectively opposite to these angles by the letters a, b, c. It must be understood that a, b, c are *numerical quantities* expressing the number of units of length contained in the three sides.

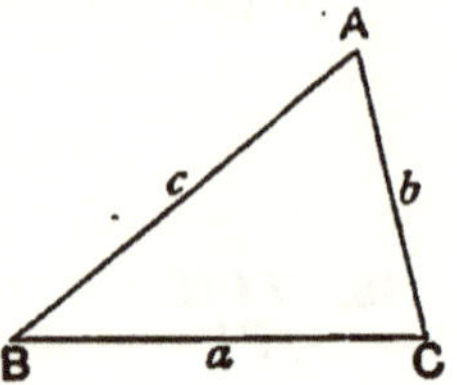

43. We know from Geometry that it is always possible to construct a triangle when any three parts are given, provided that one at least of the parts is a side. Similarly, if the values of suitable parts of a triangle be given, we can by Trigonometry find the remaining parts. The process by which this is effected is called the **Solution of the triangle.**

The *general* solution of triangles will be discussed at a later stage; in this chapter we shall confine our attention to right-angled triangles.

44. From Euc. I. 47, we know that when a triangle is right-angled, if any two sides are given the third can be found. Thus in the figure of the next article, where ABC is a triangle right-angled at A, we have $a^2=b^2+c^2$; whence if any two of the three quantities a, b, c are given, the third may be determined.

Again, the two acute angles are complementary, so that if one is given the other is also known.

Hence in the solution of right-angled triangles there are really only two cases to be considered:

I. when *any two sides* are given;

II. when *one side* and *one acute angle* are given.

45. Case I. *To solve a right-angled triangle when two sides are given.*

Let ABC be a right-angled triangle, of which A is the right angle, and suppose that any two sides are given;

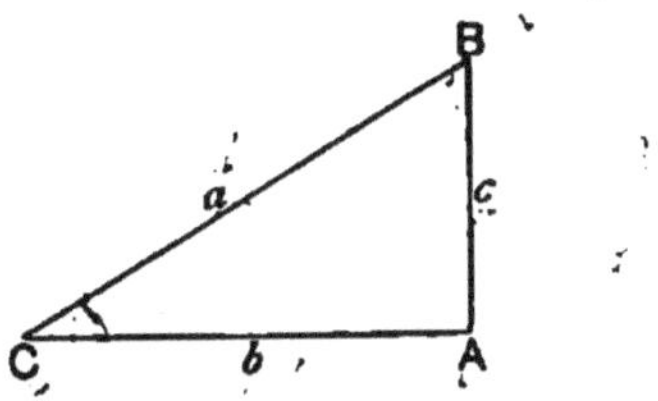

then the third side may be found from the equation

$$a^2 = b^2 + c^2.$$

Also

$$\cos C = \frac{b}{a}, \text{ and } B = 90^\circ - C;$$

whence C and B may be obtained.

Example. Given $B = 90^\circ$, $a = 20$, $b = 40$, solve the triangle.

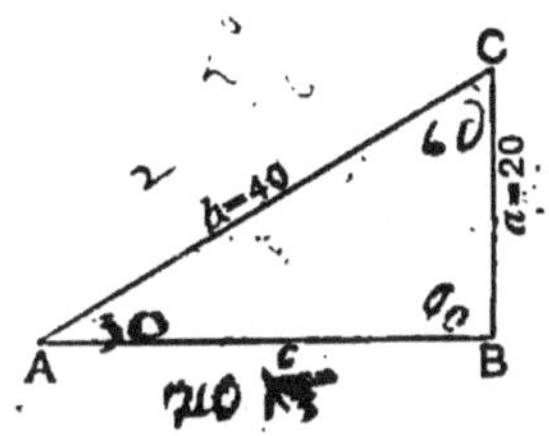

Here $\quad c^2 = b^2 - a^2$

$$= 1600 - 400 = 1200;$$

$$\therefore\ c = 20\sqrt{3}.$$

Also $\sin A = \dfrac{a}{b} = \dfrac{20}{40} = \dfrac{1}{2};$

$$\therefore\ A = 30^\circ.$$

And $C = 90^\circ - A = 90^\circ - 30^\circ = 60^\circ.$

The solution of a trigonometrical problem may often be obtained in more than one way. In the present case the triangle can be solved without making use of Euc. I. 47.

Another solution may be given as follows:

$$\cos C = \frac{a}{b} = \frac{20}{40} = \frac{1}{2};$$

$$\therefore\ C = 60^\circ.$$

And $\quad A = 90^\circ - C = 90^\circ - 60^\circ = 30^\circ.$

Also $\quad \dfrac{c}{40} = \cos A = \cos 30^\circ = \dfrac{\sqrt{3}}{2};$

$$\therefore\ c = 20\sqrt{3}.$$

46. CASE II. *To solve a right-angled triangle when one side and one acute angle are given.*

Let ABC be a right-angled triangle of which A is the right angle, and suppose one side b and one acute angle C are given; then

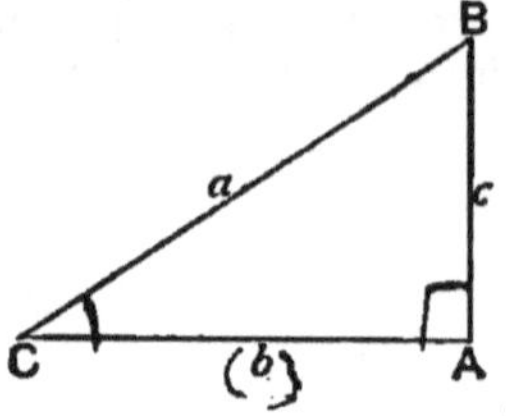

$$B=90^\circ-C, \quad \frac{a}{b}=\sec C, \quad \frac{c}{b}=\tan C;$$

whence B, a, c may be determined.

Example 1. Given $B=90^\circ$, $A=30^\circ$, $c=5$, solve the triangle.

We have $C=90^\circ-A=90^\circ-30^\circ=60^\circ$.

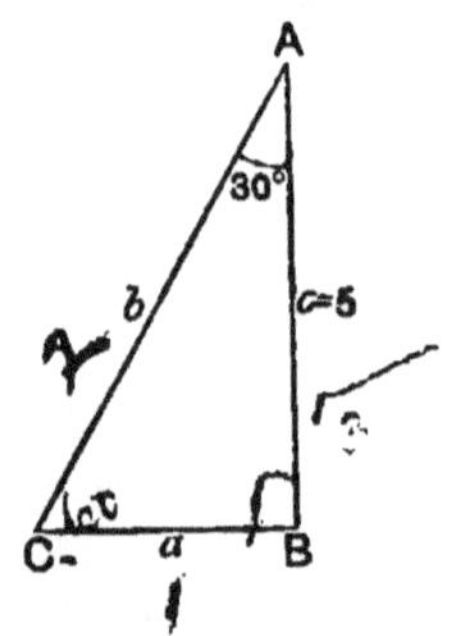

Also $\qquad \dfrac{a}{5}=\tan 30^\circ;$

$$\therefore \ a=5\tan 30^\circ=\frac{5}{\sqrt{3}}$$

$$=\frac{5}{\sqrt{3}}\times\frac{\sqrt{3}}{\sqrt{3}}=\frac{5\sqrt{3}}{3}.$$

Again, $\qquad \dfrac{b}{5}=\sec 30^\circ;$

$$\therefore \ b=5\sec 30^\circ=5\times\frac{2}{\sqrt{3}}=\frac{10}{\sqrt{3}}=\frac{10\sqrt{3}}{3}.$$

NOTE. The student should observe that in each case we write down a ratio which connects *the side we are finding with that whose value is given*, and a knowledge of the ratios of the given angle enables us to complete the solution.

Example 2. If $C=90^\circ$, $B=25^\circ 43'$, and $c=100$, solve the triangle, having given $\tan 25^\circ 43'=\cdot482$ and $\cos 25^\circ 43'=\cdot901$.

Here $A=90^\circ-B$

$=90^\circ-25^\circ 43'=64^\circ 17'$.

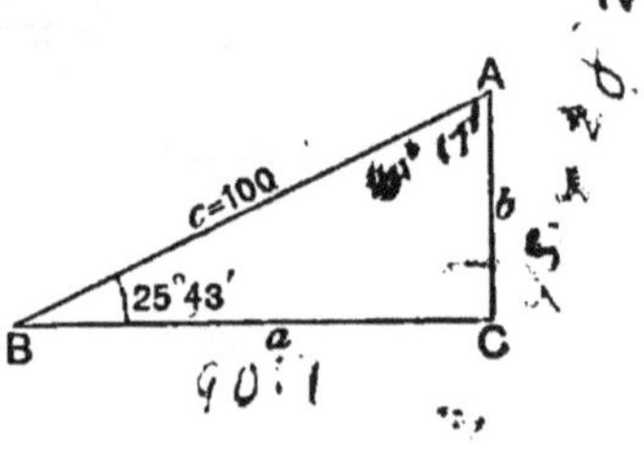

Now $\qquad \dfrac{a}{c}=\cos B;$

that is, $\dfrac{a}{100}=\cos 25^\circ 43';$

$$\therefore \ a=100\cos 25^\circ 43'$$

$$=100\times\cdot901=90\cdot1.$$

Also $\qquad \dfrac{b}{a}=\tan B$, or $b=a\tan B$;

$$\therefore \ b=90\cdot1\times\tan 25^\circ 43'=90\cdot1\times\cdot482=43\cdot4282.$$

EXAMPLES. V. a.

Solve the triangles in which the following parts are given:

1. $A=90°$, $a=4$, $b=2\sqrt{3}$.
2. $c=6$, $b=12$, $B=90°$.
3. $C=90°$, $b=12$, $a=4\sqrt{3}$.
4. $a=60$, $b=30$, $A=90°$.
5. $a=20$, $c=20$, $B=90°$.
6. $a=5\sqrt{3}$, $b=15$, $C=90°$.
7. $b=c=2$, $A=90°$.
8. $2c=b=6\sqrt{3}$, $B=90°$.
9. $C=90°$, $a=9\sqrt{3}$, $A=30°$.
10. $A=90°$, $B=30°$, $a=4$.
11. $A=60°$, $c=8$, $C=90°$.
12. $A=60°$, $C=30°$, $b=6$.
13. $B=90°$, $C=60°$, $b=100$.
14. $A=30°$, $B=60°$, $b=20\sqrt{3}$.
15. $B=C=45°$, $c=4$.
16. $2B=C=60°$, $a=8$.

17. If $C=90°$, $\cot A=·07$, $b=49$, find a.

18. If $C=90°$, $A=38°\ 19'$, $c=50$, find a; given $\sin 38°\ 19'=·62$.

19. If $a=100$, $B=90°$, $C=40°\ 51'$, find c; given $\tan 40°\ 51'=·8647$.

20. If $b=20$, $A=90°$, $C=78°\ 12'$, find a; given $\sec 78°\ 12'=4·89$.

21. If $B=90°$, $A=36°$, $c=100$, solve the triangle; given $\tan 36°=·73$, $\sec 36°=1·24$.

22. If $A=90°$, $c=37$, $a=100$, solve the triangle; given $\sin 21°\ 43'=·37$, $\cos 21°\ 43'=·93$.

23. If $A=90°$, $B=39°\ 24'$, $b=25$, solve the triangle; given $\cot 39°\ 24'=1·2174$, $\operatorname{cosec} 39°\ 24'=1·5755$.

24. If $C=90°$, $a=225$, $b=272$, solve the triangle; given $\tan 50°\ 24'=1·209$.

47. It will be found that all the varieties of the solution of right-angled triangles which can arise are either included in the two cases of Arts. 45 and 46, or in some modification of them. Sometimes the solution of a problem may be obtained by solving *two* right-angled triangles. The two examples we give as illustrations will in various forms be frequently met with in subsequent chapters.

Example 1. In the triangle ABC, the angles A and B are equal to 30° and 135° respectively, and the side AB is 100 feet; find the length of the perpendicular from C upon AB produced.

Draw CD perpendicular to AB produced, and let $CD = x$.

Then $\angle CBD = 180° - 135° = 45°$;

$$\therefore BD = CD = x.$$

Now in the right-angled triangle ADC,

$$\frac{CD}{AD} = \tan DAC = \tan 30°;$$

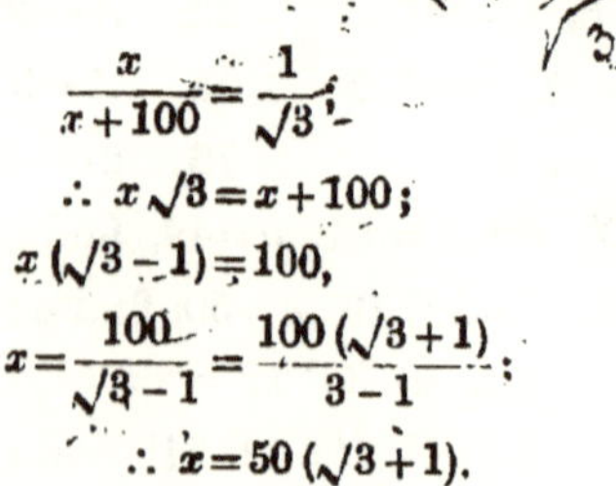

that is,

$$\frac{x}{x+100} = \frac{1}{\sqrt{3}};$$

$$\therefore x\sqrt{3} = x + 100;$$

$$x(\sqrt{3} - 1) = 100,$$

$$x = \frac{100}{\sqrt{3}-1} = \frac{100(\sqrt{3}+1)}{3-1};$$

$$\therefore x = 50(\sqrt{3}+1).$$

Thus the distance required is $50(\sqrt{3}+1)$ feet.

Example 2. In the triangle ABC, AD is drawn perpendicular to BC; solve the triangle, having given

$$AD = 5, \quad \angle ABD = 60°, \quad \angle ACD = 45°.$$

In the right-angled triangle ABD,

$$\frac{AB}{AD} = \operatorname{cosec} ABD;$$

$$\therefore AB = AD \operatorname{cosec} ABD = 5 \operatorname{cosec} 60°$$

$$= 5 \times \frac{2}{\sqrt{3}} = \frac{10}{\sqrt{3}} = \frac{10\sqrt{3}}{3}.$$

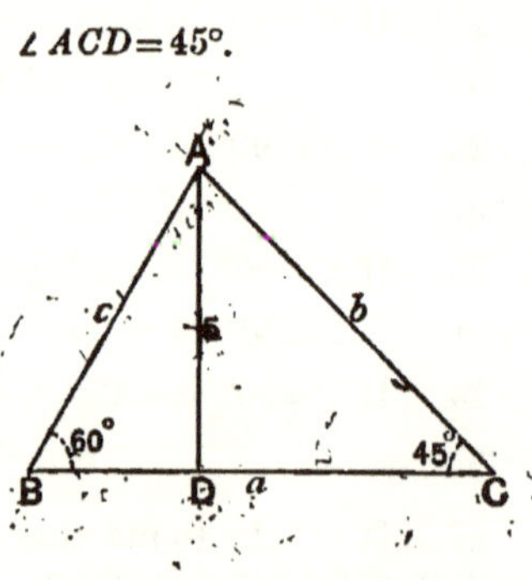

Also $\dfrac{BD}{AD} = \cot ABD$;

$$\therefore BD = AD \cot ABD$$

$$= 5 \cot 60° = \frac{5}{\sqrt{3}} = \frac{5\sqrt{3}}{3}.$$

In the right-angled triangle ADC,

$$\angle DAC = 45° = \angle DCA;$$

$$\therefore DC = DA = 5.$$

Thus $BC=CD+BD=5+\frac{5\sqrt{3}}{3}=\frac{15+5\sqrt{3}}{3}.$

And $\frac{AC}{AD}=\text{cosec}\,ACD;$

$\therefore\ AC=AD\,\text{cosec}\,ACD=5\,\text{cosec}\,45^\circ=5\sqrt{2}.$

Finally, $\angle BAC=180^\circ-60^\circ-45^\circ=75^\circ.$

Thus $a=\frac{15+5\sqrt{3}}{3},\quad b=5\sqrt{2},\quad c=\frac{10\sqrt{3}}{3},\quad A=75^\circ.$

EXAMPLES. V. b.

1. ABC is a triangle, and BD is perpendicular to AC produced: find BD, given

$$A=30^\circ,\ C=120^\circ,\ AC=20.$$

2. If BD is perpendicular to the base AC of a triangle ABC, find a and c, given

$$A=30^\circ,\ C=45^\circ,\ BD=10.$$

3. In the triangle ABC, AD is drawn perpendicular to BC making BD equal to 15 ft.: find the lengths of AB, AC, and AD, given that B and C are equal to 30° and 60° respectively.

4. In a right-angled triangle PQR, find the segments of the hypotenuse PR made by the perpendicular from Q; given

$$QR=8,\ \angle QRP=60^\circ,\ \angle QPR=30^\circ.$$

5. If PQ is drawn perpendicular to the straight line QRS, find RS, given

$$PQ=36,\ \angle RPQ=30^\circ,\ \angle SPQ=60^\circ.$$

6. If PQ is drawn perpendicular to the straight line QRS, find RS, given

$$PQ=20,\ \angle PRS=135^\circ,\ \angle PSR=30^\circ.$$

7. In the triangle ABC, the angles B and C are equal to 45° and 120° respectively; if $a=40$ find the length of the perpendicular from A on BC produced.

8. If CD is drawn perpendicular to the straight line DBA, find DC and BD, given

$$AB=59,\ \angle CBD=45^\circ,\ \angle CAB=32^\circ\,50',\quad \cot 32^\circ\,50'=1{\cdot}59.$$

CHAPTER VI.

EASY PROBLEMS.

48. THE principles explained in the previous chapters may now be applied to the solution of problems in heights and distances. It will be assumed that by the use of suitable instruments the necessary lines and angles can be measured with sufficient accuracy for the purposes required.

After the practice afforded by the examples in the last chapter, the student should be able to write down at once any side of a right-angled triangle in terms of another through the medium of the functions of either acute angle. In the present and subsequent chapters it is of great importance to acquire readiness in this respect.

For instance, from the adjoining figure, we have

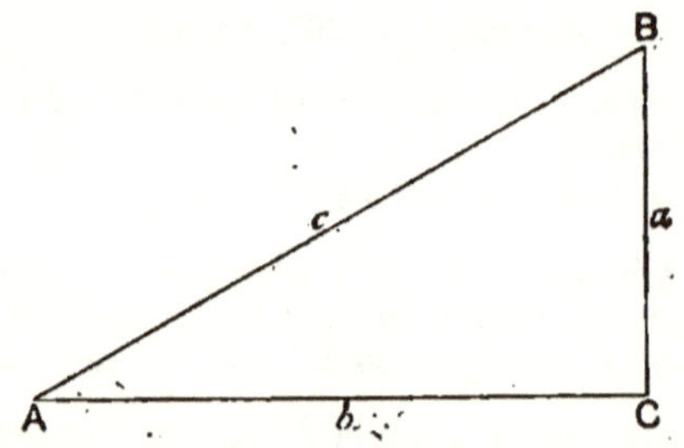

$$a=c\sin A,\quad a=c\cos B,\quad a=b\cot B,$$
$$a=b\tan A,\quad c=a\sec B,\quad b=a\tan B.$$

These relations are not to be committed to memory but in each case should be read off from the figure. There are several other similar relations connecting the parts of the above triangle, and the student should practise himself in obtaining them quickly.

Example. Q, R, T are three points in a straight line, and TP is drawn perpendicular to QT. If $PT=a$, $\angle PQT=\beta$, $\angle PRT=2\beta$, express the lengths of all the lines of the figure in terms of a and β.

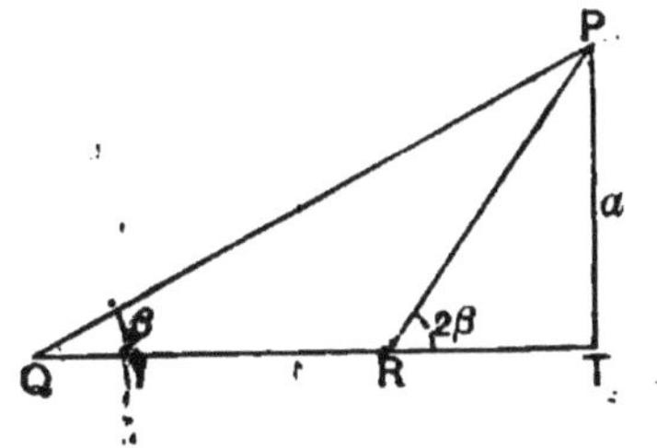

By Euc. I. 32,

$$\angle QPR = \angle PRT - \angle PQR;$$
$$\therefore\ \angle QPR = 2\beta - \beta = \beta = \angle PQR;$$
$$\therefore\ QR = PR.$$

In the right-angled triangle PRT,

$$PR = a \operatorname{cosec} 2\beta;$$
$$\therefore QR = a \operatorname{cosec} 2\beta.$$

Also $$TR = a \cot 2\beta.$$

Lastly, in the right-angled triangle PQT,

$$QT = a \cot \beta,$$
$$PQ = a \operatorname{cosec} \beta.$$

49. Angles of elevation and depression. Let OP be a horizontal line in the same vertical plane as an object Q, and let OQ be joined.

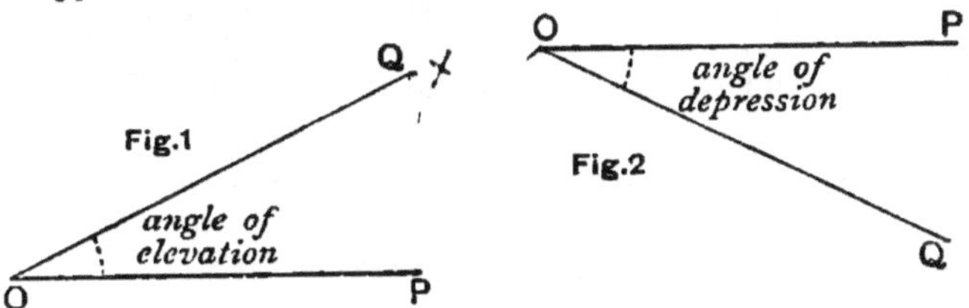

In Fig. 1, where the object Q is *above* the horizontal line OP, the angle POQ is called the **angle of elevation** of the object Q as seen from the point O.

In Fig. 2, where the object Q is *below* the horizontal line OP, the angle POQ is called the **angle of depression** of the object Q as seen from the point O.

Example I. A flagstaff stands on a horizontal plane, and from a point on the ground at a distance of 30 ft. its angle of elevation is 60°: find its height.

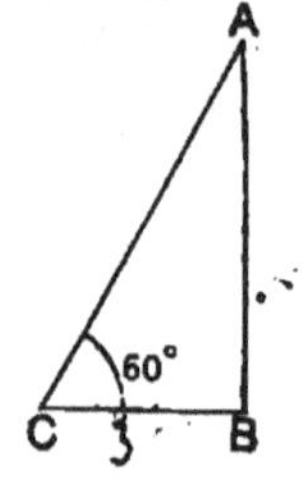

Let AB be the flagstaff, C the point of observation; then

$$AB = BC \tan 60° = 30\sqrt{3}$$
$$= 30 \times 1{\cdot}732 = 51{\cdot}96.$$

Thus the height is 51·96 ft.

EXAMPLES. VI. a.

[*The results should be expressed in a form free from surds by using the approximations* $\sqrt{2} = 1{\cdot}414$, $\sqrt{3} = 1{\cdot}732$.]

1. The angle of elevation of the top of a chimney at a distance of 300 feet is 30°: find its height.

2. From a ship's masthead 160 feet high the angle of depression of a boat is observed to be 30°: find its distance from the ship.

3. Find the angle of elevation of the sun when the shadow of a pole 6 feet high is $2\sqrt{3}$ feet long.

4. At a distance 86·6 feet from the foot of a tower the angle of elevation of the top is 30°. Find the height of the tower and the observer's distance from the top.

5. A ladder 45 feet long just reaches the top of a wall. If the ladder makes an angle of 60° with the wall, find the height of the wall, and the distance of the foot of the ladder from the wall.

6. Two masts are 60 feet and 40 feet high, and the line joining their tops makes an angle of 33° 41′ with the horizon: find their distance apart, given cot 33° 41′ = 1·5.

7. Find the distance of the observer from the top of a cliff which is 132 yards high, given that the angle of elevation is 41° 18′, and that sin 41° 18′ = ·66.

8. One chimney is 30 yards higher than another. A person standing at a distance of 100 yards from the lower observes their tops to be in a line inclined at an angle of 27° 2′ to the horizon: find their heights, given tan 27° 2′ = ·51.

Example II. From the foot of a tower the angle of elevation of the top of a column is 60°, and from the top of the tower, which is 50 ft. high, the angle of elevation is 30°: find the height of the column.

Let AB denote the column and CD the tower; draw CE parallel to DB.

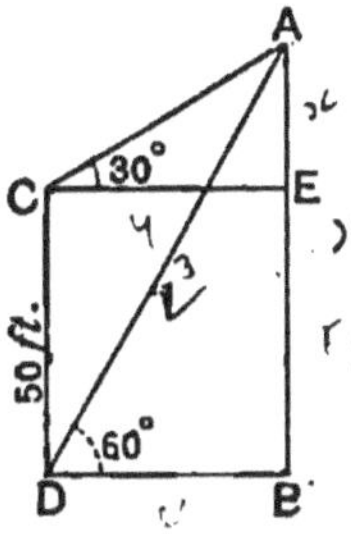

Let $AB = x$;

then $$AE = AB - BE = x - 50.$$

Let $$DB = CE = y.$$

From the right-angled triangle ADB,

$$y = x \cot 60° = \frac{x}{\sqrt{3}}.$$

From the right-angled triangle ACE,

$$y = (x - 50) \cot 30° = \sqrt{3}\,(x - 50).$$

$$\therefore \frac{x}{\sqrt{3}} = \sqrt{3}\,(x - 50),$$

$$x = 3\,(x - 50);$$

whence $$x = 75.$$

Thus the column is 75 ft. high.

9. The angle of elevation of the top of a tower is 30°; on walking 100 yards nearer the elevation is found to be 60°: find the height of the tower.

10. A flagstaff stands upon the top of a building; at a distance of 40 feet the angles of elevation of the tops of the flagstaff and building are 60° and 30°: find the length of the flagstaff.

11. The angles of elevation of a spire at two places due east of it and 200 feet apart are 45° and 30°: find the height of the spire.

12. From the foot of a post the elevation of the top of a steeple is 45°, and from the top of the post, which is 30 feet high, the elevation is 30°; find the height and distance of the steeple.

13. The height of a hill is 3300 feet above the level of a horizontal plane. From a point A on this plane the angular elevation of the top of the hill is 60°. A balloon rises from A and ascends vertically upwards at a uniform rate; after 5 minutes the angular elevation of the top of the hill to an observer in the balloon is 30°: find the rate of the balloon's ascent in miles per hour.

Example III. **From the top of a cliff 150 ft. high the angles of depression of two boats which are due South of the observer are 15° and 75°: find their distance apart, having given**

$$\cot 15° = 2 + \sqrt{3} \text{ and } \cot 75° = 2 - \sqrt{3}.$$

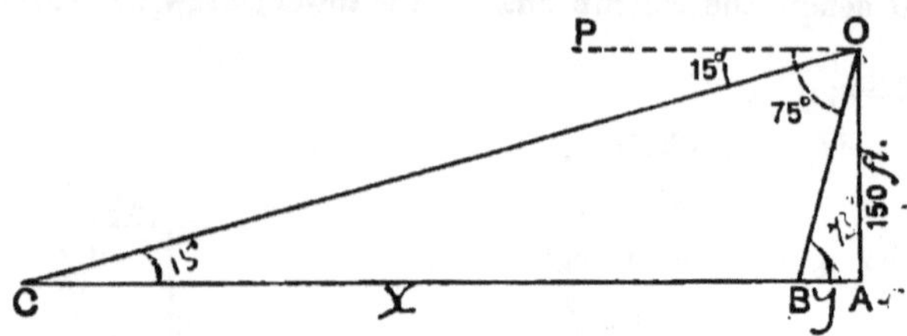

Let OA represent the cliff, B and C the boats. Let OP be a horizontal line through O; then

$$\angle POC = 15° \text{ and } \angle POB = 75°;$$
$$\therefore \angle OCA = 15° \text{ and } \angle OBA = 75°.$$

Let $CB = x$, $AB = y$; then $CA = x + y$.

From the right-angled triangle OBA,

$$y = 150 \cot 75° = 150(2 - \sqrt{3}) = 300 - 150\sqrt{3}.$$

From the right-angled triangle OCA,

$$x + y = 150 \cot 15° = 150(2 + \sqrt{3}) = 300 + 150\sqrt{3}.$$

By subtraction, $x = 300\sqrt{3} = 519{\cdot}6.$

Thus the distance between the boats is 519·6 ft.

14. From the top of a monument 100 feet high, the angles of depression of two objects on the ground due west of the monument are 45° and 30°: find the distance between them.

15. The angles of depression of the top and foot of a tower seen from a monument 96 feet high are 30° and 60°: find the height of the tower.

16. From the top of a cliff 150 feet high the angles of depression of two boats at sea, each due north of the observer, are 30° and 15°: how far are the boats apart?

17. From the top of a hill the angles of depression of two consecutive milestones on a level road running due south from the observer are 45° and 22° respectively. If $\cot 22° = 2{\cdot}475$ find the height of the hill in yards.

18. From the top of a lighthouse 80 yards above the horizon the angles of depression of two rocks due west of the observer are 75° and 15°: find their distance apart, given $\cot 75° = {\cdot}268$ and $\cot 15° = 3{\cdot}732$.

50. Trigonometrical Problems sometimes require a knowledge of the **Points of the Mariner's Compass,** which we shall now explain.

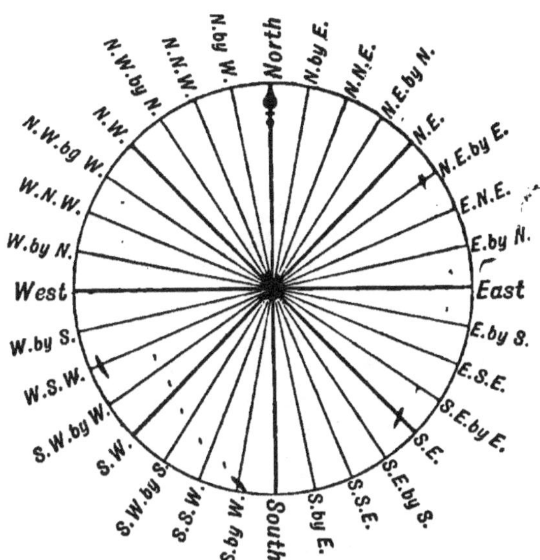

In the above figure, it will be seen that 32 points are taken at equal distances on the circumference of a circle, so that the arc between any two consecutive points subtends at the centre of the circle an angle equal to $\frac{360^\circ}{32}$, that is to $11\frac{1}{4}^\circ$.

The points North, South, East, West are called the **Cardinal Points,** and with reference to them the other *points* receive their names. The student will have no difficulty in learning these if he will carefully notice the arrangement in any one of the principal quadrants.

51. Sometimes a slightly different notation is used; thus N. $11\frac{1}{4}^\circ$ E. means a direction $11\frac{1}{4}^\circ$ east of north, and is therefore the same as N. by E. Again S.W. by S. is 3 *points* from south and may be expressed by S. $33\frac{3}{4}^\circ$ W., or since it is 5 *points* from west it can also be expressed by W. $56\frac{1}{4}^\circ$ S. In each of these cases it will be seen that the angular measurement is made from the direction which is first mentioned.

52. The angle between the directions of any two points is obtained by multiplying $11\frac{1}{4}°$ by the number of intervals between the points. Thus between S. by W. and W.S.W. there are 5 intervals and the angle is $56\frac{1}{4}°$; between N.E. by E. and S.E. there are 7 intervals and the angle is $78\frac{3}{4}°$.

53. If B lies in a certain direction with respect to A, it is said to *bear* in that direction from A; thus Birmingham *bears* N.W. of London, and from Birmingham the *bearing* of London is S.E.

Example 1. From a lighthouse L two ships A and B are observed in directions S.W. and 15° East of South respectively. At the same time B is observed from A in a S.E. direction. If LA is 4 miles find the distance between the ships.

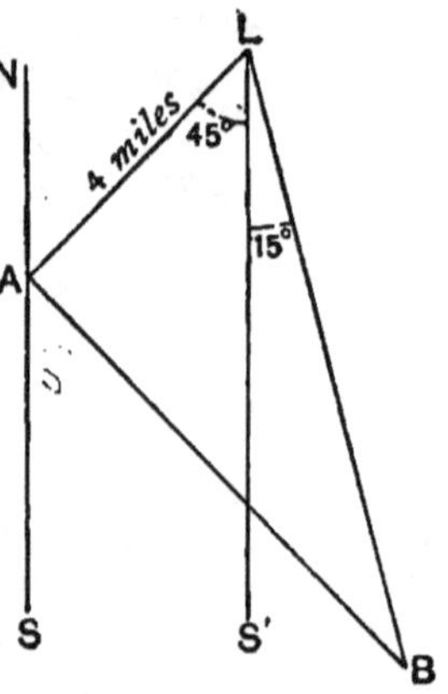

Draw LS' due South; then from the bearings of the two ships,

$$\angle ALS' = 45°, \quad \angle BLS' = 15°,$$

so that $$\angle ALB = 60°.$$

Through A draw a line NS pointing North and South; then

$$\angle NAL = \angle ALS' = 45°,$$

and $\angle BAS = 45°$, since B bears S.E. from A;

hence $\angle BAL = 180° - 45° - 45° = 90°$.

In the right-angled triangle ABL,

$$AB = AL \tan ALB = 4 \tan 60°$$
$$= 4\sqrt{3} = 6{\cdot}928.$$

Thus the distance between the ships is 6·928 miles.

Example 2. At 9 A.M. a ship which is sailing in a direction E. 40° S. at the rate of 8 miles an hour observes a fort in a direction 50° North of East. At 11 A.M. the fort is observed to bear N. 20° W.: find the distance of the fort from the ship at each observation.

Let A and C be the first and second positions of the ship; B the fort.

Through A draw lines towards the cardinal points of the compass.

From the observations made

$$\angle EAC = 40°, \quad \angle EAB = 50°, \text{ so that } \angle BAC = 90°.$$

Through C draw CN' towards the North; then $\angle BCN'=20°$, for the bearing of the fort from C is N. 20° W.

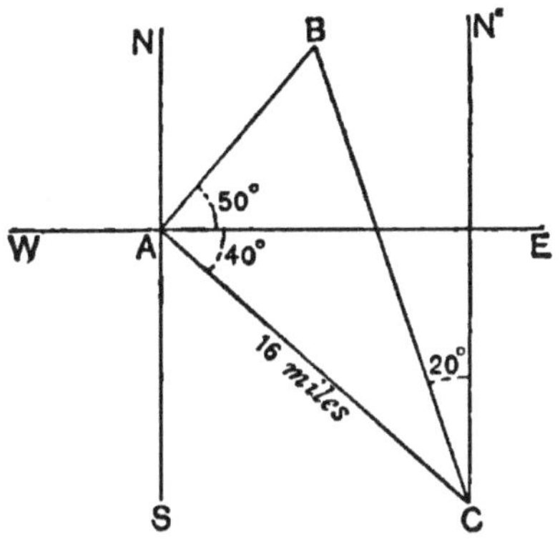

Also $\angle ACN' = \angle CAS = 90° - 40° = 50°$;

$\therefore \angle ACB = \angle ACN' - \angle BCN' = 50° - 20° = 30°$.

In the right-angled triangle ACB,

$$AB = AC \tan ACB = 16 \tan 30° = \frac{16}{\sqrt{3}} = \frac{16\sqrt{3}}{3} = 9{\cdot}237 \text{ nearly;}$$

and $BC = AC \sec ACB = 16 \sec 30° = 16 \times \frac{2}{\sqrt{3}} = \frac{32\sqrt{3}}{3} = 18{\cdot}475$ nearly.

Thus the distances are 9·237 and 18·475 miles nearly.

EXAMPLES. VI. b.

1. A person walking due E. observes two objects both in the N.E. direction. After walking 800 yards one of the objects is due N. of him, and the other lies N.W.: how far was he from the objects at first?

2. Sailing due E. I observe two ships lying at anchor due S.; after sailing 3 miles the ships bear 60° and 30° S. of W.; how far are they now distant from me?

3. Two vessels leave harbour at noon in directions W. 28° S. and E. 62° S. at the rates 10 and $10\frac{1}{2}$ miles per hour respectively. Find their distance apart at 2 p.m.

4. A lighthouse facing N. sends out a fan-shaped beam extending from N.E. to N.W. A steamer sailing due W. first sees the light when 5 miles away from the lighthouse and continues to see it for $30\sqrt{2}$ minutes. What is the speed of the steamer?

5. A ship sailing due S. observes two lighthouses in a line exactly W. After sailing 10 miles they are respectively N.W. and W.N.W.; find their distances from the position of the ship at the first observation.

6. Two vessels sail from port in directions N. 35° W. and S. 55° W. at the rates of 8 and $8\sqrt{3}$ miles per hour respectively. Find their distance apart at the end of an hour, and the bearing of the second vessel as observed from the first.

7. A vessel sailing S.S.W. is observed at noon to be E.S.E. from a lighthouse 4 miles away. At 1 p.m. the vessel is due S. of the lighthouse: find the rate at which the vessel is sailing. Given $\tan 67\frac{1}{2}° = 2·414$.

8. A, B, C are three places such that from A the bearing of C is N. 10° W., and the bearing of B is N. 50° E.; from B the bearing of C is N. 40° W. If the distance between B and C is 10 miles, find the distances of B and C from A.

9. A ship steaming due E. sights at noon a lighthouse bearing N.E., 15 miles distant; at 1.30 p.m. the lighthouse bears N.W. How many knots per day is the ship making? Given 60 knots = 69 miles.

10. At 10 o'clock forenoon a coaster is observed from a lighthouse to bear 9 miles away to N.E. and to be holding a south-easterly course; at 1 p.m. the bearing of the coaster is 15° S. of E. Find the rate of the coaster's sailing and its distance from the lighthouse at the time of the second observation.

11. The distance between two lighthouses, A and B, is 12 miles and the line joining them bears E. 15° N. At midnight a vessel which is sailing S. 15° E. at the rate of 10 miles per hour is N.E. of A and N.W. of B: find to the nearest minute when the vessel crosses the line joining the lighthouses.

12. From A to B, two stations of a railway, the line runs W.S.W. At A a person observes that two spires, whose distance apart is 1·5 miles, are in the same line which bears N.N.W. At B their bearings are N. $7\frac{1}{2}°$ E. and N. $37\frac{1}{2}°$ E. Find the rate of a train which runs from A to B in 2 minutes.

CHAPTER VII.

RADIAN OR CIRCULAR MEASURE.

54. WE shall now return to the system of measuring angles which was briefly referred to in Art. 6. In this system angles are not measured in terms of a submultiple of the right angle, as in the sexagesimal and centesimal methods, but a certain angle known as a *radian* is taken as the standard unit, in terms of which all other angles are measured.

55. DEFINITION. A **radian** is the angle subtended at the centre of any circle by an arc equal in length to the radius of the circle.

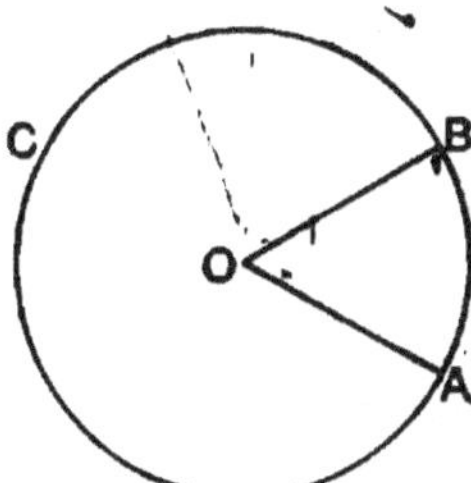

In the above figure, ABC is a circle, and O its centre. If on the circumference we measure an arc AB equal to the radius and join OA, OB, the angle AOB is a radian.

56. In any system of measurement it is essential that the unit should be always the same. In order to shew that a radian, constructed according to the above definition, is of constant magnitude, we must first establish an important property of the circle.

57. *The circumferences of circles are to one another as their radii.*

Take *any* two circles whose radii are r_1 and r_2, and in each circle let a regular polygon of n sides be described.

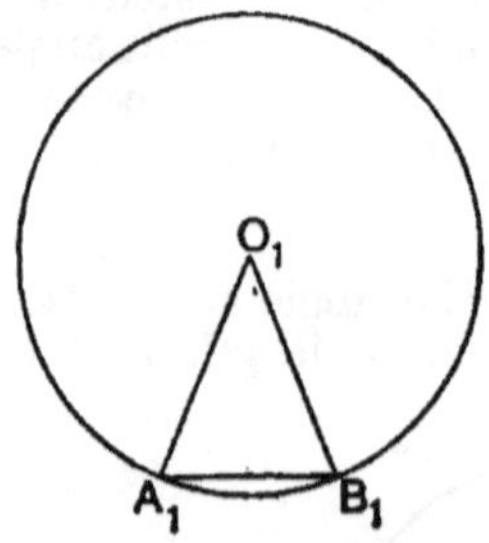

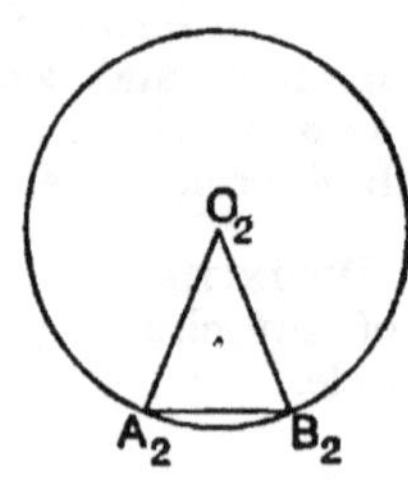

Let A_1B_1 be a side of the first, A_2B_2 a side of the second polygon, and let their lengths be denoted by a_1, a_2. Join their extremities to O_1 and O_2 the centres of the circles. We thus obtain two isosceles triangles whose vertical angles are equal, each being $\frac{1}{n}$ of four right angles.

Hence the triangles are equiangular, and therefore we have by Euc. VI. 4,

$$\frac{A_1B_1}{O_1A_1} = \frac{A_2B_2}{O_2A_2};$$

that is,

$$\frac{a_1}{r_1} = \frac{a_2}{r_2};$$

$$\therefore \frac{na_1}{r_1} = \frac{na_2}{r_2};$$

that is,

$$\frac{p_1}{r_1} = \frac{p_2}{r_2},$$

where p_1 and p_2 are the perimeters of the polygons. This is true whatever be the number of sides in the polygons. By taking n sufficiently large we can make the perimeters of the two polygons differ from the circumferences of the corresponding circles by as small a quantity as we please; so that ultimately

$$\frac{c_1}{r_1} = \frac{c_2}{r_2},$$

where c_1 and c_2 are the circumferences of the two circles.

58. It thus appears that *the ratio of the circumference of a circle to its radius is the same whatever be the size of the circle;* that is,

$$\textit{in all circles } \frac{\textit{circumference}}{\textit{diameter}} \textit{ is a constant quantity.}$$

This constant is incommensurable and is always denoted by the Greek letter π. Though its numerical value cannot be found exactly, it is shewn in a later part of the subject that it can be obtained to any degree of approximation. To ten decimal places its value is 3·1415926536. In many cases $\pi = \frac{22}{7}$, which is true to two decimal places, is a sufficiently close approximation; where greater accuracy is required the value 3·1416 may be used.

59. If c denote the circumference of the circle whose radius is r, we have

$$\frac{\text{circumference}}{\text{diameter}} = \pi;$$

$$\therefore \quad \frac{c}{2r} = \pi,$$

or

$$c = 2\pi r.$$

60. *To prove that all radians are equal.*

Draw *any* circle; let O be its centre and OA a radius. Let the arc AB be measured equal in length to OA. Join OB; then $\angle AOB$ is a radian. Produce AO to meet the circumference in C. By Euc. VI. 33, angles at the centre of a circle are proportional to the arcs on which they stand; hence

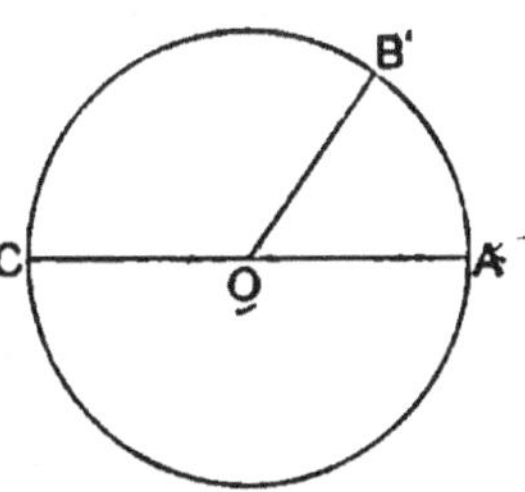

$$\frac{\angle AOB}{\text{two right angles}} = \frac{\text{arc } AB}{\text{arc } ABC}$$

$$= \frac{\text{radius}}{\text{semi-circumference}} = \frac{r}{\pi r} = \frac{1}{\pi},$$

which is constant; that is, *a radian always bears the same ratio to two right angles, and therefore is a constant angle.*

61. Since a radian is constant it is taken as a standard unit, and the *number of radians* contained in any angle is spoken of as its **radian measure** or **circular measure.** [See Art. 71.] In this system, an angle is usually denoted by a *mere number*, the unit being implied. Thus when we speak of an angle 2·5, it is understood that its radian measure is 2·5, or, in other words, that the angle contains $2\frac{1}{2}$ radians.

Where it is desirable to refer to the unit expressly, a radian may be denoted by the letter ρ, thus 3ρ represents an angle which contains three radians.

62. *To find the radian measure of a right angle.*

Let AOC be a right angle at the centre of a circle, and AOB a radian; then

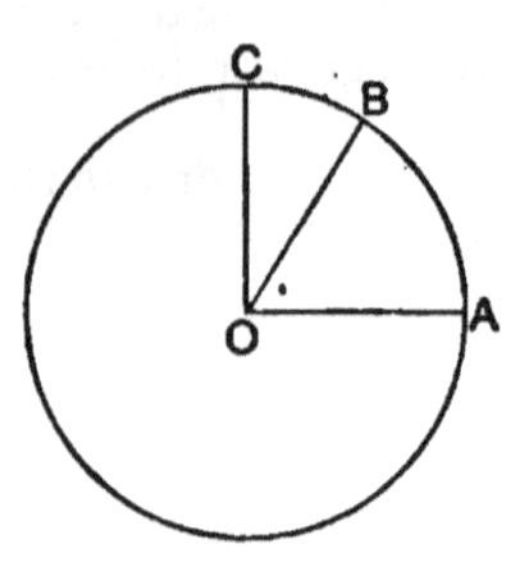

the radian measure of $\angle AOC$

$$= \frac{\angle AOC}{\angle AOB} = \frac{\text{arc } AC}{\text{arc } AB}$$

$$= \frac{\frac{1}{4}\text{(circumference)}}{\text{radius}} = \frac{\frac{1}{4}(2\pi r)}{r}$$

$$= \frac{\pi}{2};$$

that is, a right angle contains $\frac{\pi}{2}$ radians.

63. *To find the number of degrees in a radian.*

From the last article it follows that

$$\pi \textit{ radians} = 2 \textit{ right angles} = 180 \textit{ degrees}.$$

$$\therefore \text{ a radian} = \frac{180}{\pi} \text{ degrees}.$$

By division we find that $\frac{1}{\pi} = \cdot 31831$ nearly;

hence approximately, a radian $= 180 \times \cdot 31831 = 57 \cdot 2958$ degrees.

64. The formula

$$\pi \textit{ radians} = 180 \textit{ degrees}$$

connecting the sexagesimal and radian measures of an angle, is a useful result which enables us to pass readily from one system to the other.

Example. Express 75° in radian measure, and $\frac{\pi}{54}$ in sexagesimal measure.

(1) Since 180 degrees $=\pi$ radians,

$$75 \text{ degrees} = \frac{75}{180}\pi \text{ radians} = \frac{5\pi}{12} \text{ radians}.$$

Thus the radian measure is $\frac{5\pi}{12}$.

(2) Since π radians $=180$ degrees,

$$\frac{\pi}{54} \text{ radians} = \frac{180}{54} \text{ degrees}.$$

Thus the angle $=\frac{10}{3}$ degrees $=3° \, 20'$.

65. It may be well to remind the student that the symbol π always denotes a *number*, viz. 3·14159.... When the symbol stands alone, without reference to any angle, there can be no ambiguity; but even when π is used to denote an angle, it must still be remembered that π is a *number*, namely, the number of radians in two right angles.

NOTE. It is not uncommon for beginners to make statements such as "$\pi=180$" or "$\frac{\pi}{2}=90$." Without some modification this mode of expression is quite incorrect. It is true that π *radians* are equal to 180 *degrees*, but the statement '$\pi=180$' is no more correct than the statement "$20=1$" to denote the equivalence of 20 shillings and 1 sovereign.

66. *If the number of degrees and radians in an angle be represented by D and θ respectively, to prove that*

$$\frac{D}{180}=\frac{\theta}{\pi}.$$

In sexagesimal measure, the ratio of the given angle to two right angles is expressed by $\frac{D}{180}$.

In radian measure, the ratio of these same two angles is expressed by $\frac{\theta}{\pi}$.

Example 1. What is the radian measure of 45° 13′ 30″?

If D be the number of degrees in the angle, we have $D = 45{\cdot}225$.

$$\begin{array}{r} 60\,)\;30 \\ \hline 60\,)\;13{\cdot}5 \\ \hline {\cdot}225 \end{array}$$

Let θ be the number of radians in the given angle, then

$$\frac{\theta}{\pi} = \frac{45{\cdot}225}{180} = \frac{1{\cdot}005}{4};$$

$$\therefore \;\; \theta = \frac{\pi}{4} \times 1{\cdot}005 = \frac{3{\cdot}1416}{4} \times 1{\cdot}005$$

$$= {\cdot}7854 \times 1{\cdot}005 = {\cdot}789327.$$

Thus the radian measure is ·789327.

Example 2. Express in sexagesimal measure the angle whose radian measure is 1·309.

Let D be the number of degrees; then

$$\frac{D}{180} = \frac{1{\cdot}309}{\pi};$$

$$\therefore \;\; D = \frac{180 \times 1{\cdot}309}{3{\cdot}1416} = \frac{180 \times 1309 \times 10}{31416}$$

$$= \frac{180 \times 10}{24} = 75.$$

Thus the angle is 75°.

EXAMPLES. VII. a.

[*Unless otherwise stated* $\pi = 3{\cdot}1416$.

It should be noticed that $31416 = 8 \times 3 \times 7 \times 11 \times 17$.]

Express in radian measure as fractions of π:

1. 45°. 2. 30°. 3. 105°. 4. 22° 30′.
5. 18°. 6. 57° 30′. 7. 14° 24′. 8. 78° 45′.

Find numerically the radian measure of the following angles:

9. 25° 50′. 10. 37° 30′. 11. 82° 30′.
12. 68° 45′. 13. 157° 30′. 14. 52° 30′.

Express in sexagesimal measure:

15. $\frac{3\pi}{4}$. 16. $\frac{7\pi}{45}$. 17. $\frac{5\pi}{27}$. 18. $\frac{5\pi}{24}$.

19. $\cdot 3927\rho$. 20. $\cdot 5236\rho$. 21. $\cdot 6545\rho$. 22. $2\cdot 8798\rho$.

Taking $\pi = \frac{22}{7}$, find the radian measure of:

23. $36° 32' 24''$. 24. $70° 33' 36''$.

25. $116° 2' 45\cdot 6''$. 26. $171° 41' 50\cdot 4''$.

27. Taking $\frac{1}{\pi} = \cdot 31831$, shew that a radian contains 206265 seconds approximately.

28. Shew that a second is approximately equal to $\cdot 0000048$ of a radian.

67. The angles $\frac{\pi}{4}, \frac{\pi}{3}, \frac{\pi}{6}$ are the equivalents in radian measure of the angles 45°, 60°, 30° respectively.

Hence the results of Arts. 34 and 35 may be written as follows:

$$\sin\frac{\pi}{4} = \frac{1}{\sqrt{2}}, \qquad \cos\frac{\pi}{4} = \frac{1}{\sqrt{2}}, \qquad \tan\frac{\pi}{4} = 1;$$

$$\sin\frac{\pi}{3} = \frac{\sqrt{3}}{2}, \qquad \cos\frac{\pi}{3} = \frac{1}{2}, \qquad \tan\frac{\pi}{3} = \sqrt{3};$$

$$\sin\frac{\pi}{6} = \frac{1}{2}, \qquad \cos\frac{\pi}{6} = \frac{\sqrt{3}}{2}, \qquad \tan\frac{\pi}{6} = \frac{1}{\sqrt{3}}.$$

Example. Find the value of

$$3\tan^2\frac{\pi}{6} + \frac{4}{3}\cos^2\frac{\pi}{6} - \frac{1}{2}\cot^3\frac{\pi}{4} - \frac{2}{3}\sin^2\frac{\pi}{3} + \frac{1}{8}\sec^4\frac{\pi}{3}.$$

$$\text{The value} = 3\left(\frac{1}{\sqrt{3}}\right)^2 + \frac{4}{3}\left(\frac{\sqrt{3}}{2}\right)^2 - \frac{1}{2}(1)^3 - \frac{2}{3}\left(\frac{\sqrt{3}}{2}\right)^2 + \frac{1}{8}(2)^4$$

$$= \left(3 \times \frac{1}{3}\right) + \left(\frac{4}{3} \times \frac{3}{4}\right) - \frac{1}{2} - \left(\frac{2}{3} \times \frac{3}{4}\right) + \left(\frac{1}{8} \times 16\right)$$

$$= 1 + 1 - \frac{1}{2} - \frac{1}{2} + 2 = 3.$$

68. When expressed in radian measure the complement of θ is $\frac{\pi}{2}-\theta$, and corresponding to the formulæ of Art. 38 we now have relations of the form

$$\sin\left(\frac{\pi}{2}-\theta\right)=\cos\theta, \qquad \tan\left(\frac{\pi}{2}-\theta\right)=\cot\theta.$$

Example. Prove that

$$(\cot\theta+\tan\theta)\cot\left(\frac{\pi}{2}-\theta\right)=\operatorname{cosec}^2\left(\frac{\pi}{2}-\theta\right).$$

$$\begin{aligned}\text{The first side}&=(\cot\theta+\tan\theta)\tan\theta\\&=\cot\theta\tan\theta+\tan^2\theta\\&=1+\tan^2\theta=\sec^2\theta\\&=\operatorname{cosec}^2\left(\frac{\pi}{2}-\theta\right).\end{aligned}$$

69. By means of Euc. I. 32, it is easy to find the number of radians in each angle of a regular polygon.

Example. Express in radians the interior angle of a regular polygon which has n sides.

The sum of the *exterior* angles $=4$ right angles. [Euc. I. 32 Cor.]

Let θ be the number of radians in an exterior angle; then

$$n\theta=2\pi, \text{ and therefore } \theta=\frac{2\pi}{n}.$$

But interior angle $=$ two right angles $-$ exterior angle

$$=\pi-\theta=\pi-\frac{2\pi}{n}.$$

Thus each interior angle $=\dfrac{(n-2)\pi}{n}$.

EXAMPLES. VII. b.

Find the numerical value of

1. $\sin\frac{\pi}{3}\cos\frac{\pi}{6}\cot\frac{\pi}{4}$.

2. $\tan\frac{\pi}{6}\cot\frac{\pi}{3}\cos\frac{\pi}{4}$.

3. $\frac{1}{2}\cos\frac{\pi}{3}+2\operatorname{cosec}\frac{\pi}{6}$.

4. $2\sin\frac{\pi}{4}+\frac{1}{2}\sec\frac{\pi}{4}$.

Find the numerical value of

5. $\cot^2\frac{\pi}{6}+4\cos^2\frac{\pi}{4}+3\sec^2\frac{\pi}{6}$.

6. $3\tan^2\frac{\pi}{6}-\frac{1}{3}\sin^2\frac{\pi}{3}-\frac{1}{2}\operatorname{cosec}^2\frac{\pi}{4}+\frac{4}{3}\cos^2\frac{\pi}{6}$.

7. $\left(\sin\frac{\pi}{6}+\cos\frac{\pi}{6}\right)\left(\sin\frac{\pi}{3}-\cos\frac{\pi}{3}\right)\sec\frac{\pi}{3}$.

Prove the following identities:

8. $\sin\theta\sec\left(\frac{\pi}{2}-\theta\right)-\cot\theta\cot\left(\frac{\pi}{2}-\theta\right)=0.$

9. $\sin^2\left(\frac{\pi}{2}-\theta\right)\operatorname{cosec}\theta-\tan^2\left(\frac{\pi}{2}-\theta\right)\sin\theta=0.$

10. $\dfrac{\sin^2\left(\frac{\pi}{2}-\theta\right)}{\operatorname{cosec}\theta}\cdot\dfrac{\sec\theta}{\cot\left(\frac{\pi}{2}-\theta\right)}=\cos^2\theta.$

11. $\tan\theta+\tan\left(\frac{\pi}{2}-\theta\right)=\sec\theta\sec\left(\frac{\pi}{2}-\theta\right)$.

12. $\sec^2\theta+\sec^2\left(\frac{\pi}{2}-\theta\right)=(1+\tan^2\theta)\sec^2\left(\frac{\pi}{2}-\theta\right)$.

13. Find the number of radians in each exterior angle of
(1) a regular octagon, (2) a regular quindecagon.

14. Find the number of radians in each interior angle of
(1) a regular dodecagon, (2) a regular heptagon.

15. Shew that

$$\tan^2\frac{\pi}{3}-\cot^2\frac{\pi}{3}=\frac{\cos^2\frac{\pi}{6}-\cos^2\frac{\pi}{3}}{\cos^2\frac{\pi}{3}\cos^2\frac{\pi}{6}}.$$

16. Shew that the sum of the squares of

$$\sin\theta+\sin\left(\frac{\pi}{2}-\theta\right) \text{ and } \cos\theta-\cos\left(\frac{\pi}{2}-\theta\right)$$

is equal to 2.

70. *To prove that the radian measure of any angle at the centre of a circle is expressed by the fraction* $\dfrac{\textit{subtending arc}}{\textit{radius}}$.

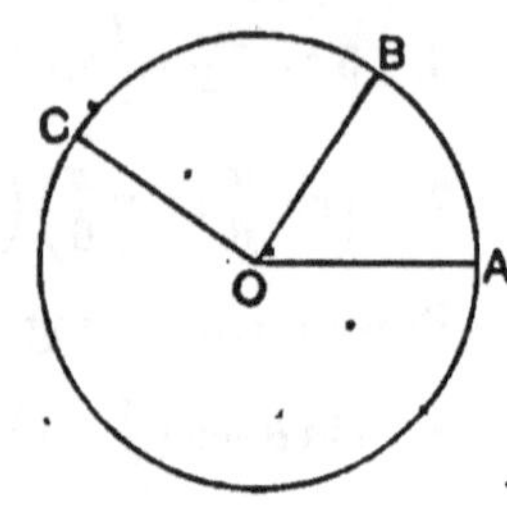

Let AOC be any angle at the centre of a circle, and AOB a radian; then

radian measure of $\angle AOC$

$$=\frac{\angle AOC}{\angle AOB}$$

$$=\frac{\text{arc } AC}{\text{arc } AB}$$

$$=\frac{\text{arc } AC}{\text{radius}},$$

since arc $AB=$radius;

that is, the radian measure of $\angle AOC=\dfrac{\text{subtending arc}}{\text{radius}}$.

71. If a be the length of the arc which subtends an angle of θ radians at the centre of a circle of radius r, we have seen in the preceding article that

$$\theta=\frac{a}{r}, \text{ and therefore } a=r\theta.$$

The fraction $\dfrac{\text{arc}}{\text{radius}}$ is usually called the *circular measure* of the angle at the centre of the circle subtended by the arc.

The *circular measure* of an angle is therefore equal to its *radian measure*, each denoting the *number of radians* contained in the angle. We have preferred to use the term *radian measure* exclusively, in order to keep prominently in view the unit of measurement, namely the radian.

NOTE. The term *circular measure* is a survival from the times when Mathematicians spoke of the trigonometrical functions of the *arc*. [See page 80.]

Example 1. Find the angle subtended by an arc of 7·5 feet at the centre of a circle whose radius is 5 yards.

Let the angle contain θ radians; then

$$\theta=\frac{\text{arc}}{\text{radius}}=\frac{7\cdot5}{15}=\frac{1}{2}.$$

Thus the angle is half a radian.

Example 2. In running a race at a uniform speed on a circular course, a man in each minute traverses an arc of a circle which subtends $2\frac{6}{7}$ radians at the centre of the course. If each lap is 792 yards, how long does he take to run a mile? $\left[\pi=\dfrac{22}{7}\right]$.

Let r yards be the radius of the circle; then

$$2\pi r = \text{circumference} = 792;$$

$$\therefore\ r=\frac{792}{2\pi}=\frac{792\times 7}{2\times 22}=126.$$

Let a yards be the length of the arc traversed in each minute; then from the formula $a=r\theta$,

$$a=126\times 2\tfrac{6}{7}=\frac{126\times 20}{7}=360;$$

that is, the man runs 360 yds. in each minute.

$$\therefore\ \text{the time}=\frac{1760}{360}\ \text{or}\ \frac{44}{9}\ \text{minutes.}$$

Thus the time is 4 min. $53\frac{1}{3}$ sec.

Example 3. Find the radius of a globe such that the distance measured along its surface between two places on the same meridian whose latitudes differ by $1^\circ 10'$ may be 1 inch, reckoning that $\pi=\dfrac{22}{7}$.

Let the adjoining figure represent a section of the globe through the meridian on which the two places P and Q lie. Let O be the centre, and denote the radius by r inches.

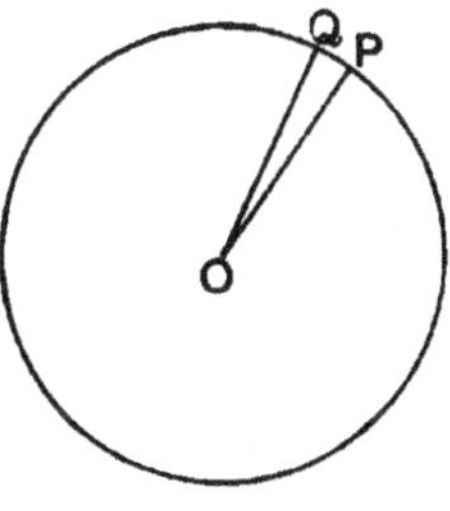

Now $\dfrac{\text{arc}\,PQ}{\text{radius}}=$ number of radians in $\angle POQ$;

but arc $PQ=1$ inch, and $\angle POQ=1^\circ 10'$;

$$\therefore\ \frac{1}{r}=\text{number of radians in } 1\tfrac{1}{6}^\circ$$

$$=1\tfrac{1}{6}\times\frac{\pi}{180}=\frac{7}{6}\times\frac{22}{7}\times\frac{1}{180}=\frac{11}{540};$$

whence $$r=\frac{540}{11}=49\tfrac{1}{11}.$$

Thus the radius is $49\frac{1}{11}$ inches.

EXAMPLES. VII. c.

1. Find the radian measure of the angle subtended by an arc of 1·6 yards at the centre of a circle whose radius is 24 feet.

2. An angle whose circular measure is ·73 subtends at the centre of a circle an arc of 219 feet; find the radius of the circle.

3. An angle at the centre of a circle whose radius is 2·5 yards is subtended by an arc of 7·5 feet; what is the angle?

4. What is the length of the arc which subtends an angle of 1·625 radians at the centre of a circle whose radius is 3·6 yards?

5. An arc of 17 yds. 1 ft. 3 in. subtends at the centre of a circle an angle of 1·9 radians; find the radius of the circle in inches.

6. The flywheel of an engine makes 35 revolutions in a second; how long will it take to turn through 5 radians? $\left[\pi = \frac{22}{7}\right]$.

7. The large hand of a clock is 2 ft. 4 in. long; how many inches does its extremity move in 20 minutes? $\left[\pi = \frac{22}{7}\right]$.

8. A horse is tethered to a stake; how long must the rope be in order that, when the horse has moved through 52·36 yards at the extremity of the rope, the angle traced out by the rope may be 75 degrees?

9. Find the length of an arc which subtends 1 minute at the centre of the earth, supposed to be a sphere of diameter 7920 miles.

10. Find the number of seconds in the angle subtended at the centre of a circle of radius 1 mile by an arc $5\frac{1}{2}$ inches long.

11. Two places on the same meridian are 145·2 miles apart; find their difference in latitude, taking $\pi = \frac{22}{7}$, and the earth's diameter as 7920 miles.

12. Find the radius of a globe such that the distance measured along its surface between two places on the same meridian whose latitudes differ by $1\frac{3}{11}^\circ$ may be 1 foot, taking $\pi = \frac{22}{7}$.

MISCELLANEOUS EXAMPLES. B.

1. Express in degrees the angle whose circular measure is ·15708.

2. If $C=90^\circ$, $A=30^\circ$, $c=110$, find b to two decimal places.

3. Find the number of degrees in the unit angle when the angle $\frac{12\pi}{25}$ is represented by $1\frac{3}{5}$.

4. What is the radius of the circle in which an arc of 1 inch subtends an angle of $1'$ at the centre?

5. Prove that

(1) $(\sin a+\cos a)(\tan a+\cot a)=\sec a+\operatorname{cosec} a$;

(2) $(\sqrt{3}+1)(3-\cot 30^\circ)=\tan^3 60^\circ-2\sin 60^\circ$.

6. Find the angle of elevation of the sun when a chimney 60 feet high throws a shadow $20\sqrt{3}$ yards long.

---

7. Prove the identities:

(1) $(\tan\theta+2)(2\tan\theta+1)=5\tan\theta+2\sec^2\theta$;

(2) $1+\dfrac{\cot^2 a}{1+\operatorname{cosec} a}=\operatorname{cosec} a$.

8. One angle of a triangle is 45° and another is $\frac{5\pi}{8}$ radians; express the third angle both in sexagesimal and radian measure.

9. The number of degrees in an angle exceeds 14 times the number of radians in it by 51. Taking $\pi=\frac{22}{7}$, find the sexagesimal measure of the angle.

10. If $B=30^\circ$, $C=90^\circ$, $b=6$, find a, c, and the perpendicular from C on the hypotenuse.

11. Shew that

(1) $\cot\theta+\cot\left(\frac{\pi}{2}-\theta\right)=\operatorname{cosec}\theta\operatorname{cosec}\left(\frac{\pi}{2}-\theta\right)$;

(2) $\operatorname{cosec}^2\theta+\operatorname{cosec}^2\left(\frac{\pi}{2}-\theta\right)=\operatorname{cosec}^2\theta\operatorname{cosec}^2\left(\frac{\pi}{2}-\theta\right)$.

12. The angle of elevation of the top of a pillar is 30°, and on approaching 20 feet nearer it is 60°: find the height of the pillar.

---

13. Shew that $\tan^2 A - \sin^2 A = \sin^4 A \sec^2 A$.

14. In a triangle the angle A is $3x$ degrees, the angle B is x grades, and the angle C is $\frac{\pi x}{300}$ radians: find the number of degrees in each of the angles.

15. Find the numerical value of

$$\sin^3 60° \cot 30° - 2 \sec^2 45° + 3 \cos 60° \tan 45° - \tan^2 60°.$$

16. Prove the identities:

(1) $(1 + \tan A)^2 + (1 + \cot A)^2 = (\sec A + \operatorname{cosec} A)^2$;

(2) $(\sec a - 1)^2 - (\tan a - \sin a)^2 = (1 - \cos a)^2$.

17. Which of the following statements is possible and which impossible?

(1) $\operatorname{cosec} \theta = \frac{a^2 + b^2}{2ab}$; (2) $2 \sin \theta = a + \frac{1}{a}$.

18. A balloon leaves the earth at the point A and rises at a uniform pace. At the end of 1·5 minutes an observer stationed at a distance of 660 feet from A finds the angular elevation of the balloon to be 60°; at what rate in miles per hour is the balloon rising?

---

19. Find the number of radians in the angles of a triangle which are in arithmetical progression, the least angle being 36°.

20. Shew that

$$\sin^2 a \sec^2 \beta + \tan^2 \beta \cos^2 a = \sin^2 a + \tan^2 \beta.$$

21. In the triangle ABC if $A = 42°$, $B = 116° 33'$, find the perpendicular from C upon AB produced; given

$$c = 55, \quad \tan 42° = \cdot 9, \quad \tan 63° 27' = 2.$$

22. Prove the identities:

(1) $\cot \alpha + \dfrac{\sin \alpha}{1+\cos \alpha} = \operatorname{cosec} \alpha$;

(2) $\operatorname{cosec} \alpha (\sec \alpha - 1) - \cot \alpha (1 - \cos \alpha) = \tan \alpha - \sin \alpha$.

23. Shew that $\left(\dfrac{1+\cot 60^\circ}{1-\cot 60^\circ}\right)^2 = \dfrac{1+\cos 30^\circ}{1-\cos 30^\circ}$.

24. A man walking N.W. sees a windmill which bears N. 15° W. In half-an-hour he reaches a place which he knows to be W. 15° S. of the windmill and a mile away from it. Find his rate of walking and his distance from the windmill at the first observation.

---

25. Find the number of radians in the complement of $\dfrac{3\pi}{8}$.

26. Solve the equations:

(1) $3 \sin \theta + 4 \cos^2\theta = 4\frac{1}{2}$; (2) $\tan \theta + \sec 30^\circ = \cot \theta$.

27. If $5 \tan \alpha = 4$, find the value of

$$\frac{5 \sin \alpha - 3 \cos \alpha}{\sin \alpha + 2 \cos \alpha}.$$

28. Prove that

$$\frac{1 - \sin A \cos A}{\cos A (\sec A - \operatorname{cosec} A)} \times \frac{\sin^2 A - \cos^2 A}{\sin^3 A + \cos^3 A} = \sin A.$$

29. Find the distance of an observer from the top of a cliff which is 195·2 yards high, given that the angle of elevation is 77° 26′, and that sin 77° 26′ = ·976.

30. A horse is tethered to a stake by a rope 27 feet long. If the horse moves along the circumference of a circle always keeping the rope tight, find how far it will have gone when the rope has traced out an angle of 70°. $\left[\pi = \dfrac{22}{7}\right]$.

CHAPTER VIII.

TRIGONOMETRICAL RATIOS OF ANGLES OF ANY MAGNITUDE.

72. In the present chapter we shall find it necessary to take account not only of the magnitude of straight lines, but also of the direction in which they are measured.

Let O be a fixed point in a horizontal line XX', then the position of any other point P in the line, whose distance from O is a given length a, will not be determined unless we know on which side of O the point P lies.

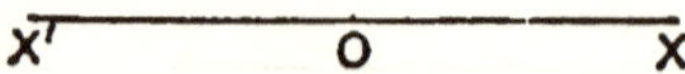

But there will be no ambiguity if it is agreed that distances measured in one direction are positive and distances measured in the opposite direction are negative.

Hence the following **Convention of Signs** is adopted:

lines measured from O to the right are positive,

lines measured from O to the left are negative.

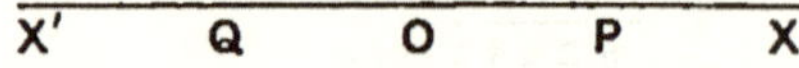

Thus in the above figure, if P and Q are two points on the line XX' at a distance a from O, their positions are indicated by the statements

$$OP = +a, \quad OQ = -a.$$

73. A similar *convention of signs* is used in the case of a plane surface.

Let O be any point in the plane; through O draw two straight lines XX' and YY' in the horizontal and vertical direction respectively, thus dividing the plane into four *quadrants*.

Then it is universally agreed to consider that

(1) *horizontal lines to the right of YY' are positive,*
horizontal lines to the left of YY' are negative;

(2) *vertical lines above XX' are positive,*
vertical lines below XX' are negative.

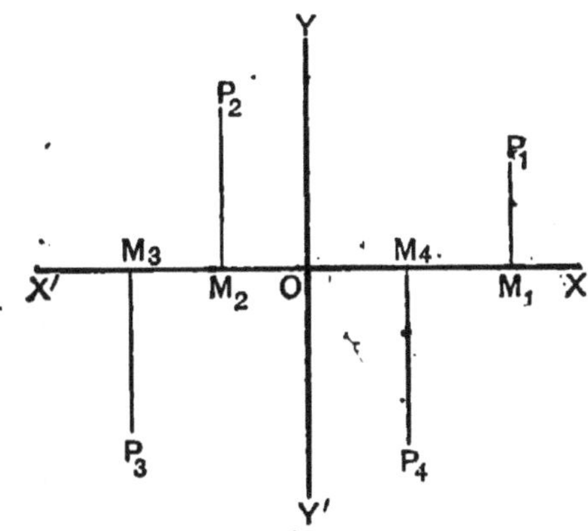

Thus OM_1, OM_4 are positive, OM_2, OM_3 are negative;
M_1P_1, M_2P_2 are positive, M_3P_3, M_4P_4 are negative.

74. Convention of Signs for Angles. In Art. 2 an angle has been defined as the amount of revolution which the radius vector makes in passing from its initial to its final position.

In the adjoining figure the straight line OP may be supposed to have arrived at its present position from the position occupied by OA by revolution about the point O in *either* of the two directions indicated by the arrows. The angle AOP may thus be regarded in two senses according as we suppose the revolution to have been in the same direction as the hands of a clock or in the opposite direction. To distinguish between these cases we adopt the following convention:

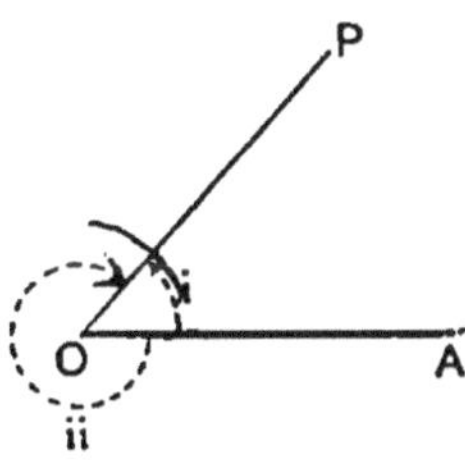

when the revolution of the radius vector is counter-clockwise the angle is positive,

when the revolution is clockwise the angle is negative.

Trigonometrical Ratios of any Angle.

75. Let XX' and YY' be two straight lines intersecting at right angles in O, and let a radius vector starting from OX revolve in either direction till it has traced out an angle A, taking up the position OP.

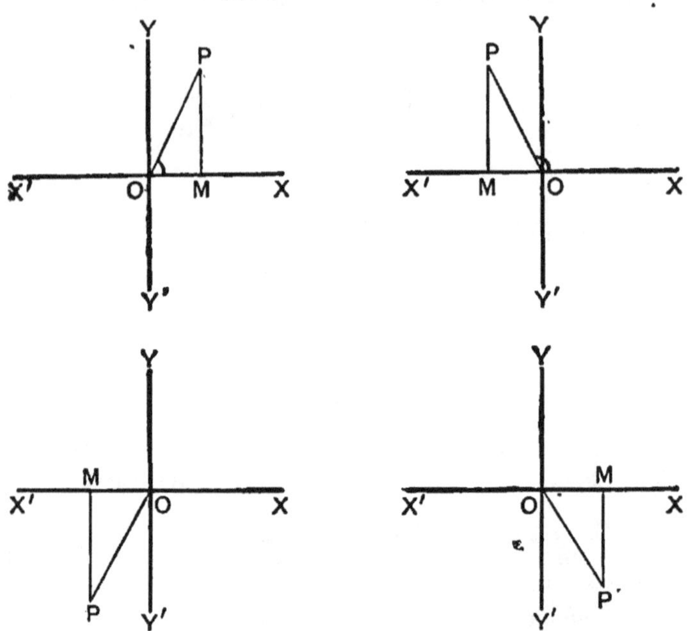

From P draw PM perpendicular to XX'; then in the right-angled triangle OPM, due regard being paid to the signs of the lines,

$$\sin A = \frac{MP}{OP}, \quad \operatorname{cosec} A = \frac{OP}{MP},$$

$$\cos A = \frac{OM}{OP}, \quad \sec A = \frac{OP}{OM},$$

$$\tan A = \frac{MP}{OM}, \quad \cot A = \frac{OM}{MP}.$$

The radius vector OP which only fixes the boundary of the angle is considered to be always positive.

From these definitions it will be seen that any trigonometrical function will be positive or negative according as the fraction which expresses its value has the numerator and denominator of the same sign or of opposite sign.

76. The four diagrams of the last article may be conveniently included in one.

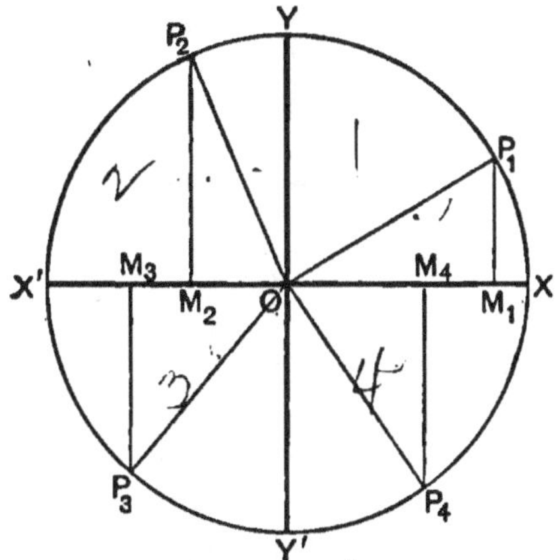

With centre O and fixed radius let a circle be described; then the diameters XX' and YY' divide the circle into four *quadrants* XOY, YOX', $X'OY'$, $Y'OX$, named *first, second, third, fourth* respectively.

Let the positions of the radius vector in the four quadrants be denoted by OP_1, OP_2, OP_3, OP_4, and let perpendiculars P_1M_1, P_2M_2, P_3M_3, P_4M_4 be drawn to XX'; then it will be seen that in the first quadrant all the lines are positive and therefore all the functions of A are positive.

In the second quadrant, OP_2 and M_2P_2 are positive, OM_2 is negative; hence $\sin A$ is positive, $\cos A$ and $\tan A$ are negative.

In the third quadrant, OP_3 is positive, OM_3 and M_3P_3 are negative; hence $\tan A$ is positive, $\sin A$ and $\cos A$ are negative.

In the fourth quadrant, OP_4 and OM_4 are positive, M_4P_4 is negative; hence $\cos A$ is positive, $\sin A$ and $\tan A$ are negative.

77. The following diagrams shew the *signs* of the trigonometrical functions in the four quadrants. It will be sufficient to consider the three principal functions only.

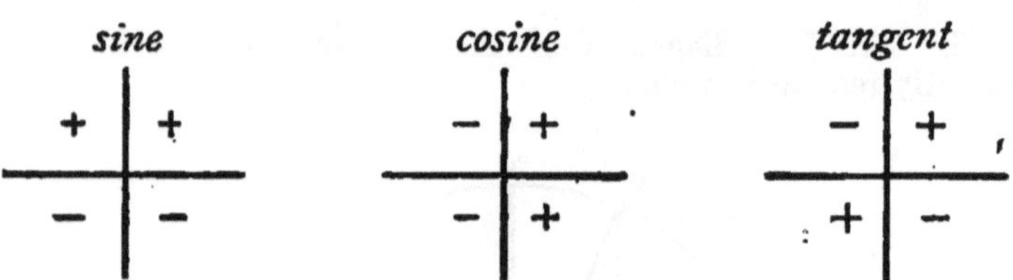

The diagram below exhibits the same results in another useful form.

sine positive *cosine negative* *tangent negative*	**all the ratios positive**
tangent positive *sine negative* *cosine negative*	**cosine positive** *sine negative* *tangent negative*

78. When an angle is increased or diminished by any multiple of four right angles, the radius vector is brought back again into the same position after one or more revolutions. There are thus an infinite number of angles which have the same boundary line. Such angles are called **coterminal angles.**

If n is *any* integer, all the angles coterminal with A may be represented by $n \,.\, 360^\circ + A$. Similarly, in radian measure all the angles coterminal with θ may be represented by $2n\pi + \theta$.

From the definitions of Art. 75, we see that the position of the boundary line is alone sufficient to determine the trigonometrical ratios of the angle; hence *all coterminal angles have the same trigonometrical ratios.*

For instance, $\sin(n \,.\, 360^\circ + 45^\circ) = \sin 45^\circ = \dfrac{1}{\sqrt{2}}$;

and $$\cos\left(2n\pi + \frac{\pi}{6}\right) = \cos\frac{\pi}{6} = \frac{\sqrt{3}}{2}.$$

Example. Draw the boundary lines of the angles 780°, −130°, −400°, and in each case state which of the trigonometrical functions are negative.

(1) Since $780 = (2 \times 360) + 60$, the radius vector has to make two complete revolutions and then turn through 60°. Thus the boundary line is in the first quadrant, so that all the functions are positive.

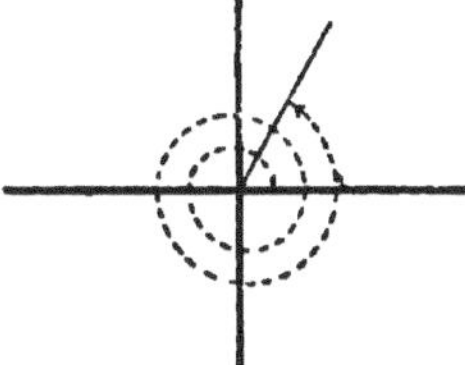

(2) Here the radius vector has to revolve through 130° in the negative direction. The boundary line is thus in the third quadrant, and since OM and MP are negative, the sine, cosine, cosecant, and secant are negative.

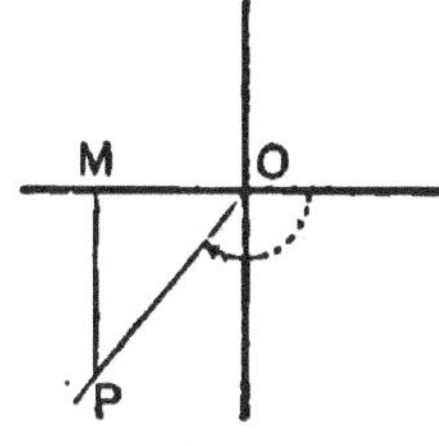

(3) Since $-400 = -(360 + 40)$, the radius vector has to make one complete revolution in the negative direction and then turn through 40°. The boundary line is thus in the fourth quadrant, and since MP is negative, the sine, tangent, cosecant, and cotangent are negative.

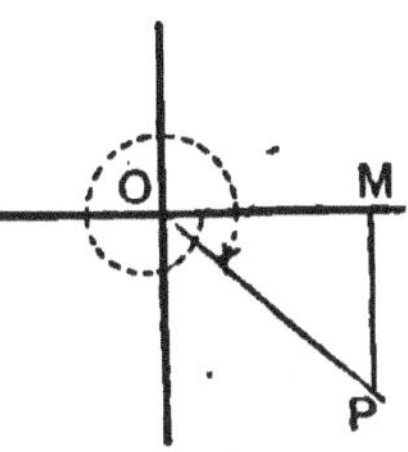

EXAMPLES. VIII. a.

State the quadrant in which the radius vector lies after describing the following angles:

1. 135°. 2. 265°. 3. −315°. 4. −120°.

5. $\frac{2\pi}{3}$. 6. $\frac{5\pi}{6}$. 7. $\frac{10\pi}{3}$. 8. $-\frac{11\pi}{4}$.

For each of the following angles state which of the three principal trigonometrical functions are positive.

9. 470°. 10. 330°. 11. 575°.

12. −230°. 13. −620°. 14. −1200°.

15. $-\frac{4\pi}{3}$. 16. $\frac{13\pi}{6}$. 17. $-\frac{13\pi}{6}$.

In each of the following cases write down the smallest positive coterminal angle, and the value of the expression.

18. $\sin 420°$. **19.** $\cos 390°$. **20.** $\tan(-315°)$.

21. $\sec 405°$. **22.** $\operatorname{cosec}(-330°)$. **23.** $\operatorname{cosec} 4380°$.

24. $\cot\frac{17\pi}{4}$. **25.** $\sec\frac{25\pi}{3}$. **26.** $\tan\left(-\frac{5\pi}{3}\right)$.

79. Since the definitions of the functions given in Art. 75 are applicable to angles of any magnitude, positive or negative, it follows that all relations derived from these definitions must be true universally. Thus we shall find that the fundamental formulæ given in Art. 29 hold in all cases; that is,

$$\sin A \times \operatorname{cosec} A = 1, \quad \cos A \times \sec A = 1, \quad \tan A \times \cot A = 1;$$

$$\tan A = \frac{\sin A}{\cos A}, \qquad \cot A = \frac{\cos A}{\sin A};$$

$$\sin^2 A + \cos^2 A = 1,$$

$$1 + \tan^2 A = \sec^2 A,$$

$$1 + \cot^2 A = \operatorname{cosec}^2 A.$$

It will be useful practice for the student to test the truth of these formulæ for different positions of the boundary line of the angle A. We shall give one illustration.

80. Let the radius vector revolve from its initial position OX till it has traced out an angle A and come into the position OP indicated in the figure. Draw PM perpendicular to XX'.

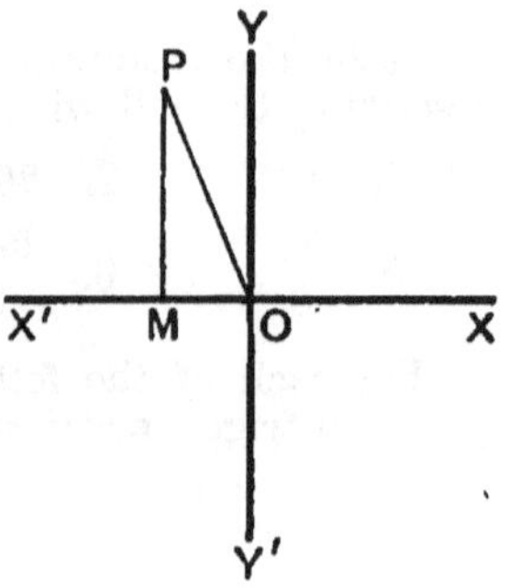

In the right-angled triangle OMP,

$$MP^2 + OM^2 = OP^2 \text{.........(1).}$$

Divide each term by OP^2; thus

$$\left(\frac{MP}{OP}\right)^2 + \left(\frac{OM}{OP}\right)^2 = 1;$$

that is, $\sin^2 A + \cos^2 A = 1$.

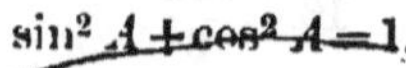

Divide each term of (1) by OM^2; thus

$$\left(\frac{MP}{OM}\right)^2+1=\left(\frac{OP}{OM}\right)^2;$$

that is, $$\tan^2 A+1=\sec^2 A.$$

Divide each term of (1) by MP^2; thus

$$1+\left(\frac{OM}{MP}\right)^2=\left(\frac{OP}{MP}\right)^2;$$

that is, $$1+\cot^2 A=\operatorname{cosec}^2 A.$$

It thus appears that the truth of these relations depends only on the statement $OP^2=MP^2+OM^2$ in the right-angled triangle OMP, and this will be the case in whatever quadrant OP lies.

NOTE. OM^2 is positive, although the line OM in the figure is negative.

81. In the statement $\cos A=\sqrt{1-\sin^2 A}$, either the positive or the negative sign may be placed before the radical. The sign of the radical hitherto has always been taken positively, because we have restricted ourselves to the consideration of acute angles. It will sometimes be necessary to examine which sign must be taken before the radical in any particular case.

Example 1. Given $\cos 126° 53'=-\frac{3}{5}$, find $\sin 126° 53'$ and $\cot 126° 53'$.

Since $\sin^2 A+\cos^2 A=1$ for angles of any magnitude, we have

$$\sin A=\pm\sqrt{1-\cos^2 A}.$$

Denote $126° 53'$ by A; then the boundary line of A lies in the second quadrant, and therefore $\sin A$ is positive. Hence the sign $+$ must be placed before the radical;

$$\therefore\ \sin 126° 53'=+\sqrt{1-\frac{9}{25}}=+\sqrt{\frac{16}{25}}=\frac{4}{5};$$

$$\cot 126° 53'=\frac{\cos 126° 53'}{\sin 126° 53'}=\left(-\frac{3}{5}\right)\div\left(\frac{4}{5}\right)=-\frac{3}{4}.$$

The same results may also be obtained by the method used in the following example. The appropriate signs of the lines are shewn in the figure.

Example 2. If $\tan A = -\frac{15}{8}$, find $\sin A$ and $\cos A$.

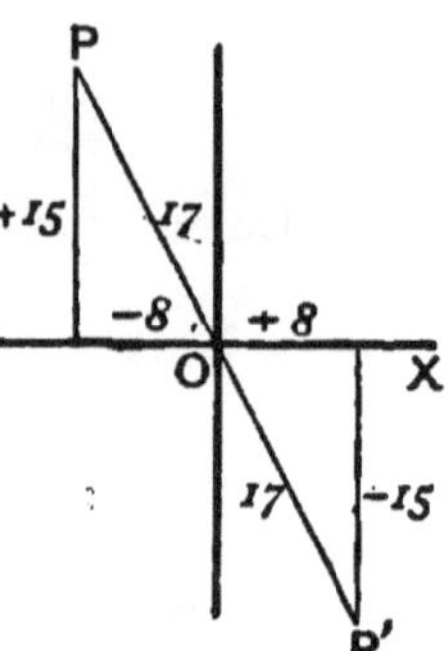

The boundary line of A will lie either in the second or in the fourth quadrant, as OP or OP'. In either position,

the radius vector $= \sqrt{(15)^2 + (8)^2}$
$= \sqrt{289} = 17.$

Hence $\sin XOP = \frac{15}{17}$, $\cos XOP = -\frac{8}{17}$;

and $\sin XOP' = -\frac{15}{17}$, $\cos XOP' = \frac{8}{17}$.

Thus corresponding to $\tan A$, there are two values of $\sin A$ and two values of $\cos A$. If however it is known in which quadrant the boundary line of A lies, $\sin A$ and $\cos A$ have each a single value.

EXAMPLES. VIII. b.

1. Given $\sin 120° = \frac{\sqrt{3}}{2}$, find $\tan 120°$.

2. Given $\tan 135° = -1$, find $\sin 135°$.

3. Find $\cos 240°$, given that $\tan 240° = \sqrt{3}$.

4. If $A = 202° \, 37'$ and $\sin A = -\frac{5}{13}$, find $\cos A$ and $\cot A$.

5. If $A = 143° \, 8'$ and $\operatorname{cosec} A = 1\frac{2}{3}$, find $\sec A$ and $\tan A$.

6. If $A = 216° \, 52'$ and $\cos A = -\frac{4}{5}$, find $\cot A$ and $\sin A$.

7. Given $\sec \frac{2\pi}{3} = -2$, find $\sin \frac{2\pi}{3}$ and $\cot \frac{2\pi}{3}$.

8. Given $\sin \frac{5\pi}{4} = -\frac{1}{\sqrt{2}}$, find $\tan \frac{5\pi}{4}$ and $\sec \frac{5\pi}{4}$.

9. If $\cos A = \frac{12}{13}$, find $\sin A$ and $\tan A$.

CHAPTER IX.

VARIATIONS OF THE TRIGONOMETRICAL FUNCTIONS.

82. A CAREFUL perusal of the following remarks will render the explanations which follow more easily intelligible.

Consider the fraction $\frac{a}{x}$ in which the numerator a has a *certain fixed value* and the denominator x is a *quantity subject to change;* then it is clear that the smaller x becomes the larger does the value of the fraction $\frac{a}{x}$ become. For instance

$$\frac{a}{\frac{1}{10}}=10a, \quad \frac{a}{\frac{1}{1000}}=1000a, \quad \frac{a}{\frac{1}{10000000}}=10000000a.$$

By making the denominator x sufficiently small the value of the fraction $\frac{a}{x}$ can be made as large as we please; that is, as x approaches to the value 0, the fraction $\frac{a}{x}$ becomes infinitely great.

The symbol ∞ is used to express a quantity infinitely great, or more shortly *infinity*, and the above statement is concisely written

$$\textit{when } x=0, \textit{ the limit of } \frac{a}{x}=\infty.$$

Again, if x is a quantity which gradually increases and finally becomes infinitely large, the fraction $\frac{a}{x}$ becomes infinitely small; that is,

$$\textit{when } x=\infty, \textit{ the limit of } \frac{a}{x}=0.$$

83. Definition. If y is a function of x, and if when x approaches nearer and nearer to the fixed quantity a, the value of y approaches nearer and nearer to the fixed quantity b and can be made to differ from it by as little as we please, then b is called the **limiting value** or the **limit** of y when $x=a$.

84. Trigonometrical Functions of 0°.

Let XOP be an angle traced out by a radius vector OP of fixed length.

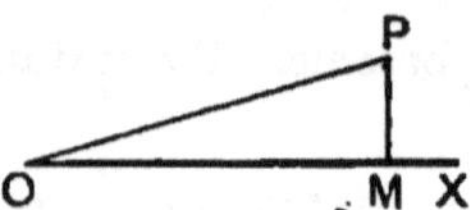

Draw PM perpendicular to OX; then

$$\sin POM=\frac{MP}{OP}.$$

If we suppose the angle POM to be gradually decreasing, MP will also gradually decrease, and if OP ultimately come into coincidence with OM the angle POM vanishes and $MP=0$.

Hence $$\sin 0^\circ=\frac{0}{OP}=0.$$

Again, $\cos POM=\frac{OM}{OP}$; but when the angle POM vanishes OP becomes coincident with OM.

Hence $$\cos 0^\circ=\frac{OM}{OM}=1.$$

Also when the angle POM vanishes,

$$\tan 0^\circ=\frac{0}{OM}=0.$$

And $$\operatorname{cosec} 0^\circ=\frac{1}{\sin 0^\circ}=\frac{1}{0}=\infty;$$

$$\sec 0^\circ=\frac{1}{\cos 0^\circ}=\frac{1}{1}=1;$$

$$\cot 0^\circ=\frac{1}{\tan 0^\circ}=\frac{1}{0}=\infty.$$

85. Trigonometrical Functions of 90° or $\frac{\pi}{2}$.

Let XOP be an angle traced out by a radius vector of fixed length.

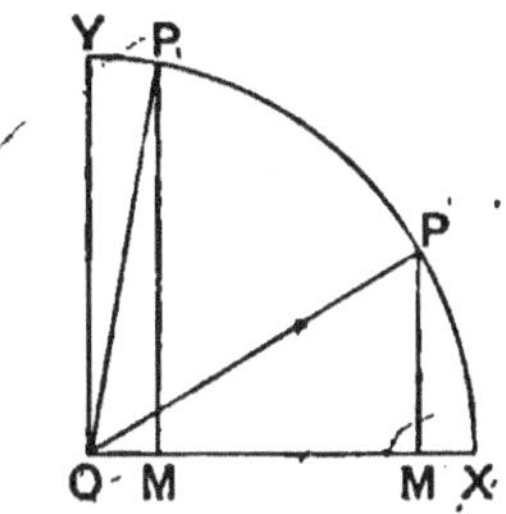

Draw PM perpendicular to OX, and OY perpendicular to OX.

By definition,

$$\sin POM = \frac{MP}{OP}, \qquad \cos POM = \frac{OM}{OP}, \qquad \tan POM = \frac{MP}{OM}.$$

If we suppose the angle POM to be gradually increasing, MP will gradually increase and OM decrease. When OP comes into coincidence with OY the angle POM becomes equal to 90°, and OM vanishes, while MP becomes equal to OP.

Hence
$$\sin 90° = \frac{OP}{OP} = 1;$$

$$\cos 90° = \frac{0}{OP} = 0;$$

$$\tan 90° = \frac{MP}{OM} = \frac{OP}{0} = \infty.$$

And
$$\cot 90° = \frac{1}{\tan 90°} = \frac{1}{\infty} = 0;$$

$$\sec 90° = \frac{1}{\cos 90°} = \frac{1}{0} = \infty;$$

$$\operatorname{cosec} 90° = \frac{1}{\sin 90°} = 1.$$

86. *To trace the changes in sign and magnitude of* sin A *as* A *increases from* 0° *to* 360°.

Let XX' and YY' be two straight lines intersecting at right angles in O.

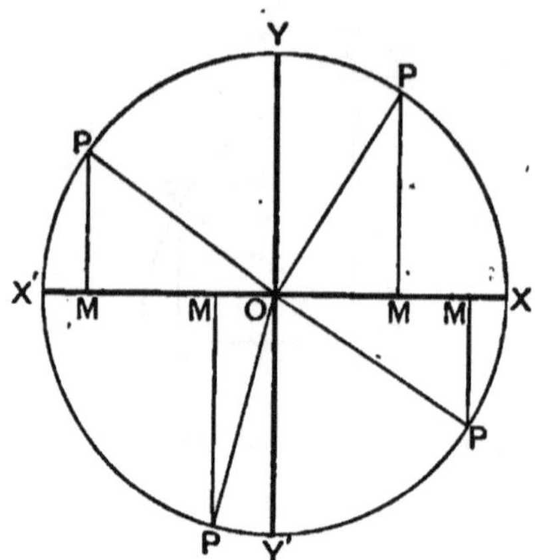

With centre O and any radius OP describe a circle, and suppose the angle A to be traced out by the revolution of OP through the four quadrants starting from OX.

Draw PM perpendicular to OX and let $OP=r$; then

$$\sin A = \frac{MP}{r},$$

and since r does not alter in sign or magnitude, we have only to consider the changes of MP as P moves round the circle.

When $A=0°$, $MP=0$, and $\sin 0° = \frac{0}{r} = 0$.

In the first quadrant, MP is positive and increasing;

$\therefore$ sin A is positive and increasing.

When $A=90°$, $MP=r$, and $\sin 90° = \frac{r}{r} = 1$.

In the second quadrant, MP is positive and decreasing;

$\therefore$ sin A is positive and decreasing.

When $A=180°$, $MP=0$, and $\sin 180° = \frac{0}{r} = 0$.

In the third quadrant, MP is negative and increasing;

$\therefore$ sin A is negative and increasing.

When $A = 270°$, MP is equal to r, but is negative; hence

$$\sin 270° = -\frac{r}{r} = -1.$$

In the fourth quadrant, MP is negative and decreasing;

$\therefore$ sin A is negative and decreasing.

When $A = 360°$, $MP = 0$, and $\sin 360° = \frac{0}{r} = 0$.

87. The results of the previous article are concisely shewn in the following diagram:

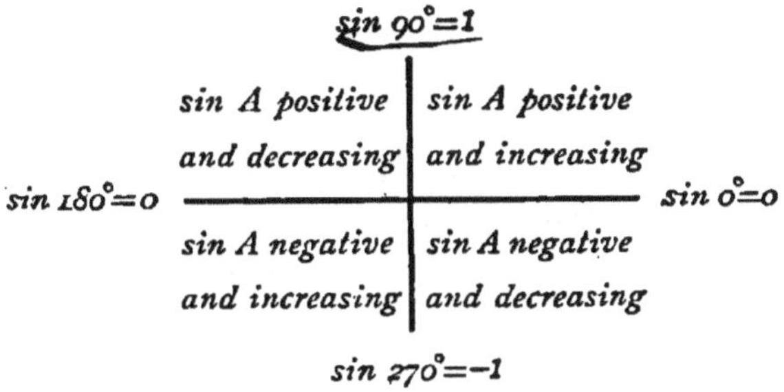

sin 90° = 1

	sin A positive and decreasing	*sin A positive and increasing*	
sin 180° = 0			*sin 0° = 0*
	sin A negative and increasing	*sin A negative and decreasing*	

sin 270° = −1

88. We leave as an exercise to the student the investigation of the changes in sign and magnitude of cos A as A increases from 0° to 360°. The following diagram exhibits these changes.

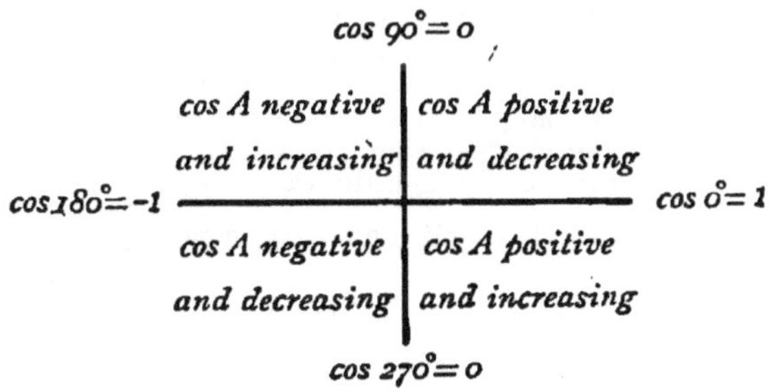

cos 90° = 0

	cos A negative and increasing	*cos A positive and decreasing*	
cos 180° = −1			*cos 0° = 1*
	cos A negative and decreasing	*cos A positive and increasing*	

cos 270° = 0

89. *To trace the changes in sign and magnitude of* tan A *as* A *increases from* 0° *to* 360°.

With the figure of Art. 86, $\tan A = \frac{MP}{OM}$, and its changes will therefore depend on those of MP and OM.

When $A=0°$, $MP=0$, $OM=r$; $\therefore \tan 0° = \frac{0}{r} = 0$.

In the first quadrant,

MP is positive and increasing,
OM is positive and decreasing;
$\therefore$ tan A is positive and increasing.

When $A=90°$, $MP=r$, $OM=0$; $\therefore \tan 90° = \frac{r}{0} = \infty$.

In the second quadrant,

MP is positive and decreasing,
OM is negative and increasing;
$\therefore$ tan A is negative and decreasing.

When $A=180°$, $MP=0$; $\therefore \tan 180° = 0$.

In the third quadrant,

MP is negative and increasing,
OM is negative and decreasing;
$\therefore$ tan A is positive and increasing.

When $A=270°$, $OM=0$; $\therefore \tan 270° = \infty$.

In the fourth quadrant,

MP is negative and decreasing,
OM is positive and increasing;
$\therefore$ tan A is negative and decreasing.

When $A=360°$, $MP=0$; $\therefore \tan 360° = 0$.

NOTE. When the numerator of a fraction changes continually from a small positive to a small negative quantity the fraction changes sign by passing through the value 0. When the denominator changes continually from a small positive to a small negative quantity the fraction changes sign by passing through the value ∞. For instance, as A passes through the value 90°, OM changes from a small positive to a small negative quantity, hence $\frac{OM}{OP}$, that is cos A, changes sign by passing through the value 0, while $\frac{PM}{OM}$, that is tan A, changes sign by passing through the value ∞.

90. The results of Art. 89 are shewn in the following diagram:

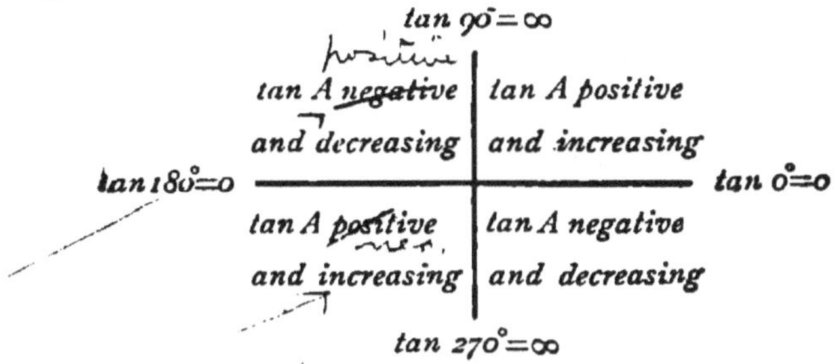

$\tan 90^\circ = \infty$

	tan A negative and decreasing	*tan A positive and increasing*	
$\tan 180^\circ = 0$			$\tan 0^\circ = 0$
	tan A positive and increasing	*tan A negative and decreasing*	

$\tan 270^\circ = \infty$

The student will now have no difficulty in tracing the variations in sign and magnitude of the other functions.

91. In Arts. 86 and 89 we have seen that the variations of the trigonometrical functions of the angle XOP depend on the position of P as P moves round the circumference of the circle. On this account the trigonometrical functions of an angle are called **circular functions**. This name is one that we shall use frequently.

EXAMPLES. IX.

Trace the changes in sign and magnitude of

1. $\cot A$, between 0° and 360°.
2. $\operatorname{cosec} \theta$, between 0 and π.
3. $\cos \theta$, between π and 2π.
4. $\tan A$, between -90° and -270°.
5. $\sec \theta$, between $\frac{\pi}{2}$ and $\frac{3\pi}{2}$.

Find the value of

6. $\cos 0^\circ \sin^2 270^\circ - 2 \cos 180^\circ \tan 45^\circ$.
7. $3 \sin 0^\circ \sec 180^\circ + 2 \operatorname{cosec} 90^\circ - \cos 360^\circ$.
8. $2 \sec^2 \pi \cos 0 + 3 \sin^3 \frac{3\pi}{2} - \operatorname{cosec} \frac{\pi}{2}$.
9. $\tan \pi \cos \frac{3\pi}{2} + \sec 2\pi - \operatorname{cosec} \frac{3\pi}{2}$.

Note on the old definitions of the Trigonometrical Functions.

Formerly, Mathematicians considered the trigonometrical functions with reference to the *arc* of a given circle, and did not regard them as *ratios* but as the *lengths* of certain straight lines drawn in relation to this arc.

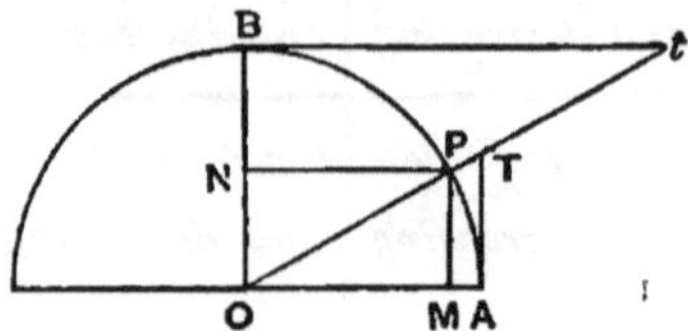

Let OA and OB be two radii of a circle at right angles, and let P be any point on the circumference. Draw PM and PN perpendicular to OA and OB respectively, and let the tangents at A and B meet OP produced in T and t respectively.

The *lines* PM, AT, OT, AM were named respectively the sine, tangent, secant, versed-sine of the arc AP, and PN, Bt, Ot, BN, which are the sine, tangent, secant, versed-sine of the complementary arc BP, were named respectively the cosine, cotangent, cosecant, coversed-sine of the arc AP.

As thus defined each trigonometrical function of the *arc* is equal to the corresponding function of the *angle*, which it subtends at the centre of the circle, multiplied by the radius. Thus

$$\frac{AT}{OA} = \tan POA\text{; that is, } AT = OA \times \tan POA\text{;}$$

and
$$\frac{Ot}{OB} = \sec BOP = \operatorname{cosec} POA\text{; that is, } Ot = OB \times \operatorname{cosec} POA.$$

The values of the functions of the arc therefore depended on the length of the radius of the circle as well as on the angle subtended by the arc at the centre of the circle, so that in Tables of the functions it was necessary to state the magnitude of the radius.

The names of the trigonometrical functions and the abbreviations for them now in use were introduced by different Mathematicians chiefly towards the end of the sixteenth and during the seventeenth century, but were not generally employed until their re-introduction by Euler. The development of the science of Trigonometry may be considered to date from the publication in 1748 of Euler's *Introductio in analysin Infinitorum.*

The reader will find some interesting information regarding the progress of Trigonometry in Ball's *Short History of Mathematics.*

MISCELLANEOUS EXAMPLES. C.

1. Draw the boundary lines of the angles whose tangent is equal to $-\frac{3}{4}$, and find the cosine of these angles.

2. Shew that

$$\cos A\,(2\sec A+\tan A)(\sec A-2\tan A)=2\cos A-3\tan A.$$

3. Given $C=90^\circ$, $b=10{\cdot}5$, $c=21$, solve the triangle.

4. If $\sec A=-\frac{25}{7}$, and A lies between 180° and 270°, find $\cot A$.

5. The latitude of Bombay is 19° N.: find its distance from the equator, taking the diameter of the earth to be 7920 miles.

6. From the top of a cliff 200 ft. high, the angles of depression of two boats due east of the observer are 34° 30′ and 18° 40′: find their distance apart, given

$$\cot 34^\circ\,30'=1{\cdot}455, \qquad \cot 18^\circ\,40'=2{\cdot}96.$$

7. If A lies between 180° and 270°, and $3\tan A=4$, find the value of $2\cot A-5\cos A+\sin A$.

8. Find, correct to three decimal places, the radius of a circle in which an arc 15 inches long subtends at the centre an angle of 71° 36′ 3·6″.

9. Shew that

$$\frac{\tan^3\theta}{1+\tan^2\theta}+\frac{\cot^3\theta}{1+\cot^2\theta}=\frac{1-2\sin^2\theta\cos^2\theta}{\sin\theta\cos\theta}.$$

10. The angle of elevation of the top of a tower is 68° 11′, and a flagstaff 24 ft. high on the summit of the tower subtends an angle of 2° 10′ at the observer's eye. Find the height of the tower, given

$$\tan 70^\circ\,21'=2{\cdot}8, \qquad \cot 68^\circ\,11'={\cdot}4.$$

CHAPTER X.

CIRCULAR FUNCTIONS OF CERTAIN ALLIED ANGLES.

92. Circular Functions of 180° – A.

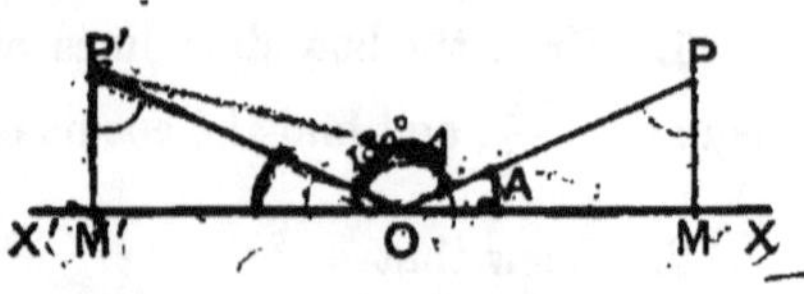

Take any straight line XOX', and let a radius vector starting from OX revolve until it has traced the angle A, taking up the position OP.

Again, let the radius vector starting from OX revolve through 180° into the position OX' and then *back again* through an angle A taking up the final position OP'. Thus XOP' is the angle $180° - A$.

From P and P' draw PM and $P'M'$ perpendicular to XX'; then by Euc. I. 26 the triangles OPM and $OP'M'$ are geometrically equal.

By definition,

$$\sin(180° - A) = \frac{M'P'}{OP'};$$

but $M'P'$ is equal to MP in magnitude and is of the same sign;

$$\therefore \sin(180° - A) = \frac{MP}{OP} = \sin A.$$

Again, $$\cos(180° - A) = \frac{OM'}{OP'};$$

and OM' is equal to OM in magnitude, but is of opposite sign;

$$\therefore \cos(180° - A) = \frac{-OM}{OP} = -\frac{OM}{OP} = -\cos A.$$

Also $$\tan(180° - A) = \frac{M'P'}{OM'} = \frac{MP}{-OM} = -\frac{MP}{OM} = -\tan A.$$

93. In the last article, for the sake of simplicity w[illegible] supposed the angle A to be less than a right angle, but[illegible] formulæ of this chapter may be shewn to be true for angles of any magnitude. A general proof of one case is given in Art. 102, and the same method may be applied to all the other cases.

94. If the angles are expressed in radian measure, the formulæ of Art. 92 become

$$\sin(\pi-\theta)=\sin\theta,$$

$$\cos(\pi-\theta)=-\cos\theta,$$

$$\tan(\pi-\theta)=-\tan\theta.$$

Example 1. Find the sine and cosine of 120°.

$$\sin 120^\circ=\sin(180^\circ-60^\circ)=\sin 60^\circ=\frac{\sqrt{3}}{2}.$$

$$\cos 120^\circ=\cos(180^\circ-60^\circ)=-\cos 60^\circ=-\frac{1}{2}.$$

Example 2. Find the cosine and cotangent of $\frac{5\pi}{6}$.

$$\cos\frac{5\pi}{6}=\cos\left(\pi-\frac{\pi}{6}\right)=-\cos\frac{\pi}{6}=-\frac{\sqrt{3}}{2}.$$

$$\cot\frac{5\pi}{6}=\cot\left(\pi-\frac{\pi}{6}\right)=-\cot\frac{\pi}{6}=-\sqrt{3}.$$

95. Definition. When the sum of two angles is equal to two right angles each is said to be the **supplement** of the other and the angles are said to be **supplementary**. Thus if A is any angle its supplement is $180^\circ-A$.

96. The results of Art. 92 are so important in a later part of the subject that it is desirable to emphasize them. We therefore repeat them in a verbal form:

the sines of supplementary angles are equal in magnitude and are of the same sign;

the cosines of supplementary angles are equal in magnitude but are of opposite sign;

the tangents of supplementary angles are equal in magnitude but are of opposite sign.

97. Circular Functions of 180°+A.

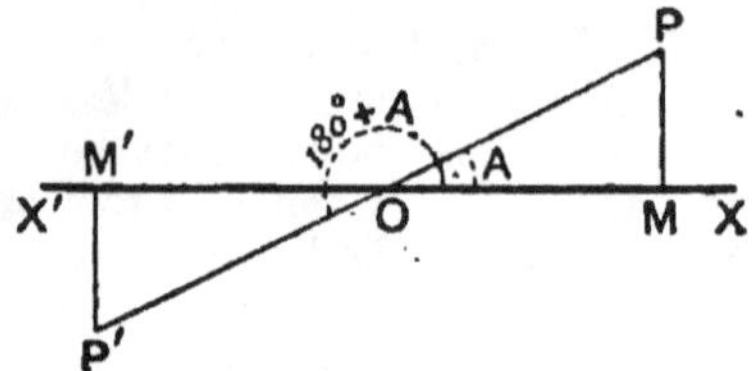

Take any straight line XOX' and let a radius vector starting from OX revolve until it has traced the angle A, taking up the position OP.

Again, let the radius vector starting from OX revolve through 180° into the position OX', and then further through an angle A, taking up the final position OP'. Thus XOP' is the angle $180°+A$.

From P and P' draw PM and $P'M'$ perpendicular to XX'; then OP and OP' are in the same straight line, and by Euc. I. 26 the triangles OPM and $OP'M'$ are geometrically equal.

By definition,

$$\sin(180°+A)=\frac{M'P'}{OP'};$$

and $M'P'$ is equal to MP in magnitude but is of opposite sign;

$$\therefore \sin(180°+A)=\frac{-MP}{OP}=-\frac{MP}{OP}=-\sin A.$$

Again, $$\cos(180°+A)=\frac{OM'}{OP'};$$

and OM' is equal to OM in magnitude but is of opposite sign;

$$\therefore \cos(180°+A)=\frac{-OM}{OP}=-\frac{OM}{OP}=-\cos A.$$

Also $$\tan(180°+A)=\frac{M'P'}{OM'}=\frac{-MP}{-OM}=\frac{MP}{OM}=\tan A.$$

Expressed in radian measure, the above formulæ are written

$\sin(\pi+\theta)=-\sin\theta,\quad \cos(\pi+\theta)=-\cos\theta,\quad \tan(\pi+\theta)=\tan\theta.$

In these results we may draw especial attention to the fact that an angle may be increased or diminished by two right angles as often as we please without altering the value of the tangent.

Example. **Find the value of cot 210°.**

$$\cot 210°=\cot(180°+30°)=\cot 30°=\sqrt{3}.$$

98. Circular Functions of 90°+A.

Take any straight line XOX', and let a radius vector starting from OX revolve until it has traced the angle A, taking up the position OP.

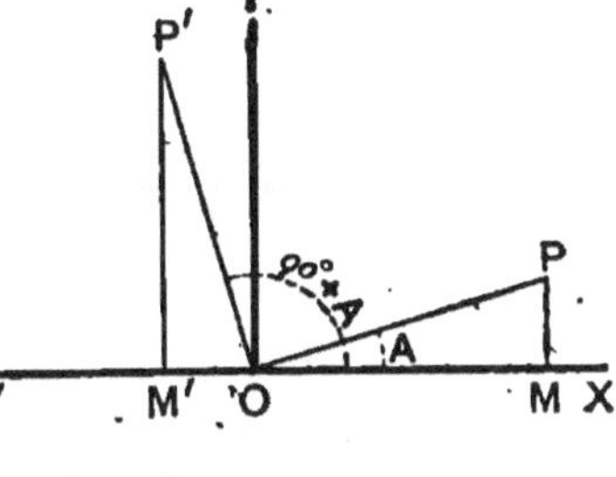

Again, let the radius vector starting from OX revolve through 90° into the position OY, and then further through an angle A, taking up the final position OP'. Thus XOP' is the angle $90°+A$.

From P and P' draw PM and $P'M'$ perpendicular to XX'; then $\angle M'P'O = \angle P'OY = A = \angle POM$.

By Euc. I. 26, the triangles OPM and $OP'M'$ are geometrically equal; hence

$M'P'$ is equal to OM in magnitude and is of the same sign,

and OM' is equal to MP in magnitude but is of opposite sign.

By definition,

$$\sin(90°+A)=\frac{M'P'}{OP'}=\frac{OM}{OP}=\cos A;$$

$$\cos(90°+A)=\frac{OM'}{OP'}=\frac{-MP}{OP}=-\frac{MP}{OP}=-\sin A;$$

$$\tan(90°+A)=\frac{M'P'}{OM'}=\frac{OM}{-MP}=-\frac{OM}{MP}=-\cot A.$$

Expressed in radian measure the above formulæ become

$$\sin\left(\frac{\pi}{2}+\theta\right)=\cos\theta,\quad \cos\left(\frac{\pi}{2}+\theta\right)=-\sin\theta,\quad \tan\left(\frac{\pi}{2}+\theta\right)=-\cot\theta.$$

Example 1. Find the value of sin 120°.

$$\sin 120°=\sin(90°+30°)=\cos 30°=\frac{\sqrt{3}}{2}.$$

Example 2. Find the values of $\tan(270°+A)$ and $\cos\left(\frac{3\pi}{2}+\theta\right)$.

$$\tan(270°+A)=\tan(180°+\overline{90°+A})=\tan(90°+A)=-\cot A;$$

$$\cos\left(\frac{3\pi}{2}+\theta\right)=\cos\left(\pi+\overline{\frac{\pi}{2}+\theta}\right)=-\cos\left(\frac{\pi}{2}+\theta\right)=\sin\theta.$$

99. Circular Functions of $-A$.

Take any straight line OX and let a radius vector starting from OX revolve until it has traced the angle A, taking up the position OP.

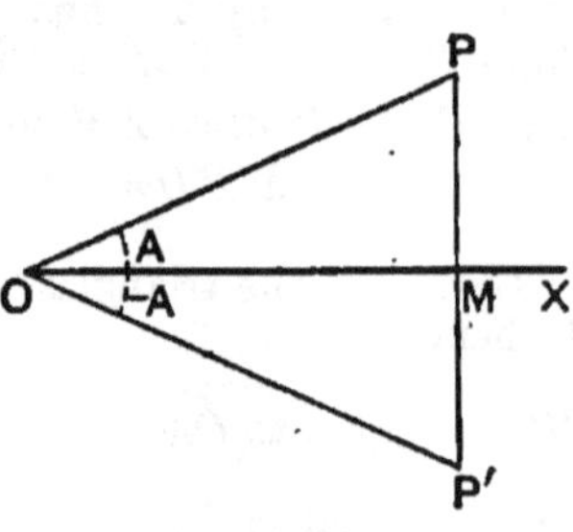

Again, let the radius vector starting from OX revolve in the *opposite* direction until it has traced the angle A, taking up the position OP'. Join PP'; then MP' is equal to MP in magnitude, and the angles at M are right angles. [Euc. I. 4.]

By definition,

$$\sin(-A)=\frac{MP'}{OP'}=\frac{-MP}{OP}=-\sin A;$$

$$\cos(-A)=\frac{OM}{OP'}=\frac{OM}{OP}=\cos A;$$

$$\tan(-A)=\frac{MP'}{OM}=\frac{-MP}{OM}=-\tan A.$$

It is especially worthy of notice that *we may change the sign of an angle without altering the value of its cosine.*

Example. Find the values of

$$\operatorname{cosec}(-210°) \text{ and } \cos(A-270°).$$

$$\operatorname{cosec}(-210°)=-\operatorname{cosec}210°=-\operatorname{cosec}(180°+30°)=\operatorname{cosec}30°=2.$$

$$\cos(A-270°)=\cos(270°-A)=\cos(180°+\overline{90°-A})$$
$$=-\cos(90°-A)=-\sin A.$$

100. If $f(A)$ denotes a function of A which is unaltered in magnitude and sign when $-A$ is written for A, then $f(A)$ is said to be an **even function** of A. In this case $f(-A)=f(A)$.

If when $-A$ is written for A, the sign of $f(A)$ is changed while the magnitude remains unaltered, $f(A)$ is said to be an **odd function** of A, and in this case $f(-A)=-f(A)$.

From the last article it will be seen that

$\cos A$ *and* $\sec A$ *are even functions of* A,

$\sin A$, $\operatorname{cosec} A$, $\tan A$, $\cot A$ *are odd functions of* A.

EXAMPLES. X. a.

Find the numerical value of

1. $\cos 135°$.
2. $\sin 150°$.
3. $\tan 240°$.
4. $\operatorname{cosec} 225°$.
5. $\sin(-120°)$.
6. $\cot(-135°)$.
7. $\cot 315°$.
8. $\cos(-240°)$.
9. $\sec(-300°)$.
10. $\tan\frac{3\pi}{4}$.
11. $\sin\frac{4\pi}{3}$.
12. $\sec\frac{2\pi}{3}$.
13. $\operatorname{cosec}\left(-\frac{\pi}{6}\right)$.
14. $\cos\left(-\frac{3\pi}{4}\right)$.
15. $\cot\left(-\frac{5\pi}{6}\right)$.

Express as functions of A:

16. $\cos(270°+A)$.
17. $\cot(270°-A)$.
18. $\sin(A-90°)$.
19. $\sec(A-180°)$.
20. $\sin(270°-A)$.
21. $\cot(A-90°)$.

Express as functions of θ:

22. $\sin\left(\theta-\frac{\pi}{2}\right)$.
23. $\tan(\theta-\pi)$.
24. $\sec\left(\frac{3\pi}{2}-\theta\right)$.

Express in the simplest form:

25. $\tan(180°+A)\sin(90°+A)\sec(90°-A)$.
26. $\cos(90°+A)+\sin(180°-A)-\sin(180°+A)-\sin(-A)$.
27. $\sec(180°+A)\sec(180°-A)+\cot(90°+A)\tan(180°+A)$.

101. In Art. 38 we have established the relations which subsist between the trigonometrical ratios of $90^\circ - A$ and those of A, when A is an acute angle. We shall now give a general proof which is applicable whatever be the magnitude of A.

102. Circular Functions of 90° – A for any value of A.

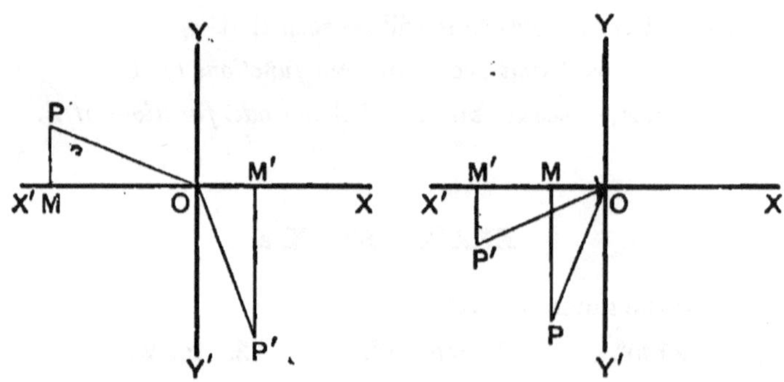

Let a radius vector starting from OX revolve until it has traced the angle A, taking up the position OP in each of the two figures.

Again, let the radius vector starting from OX revolve through 90° into the position OY and then *back again* through an angle A, taking up the final position OP' in each of the two figures.

Draw PM and $P'M'$ perpendicular to XX'; then whatever be the value of A, it will be found that $\angle OP'M' = \angle POM$, so that the triangles OMP and $OM'P'$ are geometrically equal, having MP equal to OM', and OM equal to $M'P'$, *in magnitude*.

When P is above XX', P' is to the right of YY',
and when P is below XX', P' is to the left of YY'.

When P' is above XX', P is to the right of YY',
and when P' is below XX', P is to the left of YY'.

Hence MP is equal to OM' in magnitude and is always of the same sign as OM';

and $M'P'$ is equal to OM in magnitude and is always of the same sign as OM.

By definition,

$$\sin(90^\circ - A) = \frac{M'P'}{OP'} = \frac{OM}{OP} = \cos A;$$

$$\cos(90^\circ - A) = \frac{OM'}{OP'} = \frac{MP}{OP} = \sin A;$$

$$\tan(90^\circ - A) = \frac{M'P'}{OM'} = \frac{OM}{MP} = \cot A.$$

A general method similar to the above may be applied to all the other cases of this chapter.

103. Circular Functions of n . 360° + A.

If n is any integer, $n . 360^\circ$ represents n complete revolutions of the radius vector, and therefore the boundary line of the angle $n . 360^\circ + A$ is coincident with that of A. The value of each function of the angle $n . 360^\circ + A$ is thus the same as the value of the corresponding function of A both in magnitude and in sign.

104. Since the functions of all coterminal angles are equal, there is a *recurrence* of the values of the functions each time the boundary line completes its revolution and comes round into its original position. This is otherwise expressed by saying that *the circular functions are periodic*, and 360° is said to be *the amplitude of the period.*

In radian measure, the amplitude of the period is 2π.

NOTE. In the case of the tangent and cotangent the amplitude of the period is half that of the other circular functions, being 180° or π radians. [Art. 97.]

105. Circular Functions of n . 360° − A.

If n is any integer, the boundary line of $n . 360^\circ - A$ is coincident with that of $-A$. The value of each function of $n . 360^\circ - A$ is thus the same as the value of the corresponding function of $-A$ both in magnitude and in sign; hence

$$\sin(n . 360^\circ - A) = \sin(-A) = -\sin A;$$

$$\cos(n . 360^\circ - A) = \cos(-A) = \cos A;$$

$$\tan(n . 360^\circ - A) = \tan(-A) = -\tan A.$$

106. We can always express the functions of any angle in terms of the functions of some positive acute angle. In the arrangement of the work it is advisable to follow a uniform plan.

(1) If the angle is negative, use the relations connecting the functions of $-A$ and A. [Art. 99.]

Thus $$\sin(-30^\circ) = -\sin 30^\circ = -\frac{1}{2};$$

$$\cos(-845^\circ) = \cos 845^\circ.$$

(2) If the angle is greater than 360°, by taking off multiples of 360° the angle may be replaced by a coterminal angle less than 360°. [Art. 103.]

Thus $$\tan 735^\circ = \tan(2\times 360^\circ + 15^\circ) = \tan 15^\circ.$$

(3) If the angle is still greater than 180°, use the relations connecting the functions of $180^\circ + A$ and A. [Art. 97.]

Thus $$\cot 585^\circ = \cot(360^\circ + 225^\circ) = \cot 225^\circ$$
$$= \cot(180^\circ + 45^\circ) = \cot 45^\circ = 1.$$

(4) If the angle is still greater than 90°, use the relations connecting the functions of $180^\circ - A$ and A. [Art. 92.]

Thus $$\cos 675^\circ = \cos(360^\circ + 315^\circ) = \cos 315^\circ$$
$$= \cos(180^\circ + 135^\circ) = -\cos 135^\circ$$
$$= -\cos(180^\circ - 45^\circ) = \cos 45^\circ = \frac{1}{\sqrt{2}}.$$

Example. Express $\sin(-1190^\circ)$, $\tan 1000^\circ$, $\cos(-3860^\circ)$ as functions of positive acute angles.

$$\sin(-1190^\circ) = -\sin 1190^\circ = -\sin(3\times 360^\circ + 110^\circ) = -\sin 110^\circ$$
$$= -\sin(180^\circ - 70^\circ) = -\sin 70^\circ.$$

$$\tan 1000^\circ = \tan(2\times 360^\circ + 280^\circ) = \tan 280^\circ$$
$$= \tan(180^\circ + 100^\circ) = \tan 100^\circ$$
$$= \tan(180^\circ - 80^\circ) = -\tan 80^\circ.$$

$$\cos(-3860^\circ) = \cos 3860^\circ = \cos(10\times 360^\circ + 260^\circ) = \cos 260^\circ$$
$$= \cos(180^\circ + 80^\circ) = -\cos 80^\circ.$$

107. From the investigations of this chapter we see that the number of angles which have the same circular function is unlimited. Thus if $\tan\theta=1$, θ may be any one of the angles coterminal with 45° or 225°.

Example. Draw the boundary lines of A when $\sin A=\frac{\sqrt{3}}{2}$, and write down all the angles numerically less than 360° which satisfy the equation.

Since $\sin 60°=\frac{\sqrt{3}}{2}$, if we draw OP making $\angle XOP=60°$, then OP is one position of the boundary line.

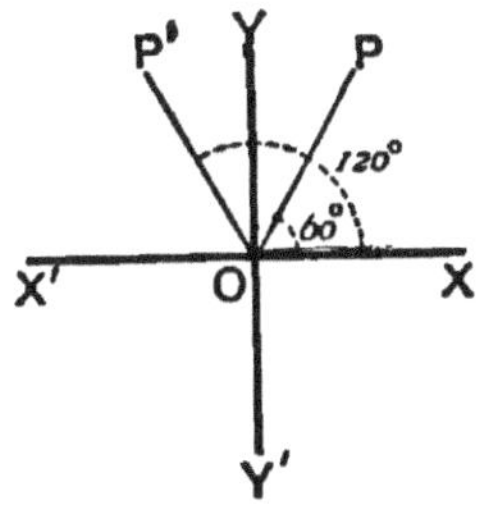

Again, $\sin 60°=\sin(180°-60°)=\sin 120°$, so that another position of the boundary line will be found by making $XOP'=120°$.

There will be no position of the boundary line in the third or fourth quadrant, since in these quadrants the sine is negative.

Thus in one complete revolution OP and OP' are the only two positions of the boundary line of the angle A.

Hence the positive angles are 60° and 120°;

and the negative angles are $-(360°-120°)$ and $-(360°-60°)$; that is, $-240°$ and $-300°$.

EXAMPLES. X. b.

Find the numerical value of

1. $\cos 480°$.	2. $\sin 960°$.	3. $\cos(-780°)$.
4. $\sin(-870°)$.	5. $\sec 900°$.	6. $\tan(-855°)$.
7. $\operatorname{cosec}(-660°)$.	8. $\cot 840°$.	9. $\operatorname{cosec}(-765°)$.
10. $\cos 4005°$.	11. $\cot 990°$.	12. $\sin 3015°$.
13. $\sec 2745°$.	14. $\cos 2400°$.	15. $\sec(-5895°)$.
16. $\sin\frac{15\pi}{4}$.	17. $\cot\frac{23\pi}{4}$.	18. $\sec\frac{7\pi}{3}$.
19. $\cot\frac{16\pi}{3}$.	20. $\sec\left(\frac{3\pi}{2}+\frac{\pi}{3}\right)$.	

Find all the angles numerically less than 360° which satisfy the equations:

21. $\cos\theta = \frac{\sqrt{3}}{2}$.   22. $\sin\theta = -\frac{1}{2}$.

23. $\tan\theta = -\sqrt{3}$.   24. $\cot\theta = -1$.

If A is less than 90°, prove geometrically

25. $\sec(A-180^\circ) = -\sec A$.

26. $\tan(270^\circ + A) = -\cot A$.

27. $\cos(A-90^\circ) = \sin A$.

28. Prove that

$$\tan A + \tan(180^\circ - A) + \cot(90^\circ + A) = \tan(360^\circ - A).$$

29. Shew that

$$\frac{\sin(180^\circ - A)}{\tan(180^\circ + A)} \cdot \frac{\cot(90^\circ - A)}{\tan(90^\circ + A)} \cdot \frac{\cos(360^\circ - A)}{\sin(-A)} = \sin A.$$

Express in the simplest form

30. $$\frac{\sin(-A)}{\sin(180^\circ + A)} - \frac{\tan(90^\circ + A)}{\cot A} + \frac{\cos A}{\sin(90^\circ + A)}.$$

31. $$\frac{\operatorname{cosec}(180^\circ - A)}{\sec(180^\circ + A)} \cdot \frac{\cos(-A)}{\cos(90^\circ + A)}.$$

32. $$\frac{\cos(90^\circ + A)\sec(-A)\tan(180^\circ - A)}{\sec(360^\circ + A)\sin(180^\circ + A)\cot(90^\circ - A)}.$$

33. Prove that $\sin\left(\frac{\pi}{2} + \theta\right)\cos(\pi - \theta)\cot\left(\frac{3\pi}{2} + \theta\right)$

$$= \sin\left(\frac{\pi}{2} - \theta\right)\sin\left(\frac{3\pi}{2} - \theta\right)\cot\left(\frac{\pi}{2} + \theta\right).$$

34. When $\alpha = \frac{11\pi}{4}$, find the numerical value of

$$\sin^2\alpha - \cos^2\alpha + 2\tan\alpha - \sec^2\alpha.$$

CHAPTER XI.

FUNCTIONS OF COMPOUND ANGLES.

108. WHEN an angle is made up by the algebraical sum of two or more angles it is called a **compound angle**; thus $A+B$, $A-B$, and $A+B-C$ are compound angles.

109. Hitherto we have only discussed the properties of the functions of single angles, such as A, B, α, θ. In the present chapter we shall prove some fundamental properties relating to the functions of compound angles. We shall begin by finding expressions for the sine, cosine, and tangent of $A+B$ and $A-B$ in terms of the functions of A and B.

It may be useful to caution the student against the prevalent mistake of supposing that a function of $A+B$ is equal to the sum of the corresponding functions of A and B, and a function of $A-B$ to the difference of the corresponding functions.

Thus $\sin(A+B)$ is not equal to $\sin A+\sin B$,

and $\quad \cos(A-B)$ is not equal to $\cos A-\cos B$.

A numerical instance will illustrate this.

Thus if $A=60°$, $B=30°$, then $A+B=90°$,

so that $$\cos(A+B)=\cos 90°=0;$$

but $$\cos A+\cos B=\cos 60°+\cos 30°=\frac{1}{2}+\frac{\sqrt{3}}{2}.$$

Hence $\cos(A+B)$ is not equal to $\cos A+\cos B$.

In like manner, $\sin(A+A)$ is not equal to $\sin A+\sin A$;

that is, $\quad \sin 2A$ is not equal to $2\sin A$.

Similarly $\quad \tan 3A$ is not equal to $3\tan A$.

110. *To prove the formulæ*

$$\sin(A+B)=\sin A\cos B+\cos A\sin B,$$

$$\cos(A+B)=\cos A\cos B-\sin A\sin B.$$

Let $\angle LOM=A$, and $\angle MON=B$; then $\angle LON=A+B$.

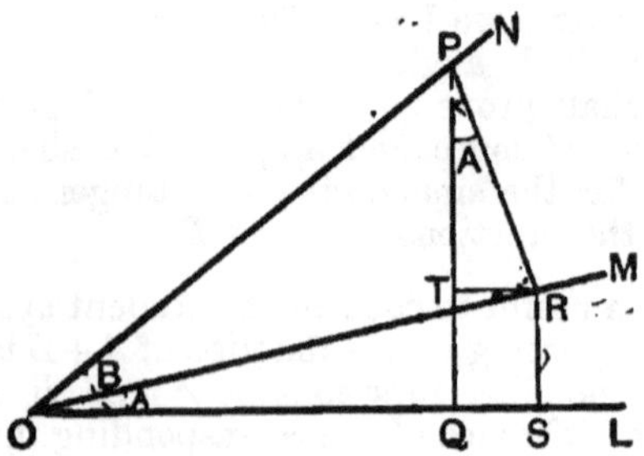

In ON, *the boundary line of the compound angle* $A+B$, take any point P, and draw PQ and PR perpendicular to OL and OM respectively; also draw RS and RT perpendicular to OL and PQ respectively.

By definition,

$$\sin(A+B)=\frac{PQ}{OP}=\frac{RS+PT}{OP}=\frac{RS}{OP}+\frac{PT}{OP}$$

$$=\frac{RS}{OR}\cdot\frac{OR}{OP}+\frac{PT}{PR}\cdot\frac{PR}{OP}$$

$$=\sin A\,.\cos B+\cos TPR\,.\sin B.$$

But $\angle TPR=90^\circ-\angle TRP=\angle TRO=\angle ROS=A$;

$$\therefore\ \sin(A+B)=\sin A\cos B+\cos A\sin B.$$

Also $$\cos(A+B)=\frac{OQ}{OP}=\frac{OS-TR}{OP}=\frac{OS}{OP}-\frac{TR}{OP}$$

$$=\frac{OS}{OR}\cdot\frac{OR}{OP}-\frac{TR}{PR}\cdot\frac{PR}{OP}$$

$$=\cos A\,.\cos B-\sin TPR\,.\sin B$$

$$=\cos A\cos B-\sin A\sin B.$$

111. *To prove the formulæ*

$$\sin(A-B)=\sin A\cos B-\cos A\sin B,$$
$$\cos(A-B)=\cos A\cos B+\sin A\sin B.$$

Let $\angle LOM=A$, and $\angle MON=B$; then $\angle LON=A-B$.

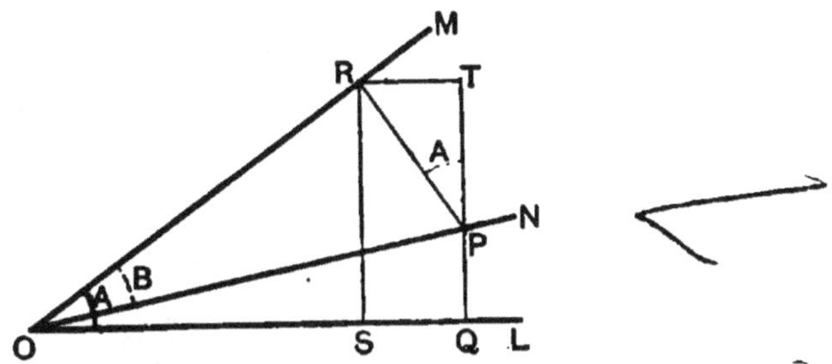

In ON, *the boundary line of the compound angle* $A-B$, take any point P, and draw PQ and PR perpendicular to OL and OM respectively; also draw RS and RT perpendicular to OL and QP respectively.

By definition,

$$\sin(A-B)=\frac{PQ}{OP}=\frac{RS-PT}{OP}=\frac{RS}{OP}-\frac{PT}{OP}$$
$$=\frac{RS}{OR}\cdot\frac{OR}{OP}-\frac{PT}{PR}\cdot\frac{PR}{OP}$$
$$=\sin A\,.\cos B-\cos TPR\,.\sin B.$$

But $\angle TPR=90^\circ-\angle TRP=\angle MRT=\angle MOL=A$;

$$\therefore\ \sin(A-B)=\sin A\cos B-\cos A\sin B.$$

Also $$\cos(A-B)=\frac{OQ}{OP}=\frac{OS+RT}{OP}=\frac{OS}{OP}+\frac{RT}{OP}$$
$$=\frac{OS}{OR}\cdot\frac{OR}{OP}+\frac{RT}{RP}\cdot\frac{RP}{OP}$$
$$=\cos A\,.\cos B+\sin TPR\,.\sin B$$
$$=\cos A\cos B+\sin A\sin B.$$

112. The *expansions* of $\sin(A \pm B)$ and $\cos(A \pm B)$ are frequently called the "Addition Formulæ." We shall sometimes refer to them as the "$A+B$" and "$A-B$" formulæ.

113. In the foregoing geometrical proofs we have supposed that the angles A, B, $A+B$ are all less than a right angle, and that $A-B$ is positive. If the angles are not so restricted some modification of the figures will be required. It is however unnecessary to consider these cases in detail, as in Chap. XXII. we shall shew by the Method of Projections that the Addition Formulæ hold universally. In the meantime the student may assume that they are always true.

Example 1. Find the value of $\cos 75°$.

$$\cos 75° = \cos(45° + 30°) = \cos 45° \cos 30° - \sin 45° \sin 30°$$

$$= \frac{1}{\sqrt{2}} \cdot \frac{\sqrt{3}}{2} - \frac{1}{\sqrt{2}} \cdot \frac{1}{2} = \frac{\sqrt{3}-1}{2\sqrt{2}}.$$

Example 2. If $\sin A = \frac{4}{5}$ and $\sin B = \frac{5}{13}$, find $\sin(A-B)$.

$$\sin(A-B) = \sin A \cos B - \cos A \sin B.$$

But $$\cos A = \sqrt{1 - \sin^2 A} = \sqrt{1 - \frac{16}{25}} = \frac{3}{5};$$

and $$\cos B = \sqrt{1 - \sin^2 B} = \sqrt{1 - \frac{25}{169}} = \frac{12}{13};$$

$$\therefore \sin(A-B) = \frac{4}{5} \cdot \frac{12}{13} - \frac{3}{5} \cdot \frac{5}{13} = \frac{33}{65}.$$

Note. Strictly speaking $\cos A = \pm \frac{3}{5}$ and $\cos B = \pm \frac{12}{13}$, so that $\sin(A-B)$ has *four* values. We shall however suppose that in similar cases only the positive value of the square root is taken.

114. *To prove that* $\sin(A+B)\sin(A-B) = \sin^2 A - \sin^2 B$.

The first side

$$= (\sin A \cos B + \cos A \sin B)(\sin A \cos B - \cos A \sin B)$$

$$= \sin^2 A \cos^2 B - \cos^2 A \sin^2 B$$

$$= \sin^2 A (1 - \sin^2 B) - (1 - \sin^2 A) \sin^2 B$$

$$= \sin^2 A - \sin^2 B.$$

EXAMPLES. XI. a.

[*The examples printed in more prominent type are important, and should be regarded as standard formulæ.*]

Prove that

1. $\sin(A+45°)=\frac{1}{\sqrt{2}}(\sin A+\cos A)$.

2. $\cos(A+45°)=\frac{1}{\sqrt{2}}(\cos A-\sin A)$.

3. $2\sin(30°-A)=\cos A-\sqrt{3}\sin A$.

4. If $\cos A=\frac{4}{5}$, $\cos B=\frac{3}{5}$, find $\sin(A+B)$ and $\cos(A-B)$.

5. If $\sin A=\frac{3}{5}$, $\cos B=\frac{12}{13}$, find $\cos(A+B)$ and $\sin(A-B)$.

6. If $\sec A=\frac{17}{8}$, $\operatorname{cosec} B=\frac{5}{4}$, find $\sec(A+B)$.

Prove that

7. $\mathbf{\sin 75°=\cos 15°=\frac{\sqrt{3}+1}{2\sqrt{2}}}$.

8. $\mathbf{\sin 15°=\cos 75°=\frac{\sqrt{3}-1}{2\sqrt{2}}}$.

9. $\mathbf{\frac{\sin(\alpha+\beta)}{\cos\alpha\cos\beta}=\tan\alpha+\tan\beta}$.

10. $\mathbf{\frac{\sin(\alpha-\beta)}{\sin\alpha\sin\beta}=\cot\beta-\cot\alpha}$.

11. $\frac{\cos(\alpha-\beta)}{\cos\alpha\sin\beta}=\cot\beta+\tan\alpha$.

12. $\mathbf{\cos(A+B)\cos(A-B)=\cos^2 A-\sin^2 B}$.

13. $\mathbf{\sin(A+B)\sin(A-B)=\cos^2 B-\cos^2 A}$.

14. $\cos(45°-A)-\sin(45°+A)=0$.

15. $\cos(45°+A)+\sin(A-45°)=0$.

16. $\cos(A-B)-\sin(A+B)=(\cos A-\sin A)(\cos B-\sin B)$.

17. $\cos(A+B)+\sin(A-B)=(\cos A+\sin A)(\cos B-\sin B)$.

Prove the following identities:

18. $2\sin(A+45^\circ)\sin(A-45^\circ)=\sin^2 A-\cos^2 A.$

19. $2\cos\left(\frac{\pi}{4}+\alpha\right)\cos\left(\frac{\pi}{4}-\alpha\right)=\cos^2\alpha-\sin^2\alpha.$

20. $2\sin\left(\frac{\pi}{4}+\alpha\right)\cos\left(\frac{\pi}{4}+\beta\right)=\cos(\alpha+\beta)+\sin(\alpha-\beta).$

21. $\frac{\sin(\beta-\gamma)}{\cos\beta\cos\gamma}+\frac{\sin(\gamma-\alpha)}{\cos\gamma\cos\alpha}+\frac{\sin(\alpha-\beta)}{\cos\alpha\cos\beta}=0.$

115. *To expand* $\tan(A+B)$ *in terms of* $\tan A$ *and* $\tan B$.

$$\tan(A+B)=\frac{\sin(A+B)}{\cos(A+B)}=\frac{\sin A\cos B+\cos A\sin B}{\cos A\cos B-\sin A\sin B}.$$

To express this fraction in terms of *tangents*, divide each term of numerator and denominator by $\cos A\cos B$;

$$\therefore \tan(A+B)=\frac{\frac{\sin A}{\cos A}+\frac{\sin B}{\cos B}}{1-\frac{\sin A}{\cos A}\cdot\frac{\sin B}{\cos B}};$$

that is,
$$\tan(A+B)=\frac{\tan A+\tan B}{1-\tan A\tan B}.$$

A geometrical proof of this result is given in Chap. XXII.

Similarly, we may prove that

$$\tan(A-B)=\frac{\tan A-\tan B}{1+\tan A\tan B}.$$

Example. Find the value of tan 75°.

$$\tan 75^\circ=\tan(45^\circ+30^\circ)=\frac{\tan 45^\circ+\tan 30^\circ}{1-\tan 45^\circ\tan 30^\circ}$$

$$=\frac{1+\frac{1}{\sqrt{3}}}{1-\frac{1}{\sqrt{3}}}=\frac{\sqrt{3}+1}{\sqrt{3}-1}$$

$$=\frac{(\sqrt{3}+1)(\sqrt{3}+1)}{3-1}=\frac{4+2\sqrt{3}}{2}$$

$$=2+\sqrt{3}.$$

116. *To expand* $\cot(A+B)$ *in terms of* $\cot A$ *and* $\cot B$.

$$\cot(A+B)=\frac{\cos(A+B)}{\sin(A+B)}=\frac{\cos A\cos B-\sin A\sin B}{\sin A\cos B+\cos A\sin B}.$$

To express this fraction in terms of *cotangents*, divide each term of numerator and denominator by $\sin A\sin B$;

$$\therefore\ \cot(A+B)=\frac{\dfrac{\cos A\cos B}{\sin A\sin B}-1}{\dfrac{\cos B}{\sin B}+\dfrac{\cos A}{\sin A}}=\frac{\cot A\cot B-1}{\cot B+\cot A}.$$

Similarly, we may prove that

$$\cot(A-B)=\frac{\cot A\cot B+1}{\cot B-\cot A}.$$

117. *To find the expansion of* $\sin(A+B+C)$.

$$\begin{aligned}\sin(A+B+C)&=\sin\{(A+B)+C\}\\&=\sin(A+B)\cos C+\cos(A+B)\sin C\\&=(\sin A\cos B+\cos A\sin B)\cos C\\&\qquad+(\cos A\cos B-\sin A\sin B)\sin C\\&=\sin A\cos B\cos C+\cos A\sin B\cos C\\&\qquad+\cos A\cos B\sin C-\sin A\sin B\sin C.\end{aligned}$$

118. *To find the expansion of* $\tan(A+B+C)$:

$$\begin{aligned}\tan(A+B+C)=\tan\{(A+B)+C\}&=\frac{\tan(A+B)+\tan C}{1-\tan(A+B)\tan C}\\&=\frac{\dfrac{\tan A+\tan B}{1-\tan A\tan B}+\tan C}{1-\dfrac{\tan A+\tan B}{1-\tan A\tan B}\cdot\tan C}\\&=\frac{\tan A+\tan B+\tan C-\tan A\tan B\tan C}{1-\tan A\tan B-\tan B\tan C-\tan C\tan A}.\end{aligned}$$

Cor. If $A+B+C=180°$, then $\tan(A+B+C)=0$; hence the numerator of the above expression must be zero.

$$\therefore\ \tan A+\tan B+\tan C=\tan A\tan B\tan C.$$

7—2

EXAMPLES. XI. b.

[*The examples printed in more prominent type are important, and should be regarded as standard formulæ.*]

1. Find $\tan(A+B)$ when $\tan A=\frac{1}{2}$, $\tan B=\frac{1}{3}$.

2. If $\tan A=\frac{4}{3}$, and $B=45°$, find $\tan(A-B)$.

3. If $\cot A=\frac{5}{7}$, $\cot B=\frac{7}{5}$, find $\cot(A+B)$ and $\tan(A-B)$.

4. If $\cot A=\frac{11}{2}$, $\tan B=\frac{7}{24}$, find $\cot(A-B)$ and $\tan(A+B)$.

5. $\mathbf{\tan(45°+A)=\frac{1+\tan A}{1-\tan A}}$.

6. $\mathbf{\tan(45°-A)=\frac{1-\tan A}{1+\tan A}}$.

7. $\cot\left(\frac{\pi}{4}-\theta\right)=\frac{\cot\theta+1}{\cot\theta-1}$.

8. $\cot\left(\frac{\pi}{4}+\theta\right)=\frac{\cot\theta-1}{\cot\theta+1}$.

9. $\mathbf{\tan 15°=2-\sqrt{3}}$.

10. $\mathbf{\cot 15°=2+\sqrt{3}}$.

11. Find the expansions of

$$\cos(A+B+C) \text{ and } \sin(A-B+C).$$

12. Express $\tan(A-B-C)$ in terms of $\tan A$, $\tan B$, $\tan C$.

13. Express $\cot(A+B+C)$ in terms of $\cot A$, $\cot B$, $\cot C$.

119. Beginners not unfrequently find a difficulty in the converse use of the $A+B$ and $A-B$ formulæ; that is, they fail to recognise when an expression is merely an expansion belonging to one of the standard forms.

Example 1. Simplify $\cos(\alpha-\beta)\cos(\alpha+\beta)-\sin(\alpha-\beta)\sin(\alpha+\beta)$.

This expression is the expansion of the cosine of the compound angle $(\alpha+\beta)+(\alpha-\beta)$, and is therefore equal to $\cos\{(\alpha+\beta)+(\alpha-\beta)\}$; that is, to $\cos 2\alpha$.

Example 2. Shew that $\frac{\tan A+\tan 2A}{1-\tan A\tan 2A}=\tan 3A$.

By Art. 115, the first side is the expansion of $\tan(A+2A)$, and is therefore equal to $\tan 3A$.

Example 3. Prove that $\cot 2A + \tan A = \operatorname{cosec} 2A$.

The first side $= \dfrac{\cos 2A}{\sin 2A} + \dfrac{\sin A}{\cos A} = \dfrac{\cos 2A \cos A + \sin 2A \sin A}{\sin 2A \cos A}$

$$= \frac{\cos (2A - A)}{\sin 2A \cos A} = \frac{\cos A}{\sin 2A \cos A}$$

$$= \frac{1}{\sin 2A} = \operatorname{cosec} 2A.$$

Example 4. Prove that

$$\cos 4\theta \cos \theta + \sin 4\theta \sin \theta = \cos 2\theta \cos \theta - \sin 2\theta \sin \theta.$$

The first side $= \cos (4\theta - \theta) = \cos 3\theta = \cos (2\theta + \theta)$

$$= \cos 2\theta \cos \theta - \sin 2\theta \sin \theta.$$

EXAMPLES. XI. c.

Prove the following identities :

1. $\cos (A+B) \cos B + \sin (A+B) \sin B = \cos A.$
2. $\sin 3A \cos A - \cos 3A \sin A = \sin 2A.$
3. $\cos 2\alpha \cos \alpha + \sin 2\alpha \sin \alpha = \cos \alpha.$
4. $\cos (30^\circ + A) \cos (30^\circ - A) - \sin (30^\circ + A) \sin (30^\circ - A) = \dfrac{1}{2}.$
5. $\sin (60^\circ - A) \cos (30^\circ + A) + \cos (60^\circ - A) \sin (30^\circ + A) = 1.$
6. $\dfrac{\cos 2\alpha}{\sec \alpha} - \dfrac{\sin 2\alpha}{\operatorname{cosec} \alpha} = \cos 3\alpha.$
7. $\dfrac{\tan (\alpha - \beta) + \tan \beta}{1 - \tan (\alpha - \beta) \tan \beta} = \tan \alpha.$
8. $\dfrac{\cot (\alpha + \beta) \cot \alpha + 1}{\cot \alpha - \cot (\alpha + \beta)} = \cot \beta.$
9. $\dfrac{\tan 4A - \tan 3A}{1 + \tan 4A \tan 3A} = \tan A.$
10. $\cot \theta - \cot 2\theta = \operatorname{cosec} 2\theta.$
11. $1 + \tan 2\theta \tan \theta = \sec 2\theta.$
12. $1 + \cot 2\theta \cot \theta = \operatorname{cosec} 2\theta \cot \theta.$
13. $\sin 2\theta \cos \theta + \cos 2\theta \sin \theta = \sin 4\theta \cos \theta - \cos 4\theta \sin \theta.$
14. $\cos 4\alpha \cos \alpha - \sin 4\alpha \sin \alpha = \cos 3\alpha \cos 2\alpha - \sin 3\alpha \sin 2\alpha.$

Functions of Multiple Angles.

120. *To express* $\sin 2A$ *in terms of* $\sin A$ *and* $\cos A$.

$$\sin 2A = \sin(A + A) = \sin A \cos A + \cos A \sin A;$$

that is,
$$\sin 2A = 2 \sin A \cos A,$$

Since A may have any value, this is a perfectly general formula for the sine of an angle in terms of the sine and cosine of the half angle. Thus if $2A$ be replaced by θ, we have

$$\sin\theta = 2\sin\frac{\theta}{2}\cos\frac{\theta}{2}.$$

Similarly,
$$\begin{aligned}\sin 4A &= 2\sin 2A \cos 2A\\ &= 4\sin A\cos A\cos 2A.\end{aligned}$$

121. *To express* $\cos 2A$ *in terms of* $\cos A$ *and* $\sin A$.

$$\cos 2A = \cos(A + A) = \cos A \cos A - \sin A \sin A;$$

that is,
$$\cos 2A = \cos^2 A - \sin^2 A \qquad\qquad (1).$$

There are two other useful forms in which $\cos 2A$ may be expressed, one involving $\cos A$ only, the other $\sin A$ only.

Thus from (1),

$$\cos 2A = \cos^2 A - (1 - \cos^2 A);$$

that is,
$$\cos 2A = 2\cos^2 A - 1 \qquad\qquad (2).$$

Again, from (1),

$$\cos 2A = (1 - \sin^2 A) - \sin^2 A;$$

that is,
$$\cos 2A = 1 - 2\sin^2 A \qquad\qquad (3).$$

From formulæ (2) and (3), we obtain by transposition

$$1 + \cos 2A = 2\cos^2 A \qquad\qquad (4),$$

and
$$1 - \cos 2A = 2\sin^2 A \qquad\qquad (5).$$

By division,
$$\frac{1 - \cos 2A}{1 + \cos 2A} = \tan^2 A \qquad\qquad (6).$$

Example. **Express $\cos 4a$ in terms of $\sin a$.**

From (3),
$$\begin{aligned}\cos 4a &= 1 - 2\sin^2 2a = 1 - 2(4\sin^2 a\cos^2 a)\\ &= 1 - 8\sin^2 a(1 - \sin^2 a)\\ &= 1 - 8\sin^2 a + 8\sin^4 a.\end{aligned}$$

These formulæ are perfectly general and may be applied to cases of any two angles, one of which is three times the other; thus

$$\cos 6a = 4\cos^3 2a - 3\cos 2a;$$

$$\sin 9A = 3\sin 3A - 4\sin^3 3A.$$

126. *To find the value of* sin 18°.

Let $A = 18°$, then $5A = 90°$, so that $2A = 90° - 3A$.

$$\therefore \sin 2A = \sin(90° - 3A) = \cos 3A;$$

$$\therefore 2\sin A\cos A = 4\cos^3 A - 3\cos A.$$

Divide by $\cos A$ (which is not equal to zero);

$$\therefore 2\sin A = 4\cos^2 A - 3 = 4(1 - \sin^2 A) - 3;$$

$$\therefore 4\sin^2 A + 2\sin A - 1 = 0;$$

$$\therefore \sin A = \frac{-2 \pm \sqrt{4+16}}{8} = \frac{-1 \pm \sqrt{5}}{4}.$$

Since 18° is an acute angle, we take the positive sign;

$$\therefore \sin 18° = \frac{\sqrt{5}-1}{4}.$$

Example. Find cos 18° and sin 54°.

$$\cos 18° = \sqrt{1 - \sin^2 18°} = \sqrt{1 - \frac{6 - 2\sqrt{5}}{16}} = \frac{\sqrt{10 + 2\sqrt{5}}}{4}.$$

Since 54° and 36° are complementary, sin 54° = cos 36°.

Now $\cos 36° = 1 - 2\sin^2 18° = 1 - \frac{2(6 - 2\sqrt{5})}{16} = \frac{\sqrt{5}+1}{4}$;

$$\therefore \sin 54° = \frac{\sqrt{5}+1}{4}.$$

EXAMPLES. XI. e.

1. If $\cos A = \frac{1}{3}$, find $\cos 3A$.

2. Find $\sin 3A$ when $\sin A = \frac{3}{5}$.

3. Given $\tan A = 3$, find $\tan 3A$.

Prove the following identities:

24. $\left(\sin\frac{A}{2}+\cos\frac{A}{2}\right)^2=1+\sin A.$

25. $\left(\sin\frac{A}{2}-\cos\frac{A}{2}\right)^2=1-\sin A.$

26. $\frac{\cos 2a}{1+\sin 2a}=\tan(45^\circ-a).$

27. $\frac{\cos 2a}{1-\sin 2a}=\cot(45^\circ-a).$

28. $\sin 8A=8\sin A\cos A\cos 2A\cos 4A.$

29. $\cos 4A=8\cos^4 A-8\cos^2 A+1.$

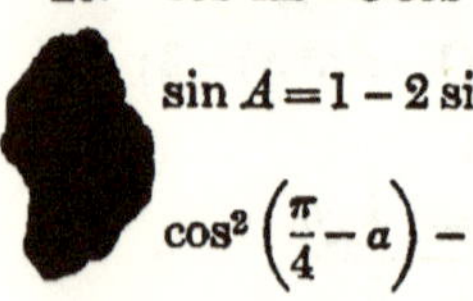

$\sin A=1-2\sin^2\left(45^\circ-\frac{A}{2}\right).$

$\cos^2\left(\frac{\pi}{4}-a\right)-\sin^2\left(\frac{\pi}{4}-a\right)=\sin 2a.$

32. $\tan(45^\circ+A)-\tan(45^\circ-A)=2\tan 2A.$

33. $\tan(45^\circ+A)+\tan(45^\circ-A)=2\sec 2A.$

125. Functions of 3A.

$$\begin{aligned}\sin 3A=\sin(2A+A)&=\sin 2A\cos A+\cos 2A\sin A\\ &=2\sin A\cos^2 A+(1-2\sin^2 A)\sin A\\ &=2\sin A(1-\sin^2 A)+(1-2\sin^2 A)\sin A;\\ &=3\sin A-4\sin^3 A.\end{aligned}$$

Similarly it may be proved that

$$\cos 3A=4\cos^3 A-3\cos A.$$

Again, $\tan 3A=\tan(2A+A)=\frac{\tan 2A+\tan A}{1-\tan 2A\tan A}$;

by putting $\tan 2A=\frac{2\tan A}{1-\tan^2 A}$,

we obtain on reduction

$$\tan 3A=\frac{3\tan A-\tan^3 A}{1-3\tan^2 A}.$$

Prove the following identities:

4. $\dfrac{\sin 3A}{\sin A} - \dfrac{\cos 3A}{\cos A} = 2.$

5. $\cot 3A = \dfrac{\cot^3 A - 3\cot A}{3\cot^2 A - 1}.$

6. $\dfrac{3\cos\alpha + \cos 3\alpha}{3\sin\alpha - \sin 3\alpha} = \cot^3\alpha.$

7. $\dfrac{\sin 3\alpha + \sin^3\alpha}{\cos^3\alpha - \cos 3\alpha} = \cot\alpha.$

8. $\dfrac{\cos^3\alpha - \cos 3\alpha}{\cos\alpha} + \dfrac{\sin^3\alpha + \sin 3\alpha}{\sin\alpha} = 3.$

9. $\sin 18° + \sin 30° = \sin 54°.$

10. $\cos 36° - \sin 18° = \dfrac{1}{2}.$

11. $\cos^2 36° + \sin^2 18° = \dfrac{3}{4}.$

12. $4\sin 18°\cos 36° = 1.$

127. The following examples further illustrate the formulæ proved in this chapter.

Example 1. Shew that $\cos^6\alpha + \sin^6\alpha = 1 - \dfrac{3}{4}\sin^2 2\alpha.$

$$\begin{aligned}\text{The first side} &= (\cos^2\alpha + \sin^2\alpha)(\cos^4\alpha + \sin^4\alpha - \cos^2\alpha\sin^2\alpha)\\ &= (\cos^2\alpha + \sin^2\alpha)^2 - 3\cos^2\alpha\sin^2\alpha\\ &= 1 - \frac{3}{4}(4\cos^2\alpha\sin^2\alpha)\\ &= 1 - \frac{3}{4}\sin^2 2\alpha.\end{aligned}$$

Example 2. Prove that $\dfrac{\cos A - \sin A}{\cos A + \sin A} = \sec 2A - \tan 2A.$

$$\text{The right side} = \frac{1}{\cos 2A} - \frac{\sin 2A}{\cos 2A} = \frac{1 - \sin 2A}{\cos 2A},$$

and since $\cos 2A = \cos^2 A - \sin^2 A = (\cos A + \sin A)(\cos A - \sin A)$, this suggests that we should multiply the numerator and denominator of the left side by $\cos A - \sin A$; thus

$$\begin{aligned}\text{the first side} &= \frac{(\cos A - \sin A)(\cos A - \sin A)}{(\cos A + \sin A)(\cos A - \sin A)}\\ &= \frac{\cos^2 A + \sin^2 A - 2\cos A\sin A}{\cos^2 A - \sin^2 A}\\ &= \frac{1 - \sin 2A}{\cos 2A} = \sec 2A - \tan 2A.\end{aligned}$$

Example 3. Shew that $\dfrac{1}{\tan 3A - \tan A} - \dfrac{1}{\cot 3A - \cot A} = \cot 2A$.

The first side $= \dfrac{1}{\dfrac{\sin 3A}{\cos 3A} - \dfrac{\sin A}{\cos A}} - \dfrac{1}{\dfrac{\cos 3A}{\sin 3A} - \dfrac{\cos A}{\sin A}}$

$$= \frac{\cos 3A \cos A}{\sin 3A \cos A - \cos 3A \sin A} - \frac{\sin 3A \sin A}{\cos 3A \sin A - \sin 3A \cos A}$$

$$= \frac{\cos 3A \cos A + \sin 3A \sin A}{\sin 3A \cos A - \cos 3A \sin A}$$

$$= \frac{\cos (3A - A)}{\sin (3A - A)} = \frac{\cos 2A}{\sin 2A} = \cot 2A.$$

NOTE. This example has been given to emphasize the fact that in identities involving the functions of $2A$ and $3A$ it is sometimes best not to substitute their equivalents in terms of functions of A.

EXAMPLES. XI. f.

Prove the following identities :

1. $\tan 2A - \sec A \sin A = \tan A \sec 2A$.
2. $\tan 2A + \cos A \operatorname{cosec} A = \cot A \sec 2A$.
3. $\dfrac{1 - \cos 2\theta + \sin 2\theta}{1 + \cos 2\theta + \sin 2\theta} = \tan \theta$.
4. $\dfrac{1 + \cos \theta + \cos \frac{\theta}{2}}{\sin \theta + \sin \frac{\theta}{2}} = \cot \dfrac{\theta}{2}$.
5. $\cos^6 a - \sin^6 a = \cos 2a \left(1 - \frac{1}{4} \sin^2 2a\right)$.
6. $4(\cos^6 \theta + \sin^6 \theta) = 1 + 3 \cos^2 2\theta$.
7. $\dfrac{\cos 3a + \sin 3a}{\cos a - \sin a} = 1 + 2 \sin 2a$.
8. $\dfrac{\cos 3a - \sin 3a}{\cos a + \sin a} = 1 - 2 \sin 2a$.
9. $\dfrac{\cos a + \sin a}{\cos a - \sin a} = \tan 2a + \sec 2a$.

Prove the following identities:

10. $\dfrac{\cot a-1}{\cot a+1}=\dfrac{1-\sin 2a}{\cos 2a}$.

11. $\dfrac{1+\sin\theta}{\cos\theta}=\dfrac{1+\tan\frac{\theta}{2}}{1-\tan\frac{\theta}{2}}$.

12. $\dfrac{\cos\theta}{1-\sin\theta}=\dfrac{\cot\frac{\theta}{2}+1}{\cot\frac{\theta}{2}-1}$.

13. $\sec A-\tan A=\tan\left(45^\circ-\dfrac{A}{2}\right)$.

14. $\tan A+\sec A=\cot\left(45^\circ-\dfrac{A}{2}\right)$.

15. $\dfrac{1+\sin\theta}{1-\sin\theta}=\tan^2\left(\dfrac{\pi}{4}+\dfrac{\theta}{2}\right)$.

16. $(2\cos A+1)(2\cos A-1)=2\cos 2A+1$.

17. $\dfrac{\sin 2A}{1+\cos 2A}\cdot\dfrac{\cos A}{1+\cos A}=\tan\dfrac{A}{2}$.

18. $\dfrac{\sin 2A}{1-\cos 2A}\cdot\dfrac{1-\cos A}{\cos A}=\tan\dfrac{A}{2}$.

19. $4\sin^3 a\cos 3a+4\cos^3 a\sin 3a=3\sin 4a$.
[*Put* $4\sin^3 a=3\sin a-\sin 3a$ *and* $4\cos^3 a=3\cos a+\cos 3a$.]

20. $\cos^3 a\cos 3a+\sin^3 a\sin 3a=\cos^3 2a$.

21. $4(\cos^3 20^\circ+\cos^3 40^\circ)=3(\cos 20^\circ+\cos 40^\circ)$.

22. $4(\cos^3 10^\circ+\sin^3 20^\circ)=3(\cos 10^\circ+\sin 20^\circ)$.

23. $\tan 3A-\tan 2A-\tan A=\tan 3A\tan 2A\tan A$.
[*Use* $\tan 3A=\tan(2A+A)$.]

24. $\dfrac{\cot\theta}{\cot\theta-\cot 3\theta}+\dfrac{\tan\theta}{\tan\theta-\tan 3\theta}=1$.

25. $\dfrac{1}{\tan 3\theta+\tan\theta}-\dfrac{1}{\cot 3\theta+\cot\theta}=\cot 4\theta$.

CHAPTER XII.

TRANSFORMATION OF PRODUCTS AND SUMS.

Transformation of products into sums or differences.

128. In the last chapter we have proved that

$$\sin A \cos B + \cos A \sin B = \sin (A + B),$$

and

$$\sin A \cos B - \cos A \sin B = \sin (A - B).$$

By addition,

$$2 \sin A \cos B = \sin (A + B) + \sin (A - B) \text{ ....(1)};$$

by subtraction

$$2 \cos A \sin B = \sin (A + B) - \sin (A - B) \text{ ....(2)}.$$

These formulæ enable us to express the product of a sine and cosine as the sum or difference of two sines.

Again, $\cos A \cos B - \sin A \sin B = \cos (A + B),$

and $\cos A \cos B + \sin A \sin B = \cos (A - B).$

By addition,

$$2 \cos A \cos B = \cos (A + B) + \cos (A - B) \text{ ....(3)};$$

by subtraction,

$$2 \sin A \sin B = \cos (A - B) - \cos (A + B) \text{ ....(4)}.$$

These formulæ enable us to express

(i) the product of two cosines as the sum of two cosines;

(ii) the product of two sines as the difference of two cosines.

129. In each of the four formulæ of the previous article it should be noticed that on the left side we have any two angles A and B, and on the right side the sum and difference of these angles.

r practical purposes the following verbal statements of the s are more useful.

$$\left\{\begin{array}{l} 2\sin A\cos B=\sin(\textit{sum})+\sin(\textit{difference});\\ 2\cos A\sin B=\sin(\textit{sum})-\sin(\textit{difference});\\ 2\cos A\cos B=\cos(\textit{sum})+\cos(\textit{difference});\\ 2\sin A\sin B=\cos(\textit{difference})-\cos(\textit{sum}).\end{array}\right.$$

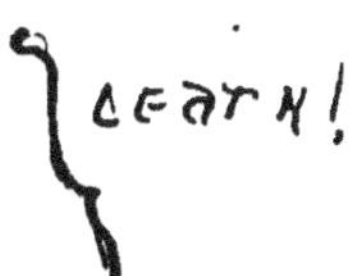

B. In the last of these formulæ, *the difference precedes m.*

ample 1. $2\sin 7A\cos 4A=\sin(\textit{sum})+\sin(\textit{difference})$

$$=\sin 11A+\sin 3A.$$

ample 2. $2\cos 3\theta\sin 6\theta=\sin(3\theta+6\theta)-\sin(3\theta-6\theta)$

$$=\sin 9\theta-\sin(-3\theta)$$
$$=\sin 9\theta+\sin 3\theta.$$

ample 3. $\cos\frac{3A}{2}\cos\frac{5A}{2}=\frac{1}{2}\left\{\cos\left(\frac{3A}{2}+\frac{5A}{2}\right)+\cos\left(\frac{3A}{2}-\frac{5A}{2}\right)\right\}$

$$=\frac{1}{2}\{\cos 4A+\cos(-A)\}$$
$$=\frac{1}{2}(\cos 4A+\cos A).$$

ample 4. $2\sin 75^\circ\sin 15^\circ=\cos(75^\circ-15^\circ)-\cos(75^\circ+15^\circ)$

$$=\cos 60^\circ-\cos 90^\circ$$
$$=\frac{1}{2}-0$$
$$=\frac{1}{2}.$$

0. After a little practice the student will be able to omit of the steps and find the equivalent very rapidly.

ample 1. $2\cos\left(\frac{\pi}{4}+\theta\right)\cos\left(\frac{\pi}{4}-\theta\right)=\cos\frac{\pi}{2}+\cos 2\theta=\cos 2\theta.$

ample 2. $\sin(a-2\beta)\cos(a+2\beta)=\frac{1}{2}\{\sin 2a+\sin(-4\beta)\}$

$$=\frac{1}{2}(\sin 2a-\sin 4\beta).$$

EXAMPLES. XII. a.

Express in the form of a sum or difference

1. $2\sin 3\theta\cos\theta$.
2. $2\cos 6\theta\sin 3\theta$.
3. $2\cos 7A\cos 5A$.
4. $2\sin 3A\sin 2A$.
5. $2\cos 5\theta\sin 4\theta$.
6. $2\sin 4\theta\cos 8\theta$.
7. $2\sin 9\theta\sin 3\theta$.
8. $2\cos 9\theta\sin 7\theta$.
9. $2\cos 2\alpha\cos 11\alpha$.
10. $2\sin 5\alpha\sin 10\alpha$.
11. $\sin 4\alpha\cos 7\alpha$.
12. $\sin 3\alpha\sin\alpha$.
13. $\cos\frac{A}{2}\sin\frac{3A}{2}$.
14. $\sin\frac{5A}{2}\cos\frac{7A}{2}$.
15. $2\cos\frac{2\theta}{3}\cos\frac{5\theta}{3}$.
16. $\sin\frac{\theta}{4}\sin\frac{3\theta}{4}$.
17. $2\cos 2\beta\cos(\alpha-\beta)$.
18. $2\sin 3\alpha\sin(\alpha+\beta)$.
19. $2\sin(2\theta+\phi)\cos(\theta-2\phi)$.
20. $2\cos(3\theta+\phi)\sin(\theta-2\phi)$.
21. $\cos(60°+\alpha)\sin(60°-\alpha)$.

Transformation of sums or differences into products.

131. Since $\sin(A+B)=\sin A\cos B+\cos A\sin B$,

and $\sin(A-B)=\sin A\cos B-\cos A\sin B$;

by addition,

$$\sin(A+B)+\sin(A-B)=2\sin A\cos B \quad\text{.........(1)};$$

by subtraction,

$$\sin(A+B)-\sin(A-B)=2\cos A\sin B \quad\text{.........(2)}.$$

Again, $\cos(A+B)=\cos A\cos B-\sin A\sin B$,

and $\cos(A-B)=\cos A\cos B+\sin A\sin B$.

By addition,

$$\cos(A+B)+\cos(A-B)=2\cos A\cos B \quad\text{.........(3)};$$

by subtraction,

$$\cos(A+B)-\cos(A-B)=-2\sin A\sin B$$
$$=2\sin A\sin(-B)\quad\text{......(4)}.$$

Let $$A+B=C, \text{ and } A-B=D;$$

then $$A=\frac{C+D}{2}, \text{ and } B=\frac{C-D}{2}.$$

By substituting for A and B in the formulæ (1), (2), (3), (4), we obtain

$$\sin C+\sin D=2\sin\frac{C+D}{2}\cos\frac{C-D}{2},$$

$$\sin C-\sin D=2\cos\frac{C+D}{2}\sin\frac{C-D}{2},$$

$$\cos C+\cos D=2\cos\frac{C+D}{2}\cos\frac{C-D}{2},$$

$$\cos C-\cos D=2\sin\frac{C+D}{2}\sin\frac{D-C}{2}.$$

132. In practice, it is more convenient to quote the formulæ we have just obtained verbally as follows:

sum of two sines = 2 sin (*half-sum*) cos (*half-difference*);

difference of two sines = 2 cos (*half-sum*) sin (*half-difference*);

sum of two cosines = 2 cos (*half-sum*) cos (*half-difference*);

difference of two cosines

= 2 sin (*half-sum*) sin (*half-difference reversed*)

Example 1. $$\sin 14\theta+\sin 6\theta=2\sin\frac{14\theta+6\theta}{2}\cos\frac{14\theta-6\theta}{2}$$
$$=2\sin 10\theta\cos 4\theta.$$

Example 2. $$\sin 9A-\sin 7A=2\cos\frac{9A+7A}{2}\sin\frac{9A-7A}{2}$$
$$=2\cos 8A\sin A.$$

Example 3. $$\cos A+\cos 8A=2\cos\frac{9A}{2}\cos\left(-\frac{7A}{2}\right)$$
$$=2\cos\frac{9A}{2}\cos\frac{7A}{2}.$$

Example 4. $$\cos 70^\circ-\cos 10^\circ=2\sin 40^\circ\sin(-30^\circ)$$
$$=-2\sin 40^\circ\sin 30^\circ=-\sin 40^\circ.$$

EXAMPLES. XII. b.

Express in the form of a product

1. $\sin 8\theta + \sin 4\theta$.
2. $\sin 5\theta - \sin \theta$.
3. $\cos 7\theta + \cos 3\theta$.
4. $\cos 9\theta - \cos 11\theta$.
5. $\sin 7a - \sin 5a$.
6. $\cos 3a + \cos 8a$.
7. $\sin 3a + \sin 13a$.
8. $\cos 5a - \cos a$.
9. $\cos 2A + \cos 9A$.
10. $\sin 3A - \sin 11A$.
11. $\cos 10° - \cos 50°$.
12. $\sin 70° + \sin 50°$.

Prove that

13. $\dfrac{\cos a - \cos 3a}{\sin 3a - \sin a} = \tan 2a$.
14. $\dfrac{\sin 2a + \sin 3a}{\cos 2a - \cos 3a} = \cot \dfrac{a}{2}$.
15. $\dfrac{\cos 4\theta - \cos \theta}{\sin \theta - \sin 4\theta} = \tan \dfrac{5\theta}{2}$.
16. $\dfrac{\cos 2\theta - \cos 12\theta}{\sin 12\theta + \sin 2\theta} = \tan 5$
17. $\sin(60° + A) - \sin(60° - A) = \sin A$.
18. $\cos(30° - A) + \cos(30° + A) = \sqrt{3} \cos A$.
19. $\cos\left(\dfrac{\pi}{4} + a\right) - \cos\left(\dfrac{\pi}{4} - a\right) = -\sqrt{2} \sin a$.
20. $\dfrac{\cos(2a - 3\beta) + \cos 3\beta}{\sin(2a - 3\beta) + \sin 3\beta} = \cot a$.
21. $\dfrac{\cos(\theta - 3\phi) - \cos(3\theta + \phi)}{\sin(3\theta + \phi) + \sin(\theta - 3\phi)} = \tan(\theta + 2\phi)$.
22. $\dfrac{\sin(a + \beta) - \sin 4\beta}{\cos(a + \beta) + \cos 4\beta} = \tan \dfrac{a - 3\beta}{2}$.

133. The eight formulæ proved in this chapter are of th utmost importance and very little further progress can be mac until they have been thoroughly learnt. In the first group, th transformation is from products to sums and differences; in th second group, there is the converse transformation from sun and differences to products.

Many examples admit of solution by applying either of thes transformations, but it is absolutely necessary that the studer should master all the formulæ and apply them with equ readiness.

134. The following examples should be studied with great care.

Example 1. Prove that

$$\sin 5A + \sin 2A - \sin A = \sin 2A\,(2\cos 3A + 1).$$

The first side $= (\sin 5A - \sin A) + \sin 2A$

$= 2\cos 3A \sin 2A + \sin 2A$

$= \sin 2A\,(2\cos 3A + 1).$

Example 2. Prove that

$$\cos 2\theta \cos\theta - \sin 4\theta \sin\theta = \cos 3\theta \cos 2\theta.$$

The first side $= \frac{1}{2}(\cos 3\theta + \cos\theta) - \frac{1}{2}(\cos 3\theta - \cos 5\theta)$

$= \frac{1}{2}(\cos\theta + \cos 5\theta)$

$= \cos 3\theta \cos 2\theta.$

Example 3. Find the value of

$$\cos 20^\circ + \cos 100^\circ + \cos 140^\circ.$$

The expression $= \cos 20^\circ + (\cos 100^\circ + \cos 140^\circ)$

$= \cos 20^\circ + 2\cos 120^\circ \cos 20^\circ$

$= \cos 20^\circ + 2\left(-\frac{1}{2}\right)\cos 20^\circ$

$= \cos 20^\circ - \cos 20^\circ = 0.$

Example 4. Express as the product of four sines

$$\sin(\beta+\gamma-\alpha) + \sin(\gamma+\alpha-\beta) + \sin(\alpha+\beta-\gamma) - \sin(\alpha+\beta+\gamma).$$

The expression $= 2\sin\gamma\cos(\beta-\alpha) + 2\cos(\alpha+\beta)\sin(-\gamma)$

$= 2\sin\gamma\,\{\cos(\beta-\alpha) - \cos(\alpha+\beta)\}$

$= 2\sin\gamma\,(2\sin\beta\sin\alpha)$

$= 4\sin\alpha\sin\beta\sin\gamma.$

Example 5. Express $4\cos\alpha\cos\beta\cos\gamma$ as the sum of four cosines.

The expression $= 2\cos\alpha\,\{\cos(\beta+\gamma) + \cos(\beta-\gamma)\}$

$= 2\cos\alpha\cos(\beta+\gamma) + 2\cos\alpha\cos(\beta-\gamma)$

$= \cos(\alpha+\beta+\gamma) + \cos(\alpha-\beta-\gamma) + \cos(\alpha+\beta-\gamma) + \cos(\alpha-\beta+\gamma)$

$= \cos(\alpha+\beta+\gamma) + \cos(\beta+\gamma-\alpha) + \cos(\gamma+\alpha-\beta) + \cos(\alpha+\beta-\gamma).$

Example 6. Prove that $\sin^2 5x - \sin^2 3x = \sin 8x \sin 2x$.

First solution.

$$\begin{aligned}\sin^2 5x - \sin^2 3x &= (\sin 5x + \sin 3x)(\sin 5x - \sin 3x)\\ &= (2 \sin 4x \cos x)(2 \cos 4x \sin x)\\ &= (2 \sin 4x \cos 4x)(2 \sin x \cos x)\\ &= \sin 8x \sin 2x.\end{aligned}$$

Second solution.

$$\begin{aligned}\sin 8x \sin 2x &= \frac{1}{2}(\cos 6x - \cos 10x)\\ &= \frac{1}{2}\{1 - 2\sin^2 3x - (1 - 2\sin^2 5x)\}\\ &= \sin^2 5x - \sin^2 3x.\end{aligned}$$

Third solution.

By using the formula of Art. 114 we have at once

$$\sin^2 5x - \sin^2 3x = \sin(5x + 3x)\sin(5x - 3x) = \sin 8x \sin 2x.$$

EXAMPLES. XII. c.

Prove the following identities:

1. $\cos 3A + \sin 2A - \sin 4A = \cos 3A\,(1 - 2\sin A)$.
2. $\sin 3\theta - \sin\theta - \sin 5\theta = \sin 3\theta\,(1 - 2\cos 2\theta)$.
3. $\cos\theta + \cos 2\theta + \cos 5\theta = \cos 2\theta\,(1 + 2\cos 3\theta)$.
4. $\sin a - \sin 2a + \sin 3a = 4 \sin\frac{a}{2}\cos a \cos\frac{3a}{2}$.
5. $\sin 3a + \sin 7a + \sin 10a = 4 \sin 5a \cos\frac{7a}{2}\cos\frac{3a}{2}$.
6. $\sin A + 2\sin 3A + \sin 5A = 4\sin 3A\cos^2 A$.
7. $\dfrac{\sin 2a + \sin 5a - \sin a}{\cos 2a + \cos 5a + \cos a} = \tan 2a$.
8. $\dfrac{\sin a + \sin 2a + \sin 4a + \sin 5a}{\cos a + \cos 2a + \cos 4a + \cos 5a} = \tan 3a$.
9. $\dfrac{\cos 7\theta + \cos 3\theta - \cos 5\theta - \cos\theta}{\sin 7\theta - \sin 3\theta - \sin 5\theta + \sin\theta} = \cot 2\theta$.
10. $\cos 3A \sin 2A - \cos 4A \sin A = \cos 2A \sin A$.

Prove the following identities:

11. $\cos 5A \cos 2A - \cos 4A \cos 3A = -\sin 2A \sin A$.

12. $\sin 4\theta \cos \theta - \sin 3\theta \cos 2\theta = \sin \theta \cos 2\theta$.

13. $\cos 5^\circ - \sin 25^\circ = \sin 35^\circ$.

[*Use* $\sin 25^\circ = \cos 65^\circ$.]

14. $\sin 65^\circ + \cos 65^\circ = \sqrt{2} \cos 20^\circ$.

15. $\cos 80^\circ + \cos 40^\circ - \cos 20^\circ = 0$.

16. $\sin 78^\circ - \sin 18^\circ + \cos 132^\circ = 0$.

17. $\sin^2 5A - \sin^2 2A = \sin 7A \sin 3A$.

18. $\cos 2A \cos 5A = \cos^2 \frac{7A}{2} - \sin^2 \frac{3A}{2}$.

19. $\sin(\alpha+\beta+\gamma) + \sin(\alpha-\beta-\gamma) + \sin(\alpha+\beta-\gamma)$
$+ \sin(\alpha-\beta+\gamma) = 4 \sin\alpha \cos\beta \cos\gamma$.

20. $\cos(\beta+\gamma-\alpha) - \cos(\gamma+\alpha-\beta) + \cos(\alpha+\beta-\gamma)$
$- \cos(\alpha+\beta+\gamma) = 4 \sin\alpha \cos\beta \sin\gamma$.

21. $\sin 2\alpha + \sin 2\beta + \sin 2\gamma - \sin 2(\alpha+\beta+\gamma)$
$= 4 \sin(\beta+\gamma) \sin(\gamma+\alpha) \sin(\alpha+\beta)$.

22. $\cos\alpha + \cos\beta + \cos\gamma + \cos(\alpha+\beta+\gamma)$
$= 4 \cos\frac{\beta+\gamma}{2} \cos\frac{\gamma+\alpha}{2} \cos\frac{\alpha+\beta}{2}$.

23. $4 \sin A \sin(60^\circ + A) \sin(60^\circ - A) = \sin 3A$.

24. $4 \cos\theta \cos\left(\frac{2\pi}{3} + \theta\right) \cos\left(\frac{2\pi}{3} - \theta\right) = \cos 3\theta$.

25. $\cos\theta + \cos\left(\frac{2\pi}{3} - \theta\right) + \cos\left(\frac{2\pi}{3} + \theta\right) = 0$.

26. $\cos^2 A + \cos^2(60^\circ + A) + \cos^2(60^\circ - A) = \frac{3}{2}$.

[*Put* $2\cos^2 A = 1 + \cos 2A$.]

27. $\sin^2 A + \sin^2(120^\circ + A) + \sin^2(120^\circ - A) = \frac{3}{2}$.

28. $\cos 20^\circ \cos 40^\circ \cos 80^\circ = \frac{1}{8}$.

29. $\sin 20^\circ \sin 40^\circ \sin 80^\circ = \frac{1}{8}\sqrt{3}$.

135. Many [illegible] identities can be established connecting the functions of the three angles A, B, C, which satisfy the relation $A+B+C=180°$. In proving these it will be necessary to keep clearly in view the properties of complementary and supplementary angles. [Arts. 39 and 96.]

From the given relation, the sum of any two of the angles is the supplement of the third; thus

$$\sin(B+C)=\sin A, \qquad \cos(A+B)=-\cos C,$$
$$\tan(C+A)=-\tan B, \qquad \cos B=-\cos(C+A),$$
$$\sin C=\sin(A+B), \qquad \cot A=-\cot(B+C).$$

Again, $\frac{A}{2}+\frac{B}{2}+\frac{C}{2}=90°$, so that each half angle is the complement of the sum of the other two; thus

$$\cos\frac{A+B}{2}=\sin\frac{C}{2}, \quad \sin\frac{C+A}{2}=\cos\frac{B}{2}, \quad \tan\frac{B+C}{2}=\cot\frac{A}{2},$$
$$\cos\frac{C}{2}=\sin\frac{A+B}{2}, \quad \sin\frac{A}{2}=\cos\frac{B+C}{2}, \quad \tan\frac{B}{2}=\cot\frac{C+A}{2}.$$

Example 1. If $A+B+C=180°$, prove that

$$\sin 2A+\sin 2B+\sin 2C=4\sin A\sin B\sin C.$$

The first side $=2\sin(A+B)\cos(A-B)+2\sin C\cos C$

$$=2\sin C\cos(A-B)+2\sin C\cos C$$
$$=2\sin C\{\cos(A-B)+\cos C\}$$
$$=2\sin C\{\cos(A-B)-\cos(A+B)\}$$
$$=2\sin C\times 2\sin A\sin B$$
$$=4\sin A\sin B\sin C.$$

Example 2. If $A+B+C=180°$, prove that

$$\tan A+\tan B+\tan C=\tan A\tan B\tan C.$$

Since $A+B$ is the supplement of C, we have

$$\tan(A+B)=-\tan C;$$

$$\therefore \frac{\tan A+\tan B}{1-\tan A\tan B}=-\tan C;$$

whence by multiplying up and rearranging,

$$\tan A+\tan B+\tan C=\tan A\tan B\tan C.$$

Example 3. If $A+B+C=180°$, prove that

$$\cos A+\cos B+\cos C=1+4\sin\frac{A}{2}\sin\frac{B}{2}\sin\frac{C}{2}.$$

The first side $=2\cos\frac{A+B}{2}\cos\frac{A-B}{2}+\cos C$

$$=2\sin\frac{C}{2}\cos\frac{A-B}{2}+1-2\sin^2\frac{C}{2}$$

$$=1+2\sin\frac{C}{2}\left(\cos\frac{A-B}{2}-\sin\frac{C}{2}\right)$$

$$=1+2\sin\frac{C}{2}\left(\cos\frac{A-B}{2}-\cos\frac{A+B}{2}\right)$$

$$=1+2\sin\frac{C}{2}\left(2\sin\frac{A}{2}\sin\frac{B}{2}\right)$$

$$=1+4\sin\frac{A}{2}\sin\frac{B}{2}\sin\frac{C}{2}.$$

EXAMPLES. XII. d.

If $A+B+C=180°$, prove that

1. $\sin 2A-\sin 2B+\sin 2C=4\cos A\sin B\cos C$.

2. $\sin 2A-\sin 2B-\sin 2C=-4\sin A\cos B\cos C$.

3. $\sin A+\sin B+\sin C=4\cos\frac{A}{2}\cos\frac{B}{2}\cos\frac{C}{2}$.

4. $\sin A+\sin B-\sin C=4\sin\frac{A}{2}\sin\frac{B}{2}\cos\frac{C}{2}$.

5. $\cos A-\cos B+\cos C=4\cos\frac{A}{2}\sin\frac{B}{2}\cos\frac{C}{2}-1$.

6. $\dfrac{\sin B+\sin C-\sin A}{\sin A+\sin B+\sin C}=\tan\frac{B}{2}\tan\frac{C}{2}$.

7. $\tan\frac{B}{2}\tan\frac{C}{2}+\tan\frac{C}{2}\tan\frac{A}{2}+\tan\frac{A}{2}\tan\frac{B}{2}=1$.

[*Use* $\tan\frac{A+B}{2}=\cot\frac{C}{2}$, *and therefore* $\tan\frac{A+B}{2}\tan\frac{C}{2}=1$.]

If $A+B+C=180°$, prove that

8. $\dfrac{1+\cos A-\cos B+\cos C}{1+\cos A+\cos B-\cos C}=\tan\dfrac{B}{2}\cot\dfrac{C}{2}$.

9. $\cos 2A+\cos 2B+\cos 2C+4\cos A\cos B\cos C+1=0$.

10. $\cot B\cot C+\cot C\cot A+\cot A\cot B=1$.

11. $(\cot B+\cot C)(\cot C+\cot A)(\cot A+\cot B)$
$=\operatorname{cosec} A\operatorname{cosec} B\operatorname{cosec} C$.

12. $\cos^2 A+\cos^2 B+\cos^2 C+2\cos A\cos B\cos C=1$.
[*Use* $2\cos^2 A=1+\cos 2A$.]

13. $\sin^2\dfrac{A}{2}+\sin^2\dfrac{B}{2}+\sin^2\dfrac{C}{2}=1-2\sin\dfrac{A}{2}\sin\dfrac{B}{2}\sin\dfrac{C}{2}$.

14. $\cos^2 2A+\cos^2 2B+\cos^2 2C=1+2\cos 2A\cos 2B\cos 2C$.

15. $\dfrac{\cot B+\cot C}{\tan B+\tan C}+\dfrac{\cot C+\cot A}{\tan C+\tan A}+\dfrac{\cot A+\cot B}{\tan A+\tan B}=1$.

16. $\dfrac{\tan A+\tan B+\tan C}{(\sin A+\sin B+\sin C)^2}=\dfrac{\tan\dfrac{A}{2}\tan\dfrac{B}{2}\tan\dfrac{C}{2}}{2\cos A\cos B\cos C}$.

136. The following examples further illustrate the formulæ proved in this and the preceding chapter.

Example 1. Prove that $\cot(A+15°)-\tan(A-15°)=\dfrac{4\cos 2A}{2\sin 2A+1}$.

$$\text{The first side}=\frac{\cos(A+15°)}{\sin(A+15°)}-\frac{\sin(A-15°)}{\cos(A-15°)}$$

$$=\frac{\cos(A+15°)\cos(A-15°)-\sin(A+15°)\sin(A-15°)}{\sin(A+15°)\cos(A-15°)}$$

$$=\frac{\cos\{(A+15°)+(A-15°)\}}{\sin(A+15°)\cos(A-15°)}$$

$$=\frac{2\cos 2A}{2\sin(A+15°)\cos(A-15°)}=\frac{2\cos 2A}{\sin 2A+\sin 30°}$$

$$=\frac{4\cos 2A}{2\sin 2A+1}.$$

NOTE. In dealing with expressions which involve numerical angles it is usually advisable to effect some simplification before substituting the known values of the functions of the angles, especially if these contain surds.

Example 2. Prove that

$$\cos\frac{A}{2}+\cos\frac{B}{2}+\cos\frac{C}{2}=4\cos\frac{\pi-A}{4}\cos\frac{\pi-B}{4}\cos\frac{\pi-C}{4}.$$

The second side $=2\cos\frac{\pi-A}{4}\left[\cos\frac{2\pi-(B+C)}{4}+\cos\frac{B-C}{4}\right]$

$$=2\cos\frac{\pi-A}{4}\cos\frac{\pi+A}{4}+2\cos\frac{\pi-A}{4}\cos\frac{B-C}{4}$$

$$=\left(\cos\frac{\pi}{2}+\cos\frac{A}{2}\right)+2\cos\frac{B+C}{4}\cos\frac{B-C}{4}$$

$$=\cos\frac{A}{2}+\cos\frac{B}{2}+\cos\frac{C}{2}.$$

EXAMPLES. XII. e.

Prove the following identities:

1. $\cos(\alpha+\beta)\sin(\alpha-\beta)+\cos(\beta+\gamma)\sin(\beta-\gamma)$
$+\cos(\gamma+\delta)\sin(\gamma-\delta)+\cos(\delta+\alpha)\sin(\delta-\alpha)=0.$

2. $\dfrac{\sin(\beta-\gamma)}{\sin\beta\sin\gamma}+\dfrac{\sin(\gamma-\alpha)}{\sin\gamma\sin\alpha}+\dfrac{\sin(\alpha-\beta)}{\sin\alpha\sin\beta}=0.$

3. $\dfrac{\sin\alpha+\sin\beta+\sin(\alpha+\beta)}{\sin\alpha+\sin\beta-\sin(\alpha+\beta)}=\cot\dfrac{\alpha}{2}\cot\dfrac{\beta}{2}.$

4. $\sin\alpha\cos(\beta+\gamma)-\sin\beta\cos(\alpha+\gamma)=\cos\gamma\sin(\alpha-\beta).$

5. $\cos\alpha\cos(\beta+\gamma)-\cos\beta\cos(\alpha+\gamma)=\sin\gamma\sin(\alpha-\beta).$

6. $(\cos A-\sin A)(\cos 2A-\sin 2A)=\cos A-\sin 3A.$

7. If $\tan\theta=\dfrac{b}{a}$, prove that $a\cos 2\theta+b\sin 2\theta=a.$

[*See Art.* 124.]

8. Prove that $\sin 2A+\cos 2A=\dfrac{(1+\tan A)^2-2\tan^2 A}{1+\tan^2 A}.$

9. Prove that $\sin 4A=\dfrac{4\tan A\,(1-\tan^2 A)}{(1+\tan^2 A)^2}.$

10. If $A+B=45^\circ$, prove that
$(1+\tan A)(1+\tan B)=2.$

Prove the following identities :

11. $\cot(15^\circ - A) + \tan(15^\circ + A) = \dfrac{4\cos 2A}{1 - 2\sin 2A}$.

12. $\cot(15^\circ + A) + \tan(15^\circ + A) = \dfrac{4}{\cos 2A + \sqrt{3}\sin 2A}$.

13. $\tan(A + 30^\circ)\tan(A - 30^\circ) = \dfrac{1 - 2\cos 2A}{1 + 2\cos 2A}$.

14. $(2\cos A + 1)(2\cos A - 1)(2\cos 2A - 1) = 2\cos 4A + 1$.

15. $\tan(\beta - \gamma) + \tan(\gamma - \alpha) + \tan(\alpha - \beta)$
$= \tan(\beta - \gamma)\tan(\gamma - \alpha)\tan(\alpha - \beta)$.

16. $\sin(\beta - \gamma) + \sin(\gamma - \alpha) + \sin(\alpha - \beta)$
$+ 4\sin\dfrac{\beta - \gamma}{2}\sin\dfrac{\gamma - \alpha}{2}\sin\dfrac{\alpha - \beta}{2} = 0$.

17. $\cos^2(\beta - \gamma) + \cos^2(\gamma - \alpha) + \cos^2(\alpha - \beta)$
$= 1 + 2\cos(\beta - \gamma)\cos(\gamma - \alpha)\cos(\alpha - \beta)$.

18. $\cos^2\alpha + \cos^2\beta - 2\cos\alpha\cos\beta\cos(\alpha + \beta) = \sin^2(\alpha + \beta)$.

19. $\sin^2\alpha + \sin^2\beta + 2\sin\alpha\sin\beta\cos(\alpha + \beta) = \sin^2(\alpha + \beta)$.

20. $\cos 12^\circ + \cos 60^\circ + \cos 84^\circ = \cos 24^\circ + \cos 48^\circ$.

If $A + B + C = 180^\circ$, shew that

21. $\cos\dfrac{A}{2} + \cos\dfrac{B}{2} + \cos\dfrac{C}{2} = 4\cos\dfrac{B + C}{4}\cos\dfrac{C + A}{4}\cos\dfrac{A + B}{4}$.

22. $\cos\dfrac{A}{2} - \cos\dfrac{B}{2} + \cos\dfrac{C}{2} = 4\cos\dfrac{\pi + A}{4}\cos\dfrac{\pi - B}{4}\cos\dfrac{\pi + C}{4}$.

23. $\sin\dfrac{A}{2} + \sin\dfrac{B}{2} + \sin\dfrac{C}{2} = 1 + 4\sin\dfrac{\pi - A}{4}\sin\dfrac{\pi - B}{4}\sin\dfrac{\pi - C}{4}$.

If $\alpha + \beta + \gamma = \dfrac{\pi}{2}$, shew that

24. $\dfrac{\sin 2\alpha + \sin 2\beta + \sin 2\gamma}{\sin 2\alpha + \sin 2\beta - \sin 2\gamma} = \cot\alpha\cot\beta$.

25. $\tan\beta\tan\gamma + \tan\gamma\tan\alpha + \tan\alpha\tan\beta = 1$.

CHAPTER XIII.

RELATIONS BETWEEN THE SIDES AND ANGLES OF A TRIANGLE.

137. *In any triangle the sides are proportional to the sines of the opposite angles; that is,*

$$\frac{a}{\sin A}=\frac{b}{\sin B}=\frac{c}{\sin C}.$$

(1) Let the triangle ABC be acute-angled.

From A draw AD perpendicular to the opposite side; then

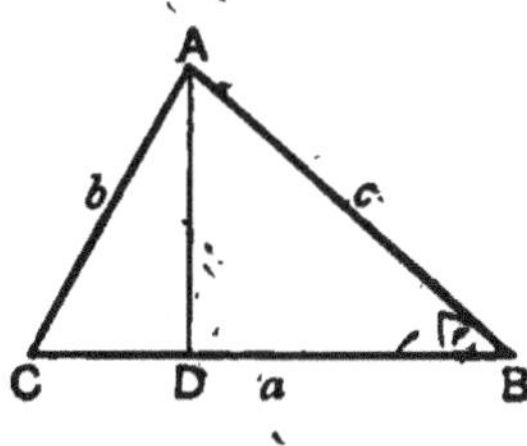

$$AD=AB\sin ABD=c\sin B,$$

and $$AD=AC\sin ACD=b\sin C;$$

$$\therefore\ b\sin C=c\sin B,$$

that is, $$\frac{b}{\sin B}=\frac{c}{\sin C}.$$

(2) Let the triangle ABC have an obtuse angle B.

Draw AD perpendicular to CB produced; then

$$AD=AC\sin ACD=b\sin C,$$

and $$AD=AB\sin ABD$$
$$=c\sin(180^\circ-B)=c\sin B;$$
$$\therefore\ b\sin C=c\sin B;$$

that is, $$\frac{b}{\sin B}=\frac{c}{\sin C}.$$

In like manner it may be proved that either of these ratios is equal to $\dfrac{a}{\sin A}$.

Thus $$\frac{a}{\sin A}=\frac{b}{\sin B}=\frac{c}{\sin C}.$$

138. *To find an expression for one side of a triangle in terms of the other two sides and the included angle.*

(1) Let ABC be an acute-angled triangle.

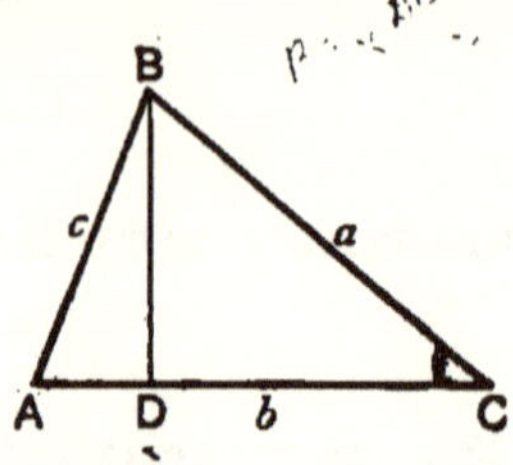

Draw BD perpendicular to AC; then by Euc. II. 13,

$$AB^2 = BC^2 + CA^2 - 2AC \cdot CD;$$

$$\therefore c^2 = a^2 + b^2 - 2b \cdot a \cos C$$

$$= a^2 + b^2 - 2ab \cos C.$$

(2) Let the triangle ABC have an obtuse angle C.

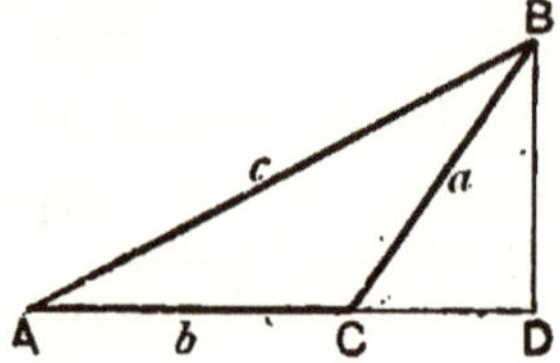

Draw BD perpendicular to AC produced; then by Euc. II. 12,

$$AB^2 = BC^2 + CA^2 + 2AC \cdot CD;$$

$$\therefore c^2 = a^2 + b^2 + 2b \cdot a \cos BCD$$

$$= a^2 + b^2 + 2ab \cos (180° - C)$$

$$= a^2 + b^2 - 2ab \cos C.$$

Hence in each case, $c^2 = a^2 + b^2 - 2ab \cos C$.

Similarly it may be shewn that

$$a^2 = b^2 + c^2 - 2bc \cos A,$$

and

$$b^2 = c^2 + a^2 - 2ca \cos B.$$

139. From the formulæ of the last article, we obtain

$$\cos A = \frac{b^2 + c^2 - a^2}{2bc}; \quad \cos B = \frac{c^2 + a^2 - b^2}{2ca}; \quad \cos C = \frac{a^2 + b^2 - c^2}{2ab}.$$

These results enable us to find the cosines of the angles when the numerical values of the sides are given.

140. *To express one side of a triangle in terms of the adjacent angles and the other two sides.*

(1) Let ABC be an acute-angled triangle.

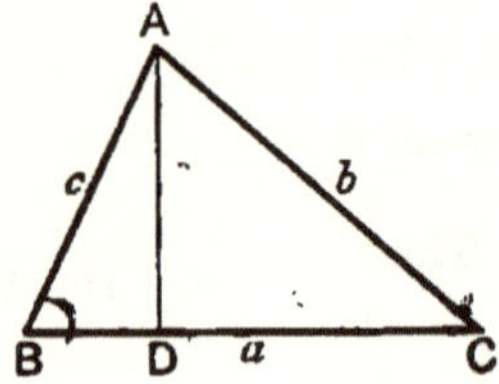

Draw AD perpendicular to BC; then

$$BC = BD + CD$$

$$= AB \cos ABD + AC \cos ACD;$$

that is, $a = c \cos B + b \cos C$.

(2) Let the triangle ABC have an obtuse angle C.

Draw AD perpendicular to BC produced; then

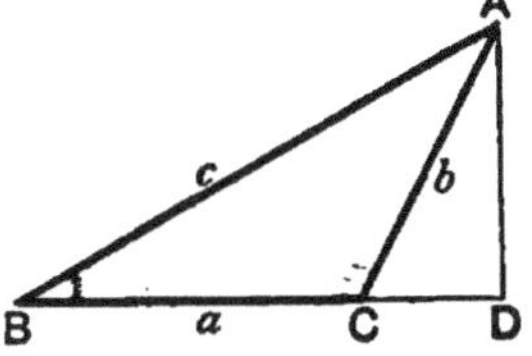

$$BC = BD - CD$$
$$= AB \cos ABD - AC \cos ACD;$$
$$\therefore\ a = c \cos B - b \cos (180° - C)$$
$$= c \cos B + b \cos C.$$

Thus in each case $a = b \cos C + c \cos B$.

Similarly it may be shewn that

$$b = c \cos A + a \cos C, \text{ and } c = a \cos B + b \cos A.$$

NOTE. The formulæ we have proved in this chapter are quite general and may be regarded as the fundamental relations subsisting between the sides and angles of a triangle. The modified forms which they assume in the case of right-angled triangles have already been considered in Chap. V.; it will therefore be unnecessary in the present chapter to make any direct reference to right-angled triangles.

141. The sets of formulæ in Arts. 137, 138, and 140 have been established independently of one another; they are however not independent, for from any one set the other two may be derived by the help of the relation $A + B + C = 180°$.

For instance, suppose we have proved as in Art. 137 that

$$\frac{a}{\sin A} = \frac{b}{\sin B} = \frac{c}{\sin C};$$

then since $\sin A = \sin (B + C) = \sin B \cos C + \sin C \cos B$;

$$\therefore\ 1 = \frac{\sin B}{\sin A} \cos C + \frac{\sin C}{\sin A} \cos B;$$

$$\therefore\ 1 = \frac{b}{a} \cos C + \frac{c}{a} \cos B;$$

$$\therefore\ a = b \cos C + c \cos B.$$

Similarly, we may prove that

$$b = c \cos A + a \cos C, \text{ and } c = a \cos B + b \cos A.$$

Multiplying these last three equations by a, b, $-c$ respectively and adding, we have

$$a^2 + b^2 - c^2 = 2ab \cos C;$$
$$\therefore\ c^2 = a^2 + b^2 - 2ab \cos C.$$

Similarly the other relations of Art. 138 may be deduced.

Solution of Triangles.

142. When any three parts of a triangle are given, provided that one at least of these is a side, the relations we have proved enable us to find the numerical values of the unknown parts. For from any equation which connects four quantities three of which are known the fourth may be found. Thus if c, a, B are given, we can find b from the formula

$$b^2 = c^2 + a^2 - 2ca \cos B;$$

and if B, C, b are given, we find c from the formula

$$\frac{c}{\sin C} = \frac{b}{\sin B}.$$

We may remark that if the three angles alone are given, the formula

$$\frac{a}{\sin A} = \frac{b}{\sin B} = \frac{c}{\sin C}$$

enables us to find the *ratios* of the sides but not their actual *lengths*, and thus the triangle cannot be completely solved. In such a case there may be an infinite number of equiangular triangles all satisfying the data of the question. [See Euc. VI. 4.]

143. CASE I. *To solve a triangle having given the three sides.*

The angles A and B may be found from the formulæ

$$\cos A = \frac{b^2 + c^2 - a^2}{2bc}, \text{ and } \cos B = \frac{c^2 + a^2 - b^2}{2ca};$$

then the angle C is known from the equation $C = 180° - A - B$.

Example 1. If $a = 7$, $b = 5$, $c = 8$, find the angles A and B, having given that $\cos 38° 11' = \frac{11}{14}$.

$$\cos A = \frac{b^2 + c^2 - a^2}{2bc} = \frac{5^2 + 8^2 - 7^2}{2 \times 5 \times 8} = \frac{40}{2 \times 5 \times 8} = \frac{1}{2};$$

$$\therefore A = 60°.$$

$$\cos B = \frac{c^2 + a^2 - b^2}{2ca} = \frac{8^2 + 7^2 - 5^2}{2 \times 8 \times 7} = \frac{88}{2 \times 8 \times 7} = \frac{11}{14};$$

$$\therefore B = 38° 11'.$$

Example 2. Find the greatest angle of the triangle whose side 6, 13, 11, having given that $\cos 84° 47' = \frac{1}{11}$.

Let $a=6$, $b=13$, $c=11$. Since the greatest angle is opposite to the greatest side, the required angle is B.

And $$\cos B = \frac{c^2+a^2-b^2}{2ca} = \frac{11^2+6^2-13^2}{2\times 11\times 6} = \frac{-12}{2\times 11\times 6};$$

$$\therefore \cos B = -\frac{1}{11} = -\cos 84° 47';$$

$$\therefore B = 180° - 84° 47' = 95° 13'.$$

Thus the required angle is 95° 13′.

144. Case II. *To solve a triangle having given two sides and the included angle.*

Let b, c, A be given; then a can be found from the formula

$$a^2 = b^2 + c^2 - 2bc \cos A.$$

We may now obtain B from either of the formulæ

$$\cos B = \frac{c^2+a^2-b^2}{2ca}, \text{ or } \sin B = \frac{b \sin A}{a};$$

then C is known from the equation $C = 180° - A - B$.

Example. If $a=3$, $b=7$, $C=98° 13'$, solve the triangle, having given $\cos 81° 47' = \frac{1}{7}$.

$$c^2 = a^2 + b^2 - 2ab \cos C$$
$$= 9 + 49 - 2\times 3\times 7 \cos 98° 13'.$$

But 98° 13′ is the supplement of 81° 47′;

$$\therefore c^2 = 58 + (2\times 3\times 7 \cos 81° 47')$$
$$= 58 + \left(2\times 3\times 7\times \frac{1}{7}\right) = 58 + 6 = 64;$$

$$\therefore c = 8.$$

$$\cos B = \frac{c^2+a^2-b^2}{2ca} = \frac{64+9-49}{2\times 8\times 3} = \frac{24}{2\times 8\times 3} = \frac{1}{2};$$

$$\therefore B = 60°.$$

$$A = 180° - 60° - 98° 13' = 21° 47'.$$

145. CASE III. *To solve a triangle having given two angl and a side.*

Let B, C, a be given.

The angle A is found from $A=180°-B-C$; and the sides and c from

$$b=\frac{a\sin B}{\sin A} \text{ and } c=\frac{a\sin C}{\sin A}.$$

Example. If $A=105°$, $C=60°$, $b=4$, solve the triangle.

$$B=180°-105°-60°=15°.$$

$$\therefore\ c=\frac{b\sin C}{\sin B}=\frac{4\sin 60°}{\sin 15°}=\frac{4\sqrt{3}}{2}\cdot\frac{2\sqrt{2}}{\sqrt{3}-1}=\frac{4\sqrt{6}}{\sqrt{3}-1}$$

$$=\frac{4\sqrt{6}(\sqrt{3}+1)}{3-1}=2\sqrt{6}(\sqrt{3}+1);$$

$$\therefore\ c=6\sqrt{2}+2\sqrt{6}.$$

$$a=\frac{b\sin A}{\sin B}=\frac{4\sin 105°}{\sin 15°}=\frac{4\sin 75°}{\sin 15°}$$

$$=4\times\frac{\sqrt{3}+1}{2\sqrt{2}}\times\frac{2\sqrt{2}}{\sqrt{3}-1}=\frac{4(\sqrt{3}+1)}{\sqrt{3}-1};$$

$$\therefore\ a=4(2+\sqrt{3}).$$

EXAMPLES. XIII. a.

1. If $a=15$, $b=7$, $c=13$, find C.
2. If $a=7$, $b=3$, $c=5$, find A.
3. If $a=5$, $b=5\sqrt{3}$, $c=5$, find the angles.
4. If $a=25$, $b=31$, $c=7\sqrt{2}$, find A.
5. The sides of a triangle are 2, $2\frac{2}{3}$, $3\frac{1}{3}$, find the greatest ang
6. Solve the triangle when $a=\sqrt{3}+1$, $b=2$, $c=\sqrt{6}$.
7. Solve the triangle when $a=\sqrt{2}$, $b=2$, $c=\sqrt{3}-1$.
8. If $a=8$, $b=5$, $c=\sqrt{19}$, find C; given $\cos 28°56'=\frac{7}{8}$.
9. If the sides are as 4 : 7 : 5, find the greatest angle;

   given $\cos 78°27'=\frac{1}{5}$.

10. If $a=2$, $b=\sqrt{3}+1$, $C=60°$, find c.

11. Given $a=3$, $c=5$, $B=120°$, find b.

12. Given $b=7$, $c=6$, $A=75° 31'$, find a; given $\cos 75° 31' = \cdot 25$.

13. If $b=8$, $c=11$, $A=93° 35'$, find a; given $\cos 86° 25' = \cdot 0625$.

14. If $a=7$, $c=3$, $B=123° 12'$, find b; given $\cos 56° 48' = \frac{23}{42}$.

15. Solve the triangle when $a=2\sqrt{6}$, $c=6-2\sqrt{3}$, $B=75°$.

16. Solve the triangle when $A=72°$, $b=2$, $c=\sqrt{5}+1$.

17. Given $A=75°$, $B=30°$, $b=\sqrt{8}$, solve the triangle.

18. If $B=60°$, $C=15°$, $b=\sqrt{6}$, solve the triangle.

19. If $A=45°$, $B=105°$, $c=\sqrt{2}$, solve the triangle.

20. Given $A=45°$, $B=60°$, shew that $c : a = \sqrt{3}+1 : 2$.

21. If $C=120°$, $c=2\sqrt{3}$, $a=2$, find b.

22. If $B=60°$, $a=3$, $b=3\sqrt{3}$, find c.

23. Given $(a+b+c)(b+c-a)=3bc$, find A.

24. Find the angles of the triangle whose sides are

$$3+\sqrt{3},\quad 2\sqrt{3},\quad \sqrt{6}.$$

25. Find the angles of the triangle whose sides are

$$\frac{\sqrt{3}+1}{2\sqrt{2}},\quad \frac{\sqrt{3}-1}{2\sqrt{2}},\quad \frac{\sqrt{3}}{2}.$$

26. Two sides of a triangle are $\frac{1}{\sqrt{6}-\sqrt{2}}$ and $\frac{1}{\sqrt{6}+\sqrt{2}}$, and the included angle is $60°$: solve the triangle.

146. When an angle of a triangle is obtained through the medium of the sine there may be ambiguity, for the sines of supplementary angles are equal in magnitude and are of the same sign, so that there are two angles less than 180° which have the same sine. When an angle is obtained through the medium of the cosine there is no ambiguity, for there is only one angle less than 180° whose cosine is equal to a given quantity.

Thus if $\sin A = \frac{1}{2}$, then $A = 30°$ or $150°$;

if $\cos A = \frac{1}{2}$, then $A = 60°$.

Example. If $C=60°$, $b=2\sqrt{3}$, $c=3\sqrt{2}$, find A.

From the equation $$\sin B=\frac{b\sin C}{c},$$

we have $$\sin B=\frac{2\sqrt{3}}{3\sqrt{2}}\cdot\frac{\sqrt{3}}{2}=\frac{1}{\sqrt{2}};$$

$$\therefore B=45° \text{ or } 135°.$$

The value $B=135°$ is inadmissible, for in this case the sum of B and C would be greater than 180°.

Thus $$A=180°-60°-45°=75°.$$

147. CASE IV. *To solve a triangle having given two sides and an angle opposite to one of them.*

Let a, b, A be given; then B is to be found from the equation

$$\sin B=\frac{b}{a}\sin A.$$

(i) If $a<b\sin A$, then $\frac{b\sin A}{a}>1$, so that $\sin B>1$, which is impossible. Thus there is no solution.

(ii) If $a=b\sin A$, then $\frac{b\sin A}{a}=1$, so that $\sin B=1$, and B has only the value 90°.

(iii) If $a>b\sin A$, then $\frac{b\sin A}{a}<1$, and two values for B may be found from $\sin B=\frac{b\sin A}{a}$. These values are supplementary, so that one angle is acute, the other obtuse.

(1) If $a<b$, then $A<B$, and therefore B may either be acute or obtuse, so that both values are admissible. This is known as **the ambiguous case.**

(2) If $a=b$, then $A=B$; and if $a>b$, then $A>B$; in either case B cannot be obtuse, and therefore only the smaller value of B is admissible.

When B is found, C is determined from $C=180°-A-B$. Finally, c may be found from the equation $c=\frac{a\sin C}{\sin A}$.

From the foregoing investigation it appears that the only case in which an ambiguous solution can arise is when the smaller of the two given sides is opposite to the given angle.

148. *To discuss the Ambiguous Case geometrically.*

Let a, b, A be the given parts. Take a line AX unlimited towards X; make $\angle XAC$ equal to A, and AC equal to b. Draw CD perpendicular to AX, then $CD = b \sin A$.

With centre C and radius equal to a describe a circle.

(i) If $a < b \sin A$, the circle will not meet AX; thus no triangle can be constructed with the given parts.

(ii) If $a = b \sin A$, the circle will *touch* AX at D; thus there is a right-angled triangle with the given parts.

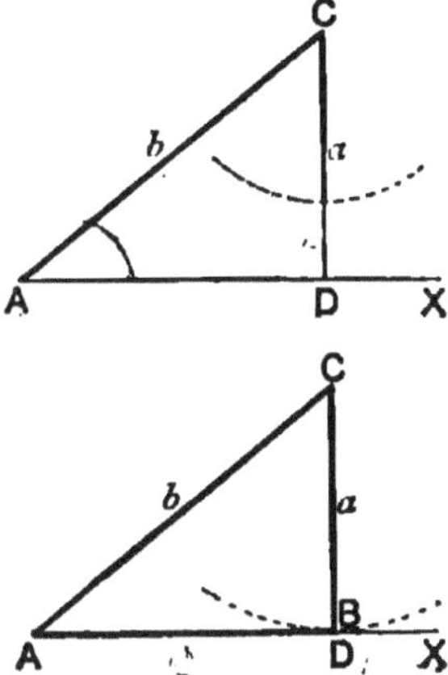

(iii) If $a > b \sin A$, the circle will cut AX in two points B_1, B_2.

(1) These points will be both on the same side of A, when $a < b$, in which case there are two solutions, namely the triangles

$$AB_1C, \quad AB_2C.$$

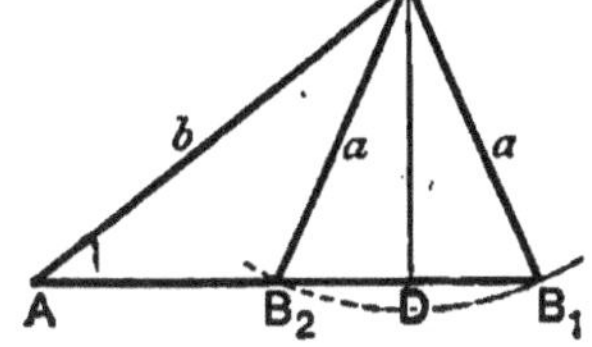

This is *the Ambiguous Case.*

(2) The points B_1, B_2 will be on opposite sides of A when $a > b$.

In this case there is only one solution, for the angle CAB_2 is *the supplement of the given angle,* and thus the triangle AB_2C does not satisfy the data.

(3) If $a = b$, the point B_2 coincides with A, so that there is only one solution.

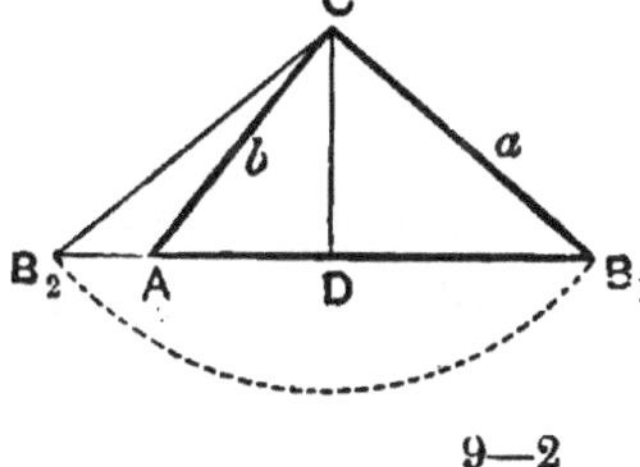

Example. Given $B=45°$, $c=\sqrt{12}$, $b=\sqrt{8}$, solve the triangle.

We have $$\sin C=\frac{c\sin B}{b}=\frac{2\sqrt{3}}{2\sqrt{2}}\cdot\frac{1}{\sqrt{2}}=\frac{\sqrt{3}}{2}.$$

$$\therefore\ C=60° \text{ or } 120°,$$

and since $b<c$, both these values are admissible. The two triangles which satisfy the data are shewn in the figure.

Denote the sides BC_1, BC_2 by a_1, a_2, and the angles BAC_1, BAC_2 by A_1, A_2 respectively.

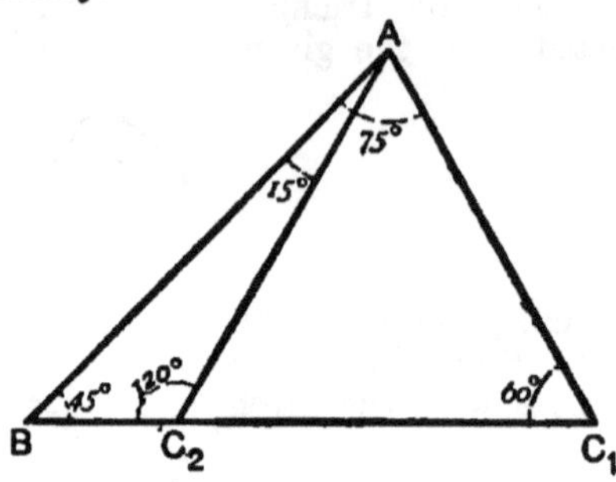

(i) In the $\triangle ABC_1$, $\angle A_1=75°$;

hence $$a_1=\frac{b\sin A_1}{\sin B}=\frac{2\sqrt{2}}{\frac{1}{\sqrt{2}}}\cdot\frac{\sqrt{3}+1}{2\sqrt{2}}=\sqrt{2}(\sqrt{3}+1).$$

(ii) In the $\triangle ABC_2$, $\angle A_2=15°$;

hence $$a_2=\frac{b\sin A_2}{\sin B}=\frac{2\sqrt{2}}{\frac{1}{\sqrt{2}}}\cdot\frac{\sqrt{3}-1}{2\sqrt{2}}=\sqrt{2}(\sqrt{3}-1).$$

Thus the complete solution is
$$\begin{cases} C=60°, \text{ or } 120°; \\ A=75°, \text{ or } 15°; \\ a=\sqrt{6}+\sqrt{2}, \text{ or } \sqrt{6}-\sqrt{2}. \end{cases}$$

EXAMPLES. XIII. b.

1. Given $a=1$, $b=\sqrt{3}$, $A=30°$, solve the triangle.

2. Given $b=3\sqrt{2}$, $c=2\sqrt{3}$, $C=45°$, solve the triangle.

3. If $C=60°$, $a=2$, $c=\sqrt{6}$, solve the triangle.

4. If $A=30°$, $a=2$, $c=5$, solve the triangle.

5. If $B=30°$, $b=\sqrt{6}$, $c=2\sqrt{3}$, solve the triangle.

6. If $B=60°$, $b=3\sqrt{2}$, $c=3+\sqrt{3}$, solve the triangle.

7. If $a=3+\sqrt{3}$, $c=3-\sqrt{3}$, $C=15°$, solve the triangle.

8. If $A=18°$, $a=4$, $b=4+\sqrt{80}$, solve the triangle.

9. If $B=135°$, $a=3\sqrt{2}$, $b=2\sqrt{3}$, solve the triangle.

149. Many relations connecting the sides and angles of a triangle may be proved by means of the formulæ we have established.

Example 1. Prove that $(b-c)\cos\frac{A}{2}=a\sin\frac{B-C}{2}$.

Let
$$k=\frac{a}{\sin A}=\frac{b}{\sin B}=\frac{c}{\sin C};$$
then
$$a=k\sin A,\quad b=k\sin B,\quad c=k\sin C;$$
$$\therefore\ (b-c)\cos\frac{A}{2}=k(\sin B-\sin C)\cos\frac{A}{2}$$
$$=2k\cos\frac{B+C}{2}\sin\frac{B-C}{2}\cos\frac{A}{2}$$
$$=2k\sin\frac{A}{2}\cos\frac{A}{2}\sin\frac{B-C}{2}$$
$$=k\sin A\sin\frac{B-C}{2}$$
$$=a\sin\frac{B-C}{2}.$$

Example 2. If $a\cos^2\frac{C}{2}+c\cos^2\frac{A}{2}=\frac{3b}{2}$, shew that the sides of the triangle are in A.P.

Since
$$2a\cos^2\frac{C}{2}+2c\cos^2\frac{A}{2}=3b,$$
$$\therefore\ a(1+\cos C)+c(1+\cos A)=3b,$$
$$\therefore\ a+c+(a\cos C+c\cos A)=3b,$$
$$\therefore\ a+c+b=3b,$$
$$\therefore\ a+c=2b.$$

Thus the sides a, b, c are in A.P.

Example 3. Prove that

$$(b^2-c^2)\cot A+(c^2-a^2)\cot B+(a^2-b^2)\cot C=0.$$

Let $k=\dfrac{a}{\sin A}=\dfrac{b}{\sin B}=\dfrac{c}{\sin C}$; then

the first side

$$=k^2\left\{(\sin^2 B-\sin^2 C)\frac{\cos A}{\sin A}+\ldots\ldots+\ldots\ldots\right\}$$

$$=k^2\left\{\sin(B+C)\sin(B-C)\frac{\cos A}{\sin A}+\ldots\ldots+\ldots\ldots\right\}. \quad \text{[Art. 114].}$$

But $\sin(B+C)=\sin A$, and $\cos A=-\cos(B+C)$;

∴ the first side

$$=-k^2\{\sin(B-C)\cos(B+C)+\ldots\ldots+\ldots\ldots\}$$

$$=-\frac{k^2}{2}\{(\sin 2B-\sin 2C)+(\sin 2C-\sin 2A)+(\sin 2A-\sin 2B)\}$$

$$=0.$$

EXAMPLES. XIII. c.

Prove the following identities :

1. $a(\sin B-\sin C)+b(\sin C-\sin A)+c(\sin A-\sin B)=0.$
2. $2(bc\cos A+ca\cos B+ab\cos C)=a^2+b^2+c^2.$
3. $a(b\cos C-c\cos B)=b^2-c^2.$
4. $(b+c)\cos A+(c+a)\cos B+(a+b)\cos C=a+b+c.$
5. $2\left(a\sin^2\dfrac{C}{2}+c\sin^2\dfrac{A}{2}\right)=c+a-b.$
6. $\dfrac{\cos B}{\cos C}=\dfrac{c-b\cos A}{b-c\cos A}.$
7. $\tan A=\dfrac{a\sin C}{b-a\cos C}.$
8. $(b+c)\sin\dfrac{A}{2}=a\cos\dfrac{B-C}{2}.$
9. $\dfrac{a+b}{c}\sin^2\dfrac{C}{2}=\dfrac{\cos A+\cos B}{2}.$
10. $a\sin(B-C)+b\sin(C-A)+c\sin(A-B)=0.$
11. $\dfrac{\sin(A-B)}{\sin(A+B)}=\dfrac{a^2-b^2}{c^2}.$
12. $\dfrac{c\sin(A-B)}{b\sin(C-A)}=\dfrac{a^2-b^2}{c^2-a^2}.$

[*All articles and examples marked with an asterisk may be omitted on the first reading of the subject.*]

*150. The *ambiguous case* may also be discussed by first finding the third side.

As before, let a, b, A be given, then

$$\cos A = \frac{b^2+c^2-a^2}{2bc};$$

$$\therefore \; c^2 - 2b\cos A \,.\, c + b^2 - a^2 = 0.$$

By solving this quadratic equation in c, we obtain

$$c = b\cos A \pm \sqrt{b^2\cos^2 A + a^2 - b^2}$$
$$= b\cos A \pm \sqrt{a^2 - b^2\sin^2 A}.$$

(i) When $a < b\sin A$, the quantity under the radical is negative, and the values of c are impossible; so that there is no solution.

(ii) When $a = b\sin A$, the quantity under the radical is zero, and $c = b\cos A$. Since $\sin A < 1$, it follows that $a < b$, and therefore $A < B$. Hence the triangle is impossible unless the given angle A is acute, in which case c is positive and there is one solution.

(iii) When $a > b\sin A$, there are three cases to consider.

(1) Suppose $a < b$, then $A < B$, and as before the triangle is impossible unless A is acute. In this case $b\cos A$ is positive.

Also $\sqrt{a^2 - b^2\sin^2 A}$ is real and $< \sqrt{b^2 - b^2\sin^2 A}$;

that is $\sqrt{a^2 - b^2\sin^2 A} < b\cos A$;

hence both values of c are real and positive, so that there are two solutions.

(2) Suppose $a > b$, then $\sqrt{a^2 - b^2\sin^2 A} > \sqrt{b^2 - b^2\sin^2 A}$;

that is $\sqrt{a^2 - b^2\sin^2 A} > b\cos A$;

hence one value of c is positive and one value is negative, whether A is acute or obtuse, and in each case there is only one solution.

(3) Suppose $a = b$, then $\sqrt{a^2 - b^2\sin^2 A} = b\cos A$;

$$\therefore \; c = 2b\cos A \text{ or } 0;$$

hence there is only one solution when A is acute, and when A is obtuse the triangle is impossible.

Example. If b, c, B are given, and if $b<c$, shew that

$$(a_1-a_2)^2+(a_1+a_2)^2\tan^2 B=4b^2,$$

where a_1, a_2 are the two values of the third side.

From the formula $\cos B=\dfrac{c^2+a^2-b^2}{2ca}$,

we have $$a^2-2c\cos B\,.\,a+c^2-b^2=0.$$

But the roots of this equation are a_1 and a_2; hence by the theory of quadratic equations

$$a_1+a_2=2c\cos B \text{ and } a_1a_2=c^2-b^2.$$

$$\therefore\ (a_1-a_2)^2=(a_1+a_2)^2-4a_1a_2$$
$$=4c^2\cos^2 B-4(c^2-b^2).$$

$$\therefore\ (a_1-a_2)^2+(a_1+a_2)^2\tan^2 B=4c^2\cos^2 B-4(c^2-b^2)+4c^2\cos^2 B\tan^2 B$$
$$=4c^2(\cos^2 B+\sin^2 B)-4c^2+4b^2$$
$$=4c^2-4c^2+4b^2$$
$$=4b^2.$$

*EXAMPLES. XIII. d.

1. In a triangle in which each base angle is double of the third angle the base is 2 : solve the triangle.

2. If $B=45°$, $C=75°$, and the perpendicular from A on BC is 3, solve the triangle.

3. If $a=2$, $b=4-2\sqrt{3}$, $c=3\sqrt{2}-\sqrt{6}$, solve the triangle.

4. If $A=18°$, $b-a=2$, $ab=4$, find the other angles.

5. Given $B=30°$, $c=150$, $b=50\sqrt{3}$, shew that of the two triangles which satisfy the data one will be isosceles and the other right-angled.

   Find the third side in the greater of these triangles. Would the solution be ambiguous if the data had been $B=30°$, $c=150$, $b=75$?

6. If $A=36°$, $a=4$, and the perpendicular from C upon AB is $\sqrt{5}-1$, find the other angles.

7. If the angles adjacent to the base of a triangle are $22\tfrac{1}{2}°$ and $112\tfrac{1}{2}°$, shew that the altitude is half the base.

8. If $a=2b$ and $A=3B$, find the angles and express c in terms of a.

9. The sides of a triangle are $2x+3$, x^2+3x+3, x^2+2x: shew that the greatest angle is 120°.

Shew that in any triangle

10. $(b-a)\cos C+c(\cos B-\cos A)=c\sin\frac{A-B}{2}\operatorname{cosec}\frac{A+B}{2}$.

11. $a\sin\left(\frac{A}{2}+B\right)=(b+c)\sin\frac{A}{2}$.

12. $\sin\left(B+\frac{C}{2}\right)\cos\frac{C}{2}=\frac{a+b}{b+c}\cos\frac{A}{2}\cos\frac{B-C}{2}$.

13. $\frac{1+\cos(A-B)\cos C}{1+\cos(A-C)\cos B}=\frac{a^2+b^2}{a^2+c^2}$.

14. If $c^4-2(a^2+b^2)c^2+a^4+a^2b^2+b^4=0$, prove that C is 60° or 120°.

15. If a, b, A are given, and if c_1, c_2 are the values of the third side in the ambiguous case, prove that if $c_1>c_2$,

(1) $c_1-c_2=2a\cos B_1$.

(2) $\cos\frac{C_1-C_2}{2}=\frac{b\sin A}{a}$.

(3) $c_1^2+c_2^2-2c_1c_2\cos 2A=4a^2\cos^2 A$.

(4) $\sin\frac{C_1+C_2}{2}\sin\frac{C_1-C_2}{2}=\cos A\cos B_1$.

16. If $A=45°$, and c_1, c_2 be the two values of the ambiguous side, shew that

$$\cos B_1CB_2=\frac{2c_1c_2}{c_1^2+c_2^2}.$$

17. If $\cos A+2\cos C:\cos A+2\cos B=\sin B:\sin C$, prove that the triangle is either isosceles or right-angled.

18. If a, b, c are in A. P., shew that

$\cot\frac{A}{2}$, $\cot\frac{B}{2}$, $\cot\frac{C}{2}$ are also in A. P.

19. Shew that

$$\frac{a^2\sin(B-C)}{\sin B+\sin C}+\frac{b^2\sin(C-A)}{\sin C+\sin A}+\frac{c^2\sin(A-B)}{\sin A+\sin B}=0.$$

MISCELLANEOUS EXAMPLES. D.

1. Prove that (1) $\tan 2\theta \cot \theta - 1 = \sec 2\theta$;
(2) $\sin a - \cot \theta \cos a = -\operatorname{cosec} \theta \cos (a+\theta)$.

2. If $a=48$, $b=35$, $C=60°$, find c.

3. If $\cos a = \frac{8}{17}$ and $\cos \beta = \frac{15}{17}$, find
$\tan (a+\beta)$ and $\operatorname{cosec} (a+\beta)$.

4. If $a = \frac{\pi}{21}$, find the value of $\frac{\sin 23a - \sin 7a}{\sin 2a + \sin 14a}$.

5. Prove that $\sin \theta (\cos 2\theta + \cos 4\theta + \cos 6\theta) = \sin 3\theta \cos 4\theta$.

6. If $b=\sqrt{2}$, $c=\sqrt{3}+1$, $A=45°$, solve the triangle.

7. Prove that
(1) $2 \sin^2 36° = \sqrt{5} \sin 18°$; (2) $4 \sin 36° \cos 18° = \sqrt{5}$.

8. Prove that $\frac{\sin 3a}{\sin a} + \frac{\cos 3a}{\cos a} = 4 \cos 2a$.

9. If $b=c=2$, $a=\sqrt{6}-\sqrt{2}$, solve the triangle.

10. Shew that
(1) $\cos 2a - \cot 3a \sin 2a = \tan a (\sin 2a + \cot 3a \cos 2a)$.
(2) $\cos a + \cos 2a + \cos 3a = 4 \cos a \cos \frac{a}{2} \cos \frac{3a}{2} - 1$.

11. In any triangle, prove that
(1) $b^2 \sin 2C + c^2 \sin 2B = 2bc \sin A$;
(2) $\frac{a^2 \sin (B-C)}{\sin A} + \frac{b^2 \sin (C-A)}{\sin B} + \frac{c^2 \sin (A-B)}{\sin C} = 0$.

12. If A, B, C, D are the angles of a quadrilateral, prove that

$$\frac{\tan A + \tan B + \tan C + \tan D}{\cot A + \cot B + \cot C + \cot D} = \tan A \tan B \tan C \tan D.$$

[*Use* $\tan (A+B) = \tan (360° - C - D)$.]

CHAPTER XIV.

LOGARITHMS.

151. Definition. The **logarithm** of any number to a given **base** is the index of the power to which the base must be raised in order to equal the given number. Thus if $a^x = N$, x is called the logarithm of N to the base a.

Example 1. Since $3^4 = 81$, the logarithm of 81 to base 3 is 4.

Example 2. Since $10^1 = 10$, $10^2 = 100$, $10^3 = 1000$,......
the natural numbers 1, 2, 3,... are respectively the logarithms of 10, 100, 1000,...... to base 10.

Example 3. Find the logarithm of $\cdot 008$ to base 25.

Let x be the required logarithm; then by definition,

$$25^x = \cdot 008 = \frac{8}{1000} = \frac{1}{125} = \frac{1}{5^3};$$

that is, $\qquad (5^2)^x = 5^{-3}$, or $5^{2x} = 5^{-3}$;

whence, by equating indices, $2x = -3$, and $x = -1 \cdot 5$.

152. The logarithm of N to base a is usually written $\log_a N$, so that the same meaning is expressed by the two equations

$$a^x = N, \qquad x = \log_a N.$$

From these equations it is evident that $a^{\log_a N} = N$.

Example. Find the value of $\log_{\cdot 01} \cdot 00001$.

Let $\log_{\cdot 01} \cdot 00001 = x$; then $(\cdot 01)^x = \cdot 00001$;

$$\therefore \left(\frac{1}{10^2}\right)^x = \frac{1}{100000}, \text{ or } \frac{1}{10^{2x}} = \frac{1}{10^5}.$$

$$\therefore 2x = 5, \text{ and } x = 2 \cdot 5.$$

153. When it is understood that a particular system of logarithms is in use, the suffix denoting the base is omitted.

Thus in arithmetical calculations in which 10 is the base, we usually write $\log 2$, $\log 3$,...... instead of $\log_{10} 2$, $\log_{10} 3$,......

Logarithms to the base 10 are known as **Common Logarithms**; this system was first introduced in 1615 by Briggs, a contemporary of Napier the inventor of Logarithms.

Before discussing the properties of common logarithms we shall prove some general propositions which are true for all logarithms independently of any particular base.

154. *The logarithm of* 1 *is* 0.

For $a^0 = 1$ for all values of a; therefore $\log 1 = 0$, whatever the base may be.

155. *The logarithm of the base itself is* 1.

For $a^1 = a$; therefore $\log_a a = 1$.

156. *To find the logarithm of a product.*

Let MN be the product; let a be the base of the system, and suppose

$$x = \log_a M, \qquad y = \log_a N;$$

so that $$a^x = M, \qquad a^y = N.$$

Thus the product $MN = a^x \times a^y = a^{x+y}$;

whence, by definition, $\log_a MN = x + y$

$$= \log_a M + \log_a N.$$

Similarly, $\log_a MNP = \log_a M + \log_a N + \log_a P$; and so on for any number of factors.

Example. $\log 42 = \log (2 \times 3 \times 7) = \log 2 + \log 3 + \log 7.$

157. *To find the logarithm of a fraction.*

Let $\frac{M}{N}$ be the fraction, and suppose

$$x = \log_a M, \qquad y = \log_a N;$$

so that $$a^x = M, \qquad a^y = N.$$

Thus the fraction $$\frac{M}{N} = \frac{a^x}{a^y} = a^{x-y};$$

whence, by definition, $\log_a \frac{M}{N} = x - y$

$$= \log_a M - \log_a N.$$

Example. $\log(2\tfrac{1}{7}) = \log\dfrac{15}{7} = \log 15 - \log 7$

$$= \log(3 \times 5) - \log 7 = \log 3 + \log 5 - \log 7.$$

158. *To find the logarithm of a number raised to any power, integral or fractional.*

Let $\log_a(M^p)$ be required, and suppose

$$x = \log_a M, \text{ so that } a^x = M;$$

then $$M^p = (a^x)^p = a^{px};$$

whence, by definition, $$\log_a(M^p) = px;$$

that is, $$\log_a(M^p) = p \log_a M.$$

Similarly, $$\log_a\left(M^{\frac{1}{r}}\right) = \frac{1}{r}\log_a M.$$

159. It follows from the results we have proved that

(1) the logarithm of a product is equal to the sum of the logarithms of its factors;

(2) the logarithm of a fraction is equal to the logarithm of the numerator diminished by the logarithm of the denominator;

(3) the logarithm of the pth power of a number is p times the logarithm of the number;

(4) the logarithm of the rth root of a number is $\frac{1}{r}$ of the logarithm of the number.

Thus by the use of logarithms the operations of multiplication and division may be replaced by those of addition and subtraction; the operations of involution and evolution by those of multiplication and division.

Example. Express $\log\dfrac{a^5\sqrt{b}}{\sqrt[3]{c^2}}$ in terms of $\log a$, $\log b$, $\log c$.

The expression $= \log(a^5\sqrt{b}) - \log\sqrt[3]{c^2}$

$$= \log a^5 + \log\sqrt{b} - \frac{2}{3}\log c$$

$$= 5\log a + \frac{1}{2}\log b - \frac{2}{3}\log c.$$

160. From the equation $10^x = N$, it is evident that common logarithms will not in general be integral, and that they will not always be positive.

For instance $3154 > 10^3$ and $< 10^4$;

$$\therefore \log 3154 = 3 + \text{a fraction.}$$

Again, $\cdot 06 > 10^{-2}$ and $< 10^{-1}$;

$$\therefore \log \cdot 06 = -2 + \text{a fraction.}$$

161. Definition. The integral part of a logarithm is called the **characteristic**, and the decimal part is called the **mantissa.**

The characteristic of the logarithm of any number to the base 10 can be found by inspection, as we shall now shew.

162. *To determine the characteristic of the logarithm of any number greater than unity.*

It is clear that a number with two digits in its integral part lies between 10^1 and 10^2; a number with three digits in its integral part lies between 10^2 and 10^3; and so on. Hence a number with n digits in its integral part lies between 10^{n-1} and 10^n.

Let N be a number whose integral part contains n digits; then

$$N = 10^{(n-1)+\text{a fraction}};$$

$$\therefore \log N = (n-1) + \text{a fraction.}$$

Hence the characteristic is $n-1$; that is, *the characteristic of the logarithm of a number greater than unity is less by one than the number of digits in its integral part, and is positive.*

163. *To determine the characteristic of the logarithm of a decimal fraction.*

A decimal with one cipher immediately after the decimal point, such as $\cdot 0324$, being greater than $\cdot 01$ and less than $\cdot 1$, lies between 10^{-2} and 10^{-1}; a number with two ciphers after the decimal point lies between 10^{-3} and 10^{-2}; and so on. Hence a decimal fraction with n ciphers immediately after the decimal point lies between $10^{-(n+1)}$ and 10^{-n}.

Let D be a decimal beginning with n ciphers; then

$$D = 10^{-(n+1)+\text{a fraction}};$$

$$\therefore \log D = -(n+1) + \text{a fraction.}$$

Hence the characteristic is $-(n+1)$; that is, *the characteristic of the logarithm of a decimal fraction is greater by unity than the number of ciphers immediately after the decimal point and is negative.*

164. The logarithms to base 10 of all integers from 1 to 200000 have been found and tabulated; in most Tables they are given to seven places of decimals.

The base 10 is chosen on account of two great advantages.

(1) From the results already proved it is evident that the characteristics can be written down by inspection, so that only the mantissæ have to be registered in the Tables.

(2) The mantissæ are the same for the logarithms of all numbers which have the same significant digits; so that it is sufficient to tabulate the mantissæ of the logarithms of *integers*.

This proposition we proceed to prove.

165. Let N be any number, then since multiplying or dividing by a power of 10 merely alters the position of the decimal point without changing the sequence of figures, it follows that $N \times 10^p$, and $N \div 10^q$, where p and q are any integers, are numbers whose significant digits are the same as those of N.

Now
$$\log(N \times 10^p) = \log N + p \log 10$$
$$= \log N + p \quad\text{..........(1)}.$$

Again,
$$\log(N \div 10^q) = \log N - q \log 10$$
$$= \log N - q \quad\text{..........(2)}.$$

In (1) an integer is added to $\log N$, and in (2) an integer is subtracted from $\log N$; that is, the mantissa or decimal portion of the logarithm remains unaltered.

In this and the three preceding articles the mantissæ have been supposed positive. In order to secure the advantages of Briggs' system, we arrange our work so as *always to keep the mantissa positive*, so that when the mantissa of any logarithm has been taken from the Tables the characteristic is prefixed with its appropriate sign, according to the rules already given.

166. In the case of a negative logarithm the minus sign is written *over the characteristic*, and not before it, to indicate that the characteristic alone is negative, and not the whole expression.

Thus $\bar{4}\cdot30103$, the logarithm of $\cdot0002$, is equivalent to $-4+\cdot30103$, and must be distinguished from $-4\cdot30103$, an expression in which both the integer and the decimal are negative. In working with negative logarithms an arithmetical artifice will sometimes be necessary in order to make the mantissa positive. For instance, a result such as $-3\cdot69897$, in which the whole expression is negative, may be transformed by subtracting 1 from the integral part and adding 1 to the decimal part. Thus

$$-3\cdot69897=-4+(1-\cdot69897)=\bar{4}\cdot30103.$$

Example 1. Required the logarithms of $\cdot0002432$.

In the Tables we find that 3859636 is the mantissa of $\log 2432$ (the decimal point as well as the characteristic being omitted); and, by Art. 163, the characteristic of the logarithm of the given number is -4;

$$\therefore \log \cdot0002432=\bar{4}\cdot3859636.$$

Example 2. Find the cube root of $\cdot0007$, having given

$$\log 7=\cdot8450980, \quad \log 887904=5\cdot9483660.$$

Let x be the required cube root; then

$$\log x=\frac{1}{3}\log(\cdot0007)=\frac{1}{3}(\bar{4}\cdot8450980)=\frac{1}{3}(\bar{6}+2\cdot8450980);$$

that is, $$\log x=\bar{2}\cdot9483660;$$

but $$\log 887904=5\cdot9483660;$$

$$\therefore x=\cdot0887904.$$

167. The logarithm of 5 and its powers can easily be obtained from log 2; for

$$\log 5=\log\frac{10}{2}=\log 10-\log 2=1-\log 2.$$

Example. Find the value of the logarithm of the reciprocal of $324\sqrt[5]{125}$, having given $\log 2=\cdot3010300$, $\log 3=\cdot4771213$.

Since $\log\frac{1}{a}=-\log a$, the required value

$$=-\log(324\sqrt[5]{125})=-\log(2^2\times3^4\times5^{\frac{3}{5}})$$

$$=-\left(2\log 2+4\log 3+\frac{3}{5}\log 5\right)$$

$$=-2\cdot9299272$$

$$=\bar{3}\cdot0700728.$$

$$\begin{aligned} 2\log 2 &= \cdot6020600\\ 4\log 3 &= 1\cdot9084852\\ \tfrac{6}{10}\log 5 &= \cdot4193820\\ \hline & 2\cdot9299272 \end{aligned}$$

EXAMPLES. XIV. a.

1. Find the logarithms respectively

of the numbers 1024, 81, $\cdot 125$, $\cdot 01$, $\cdot\dot{3}$, 100,

to the bases 2, $\sqrt{3}$, 4, $\cdot 001$, $\cdot\dot{1}$, $\cdot 01$.

2. Find the values of

$$\log_8 16, \quad \log_{81} 243, \quad \log_{\cdot 01} 10, \quad \log_{49} 343\sqrt{7}.$$

3. Find the numbers whose logarithms respectively

to the bases 49, $\cdot 25$, $\cdot 03$, 1, $\cdot 64$, 100, $\cdot 1$,

are 2, $\frac{1}{2}$, -2, -1, $-\frac{1}{2}$, $1\cdot 5$, -4.

4. Find the respective characteristics

of the logarithms of $\cdot 325$, 1603, 2400, 10000, 19,

to the bases 3, 11, 7, 9, 21.

5. Write down the characteristics of the common logarithms of $3\cdot 26$, $523\cdot 1$, $\cdot 03$, $1\cdot 5$, $\cdot 0002$, $3000\cdot 1$, $\cdot 1$.

6. The mantissa of log 64439 is $\cdot 8091488$, write down the logarithms of $\cdot 64439$, 6443900, $\cdot 00064439$.

7. The logarithm of $32\cdot 5$ is $1\cdot 5118834$, write down the numbers whose logarithms are

$$\cdot 5118834, \quad 2\cdot 5118834, \quad \bar{4}\cdot 5118834.$$

[*When required the following logarithms may be used*

$\log 2 = \cdot 3010300$, $\log 3 = \cdot 4771213$, $\log 7 = \cdot 8450980$.]

Find the value of

8. $\log 768$. 9. $\log 2352$. 10. $\log 35\cdot 28$.

11. $\log \sqrt{6804}$. 12. $\log \sqrt[5]{\cdot 00162}$. 13. $\log \cdot 021\dot{7}$.

14. $\log \cos 60°$. 15. $\log \sin^3 60°$. 16. $\log \sqrt[3]{\sec 45°}$.

Find the numerical value of

17. $2 \log \frac{15}{8} - \log \frac{25}{162} + 3 \log \frac{4}{9}$.

18. Evaluate $16 \log \frac{10}{9} - 4 \log \frac{25}{24} - 7 \log \frac{80}{81}$.

19. Find the seventh root of 7,

given $\log 1{\cdot}320469 = {\cdot}1207283$.

20. Find the cube root of ·00001764,

given $\log 260315 = 5{\cdot}4154995$.

21. Given $\log 3571 = 3{\cdot}5527899$, find the logarithm of

$$3{\cdot}571 \times {\cdot}03571 \times \sqrt[3]{3571}.$$

22. Given $\log 11 = 1{\cdot}0413927$, find the logarithm of

$$({\cdot}00011)^{\frac{1}{3}} \times (1{\cdot}21)^2 \times (13{\cdot}31)^{\frac{4}{3}} \div 12100000.$$

23. Find the number of digits in the integral parts of

$$\left(\frac{21}{20}\right)^{300} \text{ and } \left(\frac{126}{125}\right)^{1000}.$$

24. How many positive integers have characteristic 3 when the base is 7?

168. Suppose that we have a table of logarithms of numbers to base a and require to find the logarithms to base b.

Let N be one of the numbers, then $\log_b N$ is required.

Let $b^y = N$, so that $y = \log_b N$.

$$\therefore \log_a (b^y) = \log_a N;$$

that is,

$$y \log_a b = \log_a N;$$

$$\therefore y = \frac{1}{\log_a b} \times \log_a N,$$

or

$$\log_b N = \frac{1}{\log_a b} \times \log_a N \;\ldots\ldots\ldots\ldots\ldots\ldots\ldots (1).$$

Now since N and b are given, $\log_a N$ and $\log_a b$ are known from the Tables, and thus $\log_b N$ may be found.

Hence it appears that to transform logarithms from base a to base b we have only to multiply them all by $\frac{1}{\log_a b}$; this is a constant quantity and is given by the Tables; it is known as the *modulus*.

If in equation (1) we put a for N, we obtain

$$\log_b a = \frac{1}{\log_a b} \times \log_a a = \frac{1}{\log_a b};$$

$$\therefore \log_b a \times \log_a b = 1.$$

169. The following examples further illustrate the great use of logarithms in arithmetical work.

Example 1. Given $\log 2 = \cdot 3010300$ and $\log 4844544 = 6\cdot 6852530$, find the value of $(6\cdot 4)^{\frac{1}{10}} \times (\sqrt[4]{\cdot 256})^3 \div \sqrt{80}$.

Let x be the value of the expression; then

$$\log x = \frac{1}{10}\log\frac{64}{10} + \frac{3}{4}\log\frac{256}{1000} - \frac{1}{2}\log 80$$

$$= \frac{1}{10}(\log 2^6 - 1) + \frac{3}{4}(\log 2^8 - 3) - \frac{1}{2}(\log 2^3 + 1)$$

$$= \left(\frac{6}{10} + 6 - \frac{3}{2}\right)\log 2 - \left(\frac{1}{10} + \frac{9}{4} + \frac{1}{2}\right)$$

$$= \left(5 + \frac{1}{10}\right)\log 2 - 2\tfrac{17}{20}$$

$$= 1\cdot 5051500 + \cdot 0301030 - 2\cdot 85.$$

Thus $\log x = \bar{2}\cdot 6852530.$

But $\log 4844544 = 6\cdot 6852530,$

$$\therefore x = \cdot 04844544.$$

Example 2. Find how many ciphers there are between the decimal point and the first significant digit in $(\cdot 0504)^{10}$; having given

$$\log 2 = \cdot 301,\ \log 3 = \cdot 477,\ \log 7 = \cdot 845.$$

Denote the expression by E; then

$$\log E = 10 \log\frac{504}{10000}$$

$$= 10(\log 504 - 4)$$

$$= 10\{\log(2^3 \times 3^2 \times 7) - 4\}$$

$$= 10\{3\log 2 + 2\log 3 + \log 7 - 4)$$

$$= 10(2\cdot 702 - 4) = 10(\bar{2}\cdot 702)$$

$$= \overline{20} + 7\cdot 02 = \overline{13}\cdot 02.$$

$3\log 2 =$	$\cdot 903$
$2\log 3 =$	$\cdot 954$
$\log 7 =$	$\cdot 845$
	$2\cdot 702$

Thus the number of ciphers is 12. [Art. 163.]

Exponential equations.

170. If in an equation the unknown quantity appears as an exponent, the solution may be effected by the help of logarithms.

Example 1. Solve the equation $8^{5-3x}=12^{4-2x}$, having given

$$\log 2=\cdot 30103, \text{ and } \log 3=\cdot 47712.$$

From the given equation, by taking logarithms, we have

$$(5-3x)\log 8=(4-2x)\log 12;$$

$$\therefore 3(5-3x)\log 2 = (4-2x)(2\log 2+\log 3);$$

$$\therefore 15\log 2-8\log 2-4\log 3 = x(9\log 2-4\log 2-2\log 3);$$

$$\therefore x=\frac{7\log 2-4\log 3}{5\log 2-2\log 3}=\frac{\cdot 19873}{\cdot 55091}.$$

Thus $x=\cdot 36$ nearly.

$$\begin{array}{r} 7\log 2=2\cdot 10721 \\ 4\log 3=1\cdot 90848 \\ \hline \cdot 19873 \end{array}$$

$$\begin{array}{r} 5\log 2=1\cdot 50515 \\ 2\log 3=\ \cdot 95424 \\ \hline \cdot 55091 \end{array}$$

$$55091\ )\ 198730\ (\ 36$$
$$\begin{array}{r} 165273 \\ \hline 334570 \end{array}$$

Example 2. Given $\log 2=\cdot 30103$, solve the simultaneous equations

$$2^x . 5^y=1, \qquad 5^{x+1} . 2^y=2.$$

Take logarithms of the given equations;

$$\therefore x\log 2+y\log 5=0, \qquad (x+1)\log 5+y\log 2=\log 2.$$

For shortness, put $\log 2=a$, $\log 5=b$.

Thus $$ax+by=0,$$

and $$b(x+1)+ay=a, \text{ or } bx+ay=a-b.$$

By eliminating y, $x(a^2-b^2)=-b(a-b)$,

$$\therefore x=-\frac{b}{a+b}=-\frac{\log 5}{\log 2+\log 5}=-\frac{\log 5}{\log 10}=-\log 5=-\cdot 69897.$$

And $$y=-\frac{ax}{b}=\frac{a}{b}\log 5=a=\log 2=\cdot 30103.$$

EXAMPLES. XIV. b.

[*When required the values of* log 2, log 3, log 7 *given on p.* 145 *may be used.*]

Find the value of

1. $\left(\frac{147 \times 375}{126 \times 16}\right)^{\frac{2}{3}}$, given $\log 9 \cdot 076226 = \cdot 9579053$.

2. $\sqrt[3]{378} \times \sqrt{108} \div (\sqrt[6]{1008} \times \sqrt[3]{486})$,
given $\log 301824 = 5 \cdot 4797536$.

3. $(1080)^{\frac{1}{2}} \times (\cdot 24)^{\frac{5}{3}} \times 810$,
given $\log 2467266 = 6 \cdot 3922160$.

Calculate to two decimal places the values of

4. $\log_{20} 800$. 5. $\log_3 49$. 6. $\log_{125} 4000$.

7. Find how many ciphers there are before the first significant digits in

$$(\cdot 00378)^{\frac{40}{3}} \text{ and } (\cdot 0\dot{2}5\dot{9})^{50}.$$

8. To what base is 3 the logarithm of 11000?
given $\log 11 = 1 \cdot 0413927$ and $\log 222398 = 5 \cdot 3471309$.

Solve to two decimal places the equations:

9. $2^{x-1} = 5$. 10. $3^{x-4} = 7$. 11. $5^{1-x} = 6^{x-3}$.

12. $5^x = 2^{-y}$ and $5^{2+y} = 2^{2-x}$.

13. $2^x = 3^y$ and $2^{y+1} = 3^{x-1}$.

14. Given $\log 28 = a$, $\log 21 = b$, $\log 25 = c$, find $\log 27$ and $\log 224$ in terms of a, b, c.

15. Given $\log 242 = a$, $\log 80 = b$, $\log 45 = c$, find $\log 36$ and $\log 66$ in terms of a, b, c.

MISCELLANEOUS EXAMPLES. E.

1. Prove that

$\cos(30^\circ + A)\cos(30^\circ - A) - \cos(60^\circ + A)\cos(60^\circ - A) = \frac{1}{2}$.

2. If $A+B+C=180^\circ$, shew that

$$\frac{\sin 2A + \sin 2B + \sin 2C}{\sin A + \sin B + \sin C} = 8 \sin\frac{A}{2} \sin\frac{B}{2} \sin\frac{C}{2}.$$

3. If $a=2$, $c=\sqrt{2}$, $B=15^\circ$, solve the triangle.

4. Shew that $\cos a + \tan\frac{a}{2}\sin a = \cot\frac{a}{2}\sin a - \cos a$.

5. If $b\cos A = a\cos B$, shew that the triangle is isosceles.

6. Prove that

(1) $\sin\theta(\sin 3\theta + \sin 5\theta + \sin 7\theta + \sin 9\theta) = \sin 6\theta \sin 4\theta$;

(2) $\dfrac{\sin a + \sin 3a + \sin 5a + \sin 7a}{\cos a + \cos 3a + \cos 5a + \cos 7a} = \tan 4a$.

7. Shew that $\dfrac{\cos 3a}{\sin a} + \dfrac{\sin 3a}{\cos a} = 2\cot 2a$.

8. If $b = a(\sqrt{3}-1)$, $C=30^\circ$, find A and B.

9. Shew that $\tan 4a = \dfrac{4\tan a - 4\tan^3 a}{1 - 6\tan^2 a + \tan^4 a}$.

10. In a triangle, shew that

(1) $a^2\cos 2B + b^2\cos 2A = a^2 + b^2 - 4ab\sin A\sin B$;

(2) $4\left(bc\cos^2\frac{A}{2} + ca\cos^2\frac{B}{2} + ab\cos^2\frac{C}{2}\right) = (a+b+c)^2$.

11. If $a^4+b^4+c^4 = 2c^2(a^2+b^2)$, prove that $C=45^\circ$ or 135°.

[*Solve as a quadratic in* c^2.]

12. If in a triangle $\cos 3A + \cos 3B + \cos 3C = 1$, shew that one angle must be 120°.

CHAPTER XV.

THE USE OF LOGARITHMIC TABLES.

171. We shall now explain the use of logarithmic Tables to which reference has been made in the previous chapter.

In a book of Tables there will usually be found the *mantissæ* of the logarithms of all *integers* from 1 to 100000; the *characteristics* can be written down by inspection and are therefore omitted. [Art. 162.]

The logarithm of any number consisting of not more than 5 significant digits can be obtained directly from these Tables. For instance, suppose the logarithm of 336·34 is required. Opposite to 33634 we find the figures 5267785; this, with the decimal point prefixed, is the mantissa for the logarithms of all numbers whose significant digits are the same as 33634. We have therefore only to prefix the characteristic 2, and we obtain

$$\log 336{\cdot}34 = 2{\cdot}5267785.$$

Similarly, $$\log 33634 = 4{\cdot}5267785,$$

and $$\log {\cdot}0033634 = \bar{3}{\cdot}5267785.$$

172. Suppose now that we required log 33634·392.

Since this number contains more than 5 significant digits it cannot be obtained directly from the tables; but it lies between the two consecutive numbers 33634 and 33635, and therefore its logarithm lies between the logarithms of these two numbers. If we pass from 33634 to 33635, making an increase of 1 in the number, the corresponding increase in the logarithm as obtained from the tables is ·0000129. If now we pass from 33634 to 33634·392, making an increase of ·392 in the number, the increase in the logarithm will be $\cdot 392 \times \cdot 0000129$, provided that the increase in the logarithm is proportional to the increase in the number.

Now it can be proved that *when the increase made is small in comparison with the number, the increase in the logarithm is very nearly proportional to the increase in the number.*

This principle is known as the **Rule of Proportional Parts.**

The application of this rule will be illustrated in the examples which follow.

173. In order to make the explanations more intelligible we give here an Extract from Chambers' *Mathematical Tables.*

No.		0	1	2	3	4	5	6	7	8	9	Diff.
3361	526	4685	4814	4944	5073	5202	5331	5460	5590	5719	5848	
62		5977	6106	6235	6365	6494	6623	6752	6881	7010	7140	129
63		7269	7398	7527	7656	7785	7914	8043	8173	8302	8431	1 13
64		8560	8689	8818	8947	9076	9205	9334	9463	9593	9722	2 26
65		9851	9980	0109	0238	0367	0496	0625	0754	0883	1012	3 39
												4 52
66	527	1141	1270	1399	1528	1657	1786	1915	2044	2173	2302	5 65
67		2431	2560	2689	2818	2947	3076	3205	3334	3463	3592	6 77
68		3721	3850	3979	4108	4237	4366	4494	4623	4752	4881	7 90
69		5010	5139	5268	5397	5526	5655	5783	5912	6041	6170	8 103
70		6299	6428	6557	6686	6814	6943	7072	7201	7330	7459	9 116

174. Suppose that log 33635 is required.

In the third horizontal line we have the logarithms of numbers beginning with 3363. As the next digit is 5 we choose from this line the mantissa which stands under the column 5. We have now only to prefix the characteristic and we obtain log 33635 = 4·5267914.

Similarly, $\log 33651 = 4{\cdot}5269980$,

and $\log 33652 = 4{\cdot}5\overline{270109}$,

the transition in the mantissæ from 526... to 527... being shewn by the bar drawn over 0109. This bar is repeated over each of the subsequent logarithms as far as the end of the line, and in the next line the mantissæ begin with 527.

Example. Find log 33634·392.

$$\begin{aligned} \text{From the Tables,}\quad \log 33635 &= 4{\cdot}5267914 \\ \log 33634 &= 4{\cdot}5267785 \\ \text{difference for } 1 &= {\cdot}0000129 \end{aligned}$$

Now by the Rule of Proportional Parts, log 33634·392 will be greater than log 33634 by ·392 times the difference for 1; hence to 7 places of decimals, we have

$$\begin{array}{r|l} {\cdot}0000129 & \\ {\cdot}392 & \\ \hline & 258 \\ 11 & 61 \\ 38 & 7 \\ \hline {\cdot}0000050 & 568 \end{array}$$

$$\begin{aligned} \log 33634 &= 4{\cdot}5267785 \\ \text{proportional difference for } {\cdot}392 &= {\cdot}0000051 \\ \therefore\ \log 33634{\cdot}392 &= 4{\cdot}5267836 \end{aligned}$$

In practice, the difference for 1 is usually quoted without the ciphers; if therefore we *treat the difference* 129 *as a whole number*, on multiplying by ·392 we obtain the product 50·568, and we take the digits given by its integral part (51 approximately) as the proportional increase for ·392.

175. The method of calculating the proportional difference for ·392 which we have explained is that which must be adopted when we have nothing given but the logarithms of two consecutive numbers between which lies the number whose logarithm we are seeking.

But when the Tables are used the calculation is facilitated by means of the proportional differences standing in the column to the right. This gives the differences for *tenths* of unity.

The difference for ·392 is obtained as follows.

$${\cdot}392 \times 129 = \left(\frac{3}{10} + \frac{9}{100} + \frac{2}{1000}\right) \times 129 = 39 + 11{\cdot}6 + {\cdot}26 = 50{\cdot}86.$$

The difference for 9 quoted in the margin (really 9 *tenths*) is 116, and therefore the difference for 9 hundredths is 11·6; and similarly the difference for 2 thousandths is ·26.

In practical work, the following arrangement is adopted.

$$\begin{array}{lllr|l} & \log 33634 & & =4\cdot5267785 & \\ \text{add for} & & 3 & 39 & \\ & & 9 & 11 & 6 \\ & & 2 & & 26 \\ \hline \therefore & \log 33634\cdot392 & & =4\cdot5267836 & \end{array}$$

176. The following example is solved more concisely as a model for the student. In the column on the left we work from the data of the question; in the column on the right we obtain the logarithm by the use of the Tables independently of the two given logarithms.

Example. Find log 33·656208, having given

$$\log 33656 = 4\cdot5270625 \text{ and } \log 33657 = 4\cdot5270754.$$

$$\begin{array}{lr|l} \log 33\cdot657 = & 1\cdot5270754 & \\ \log 33\cdot656 = & 1\cdot5270625 & \\ \hline \text{diff. for } \cdot001 = & 129 & \\ & 208 & \\ \hline & 1 & 032 \\ & 25 & 80 \\ \hline \text{diff. for } \cdot000208 = & 26 & 832 \\ \log 33\cdot656 = & 1\cdot5270625 & \\ \hline \log 33\cdot656208 = & 1\cdot5270652 & \end{array}$$

From the Tables, we have

$$\begin{array}{llr|l} \log 33\cdot656 & & =1\cdot5270625 & \\ \text{add for} & 2 & 26 & \\ & 0 & 0 & \\ & 8 & 1 & 03 \end{array}$$

$$\log 33\cdot656208 = 1\cdot5270652$$

177. The Rule of Proportional Parts also enables us to find the number corresponding to a given logarithm.

Example 1. Find the number whose logarithm is $\bar{2}\cdot5274023$, having given log 3·3683 = ·5274108 and log 3·3682 = ·5273979.

Let x be the required number; then

$$\begin{array}{rl} \log x = & \bar{2}\cdot5274023 \\ \log \cdot033682 = & \bar{2}\cdot5273979 \\ \hline \text{diff.} = & 44 \end{array} \qquad \begin{array}{rl} \log \cdot033683 = & \bar{2}\cdot5274108 \\ \log \cdot033682 = & \bar{2}\cdot5273979 \\ \hline \text{diff. for } \cdot000001 = & 129 \end{array}$$

hence x lies between ·033682 and ·033683, and is greater than ·033682 by $\frac{44}{129} \times \cdot000001$, that is by ·00000034.

$$\therefore x = \cdot03368234.$$

$$\begin{array}{r} 129\ )\ 440\ (\ 34 \\ 387 \\ \hline 530 \\ 516 \\ \hline \end{array}$$

In working from the Tables, we proceed as follows.

$$
\begin{array}{lrr}
\log x & = \bar{2}\text{·}5274023 & \\
\log \text{·}033682 & = \bar{2}\text{·}5273979 & \\
 & 44 & \\
3 & 39 & \\
 & 50 & \\
4 & 52 &
\end{array}
$$

$\therefore\ x = \text{·}03368234.$

We are saved the trouble of the division, as the multiples of 129 which occur during the work are given in the approximate forms 39 and 52 in the difference column opposite to the numbers 3 and 4.

Example 2. Find the fifth root of ·0025612, having given

$\log 2\text{·}5612 = \text{·}4084435,\ \log 3\text{·}0317 = \text{·}4816862,\ \log 3\text{·}0318 = \text{·}4817005.$

Let $x = (\text{·}0025612)^{\frac{1}{5}}$; then

$$\log x = \frac{1}{5}\log(\text{·}0025612) = \frac{1}{5}(\bar{3}\text{·}4084435) = \frac{1}{5}(\bar{5} + 2\text{·}4084435);$$

$$= \bar{1}\text{·}4816887.$$

$$
\begin{array}{lr}
\log x = \bar{1}\text{·}4816887 & \log \text{·}30318 = \bar{1}\text{·}4817005 \\
\log \text{·}30317 = \bar{1}\text{·}4816862 & \log \text{·}30317 = \bar{1}\text{·}4816862 \\
\text{diff.} = 25 & \text{diff. for ·}00001 = 143
\end{array}
$$

$\therefore$ proportional increase $= \dfrac{25}{143} \times \text{·}00001 = \text{·}00000175.$

Thus $x = \text{·}30317175.$

$$
\begin{array}{r}
143\ )\ 250\ (\ 175 \\
143 \\
1070 \\
1001 \\
690 \\
715
\end{array}
$$

EXAMPLES. XV. a.

1. Find the value of log 4951634, given that
   $\log 49516 = 4\text{·}6947456,\quad \log 49517 = 4\text{·}6947543.$

2. Find log 3·4713026, having given that
   $\log 347\text{·}13 = 2\text{·}5404921,\quad \log 34714 = 4\text{·}5405047.$

3. Find log 2849614, having given that
   $\log 2\text{·}8496 = \text{·}4547839,\quad \log 2\text{·}8497 = \text{·}4547991.$

4. Find log 57·63325, having given that
log 576·33 = 2·7606712, log 5763·4 = 3·7606788.

5. Given log 60814 = 4·7840036, diff. for 1 = 72, find
log 6·081465.

6. Find the number whose logarithm is 4·7461735, given
log 55740 = 4·7461670, log 55741 = 4·7461748.

7. Find the number whose logarithm is 2·8283676, given
log 6·7354 = ·8283634, log 67355 = 4·8283698.

8. Find the number whose logarithm is $\bar{2}$·0288435, given
log 1068·6 = 3·0288152, log 1·0687 = ·0288558.

9. Find the number whose logarithm is $\bar{3}$·9184377, given
log 8·2877 = ·9184340, log 8287·8 = 3·9184392.

10. Given log 253·19 = 2·4034465, diff. for 1 = 172, find the number whose logarithm is $\bar{1}$·4034508.

11. Given log 2·0313 = ·3077741, log 2·0314 = ·3077954,
and log 1·4271 = ·1544544,
find the seventh root of 142·71.

12. Find the eighth root of 13·89492, given
log 13894 = 4·1428273, log 138·95 = 2·1428586.

13. Find the value of $\sqrt[14]{242447}$, given
log 2·4244 = ·3846043, diff. for 1 = 179.

14. Find the twentieth root of 2069138, given
log 20691 = 4·3157815, diff. for 1 = 210.

Tables of Natural and Logarithmic Functions.

178. Tables have been constructed giving the values of the trigonometrical functions of all angles between 0° and 90° at intervals of 10″. These are called the Tables of **natural sines, cosines, tangents,...** In the smaller Tables, such as Chambers', the interval is 1′.

The logarithms of the functions have also been calculated. Since many of the trigonometrical functions are less than unity their logarithms are negative, and as the characteristics are not always evident on inspection they cannot be omitted. To avoid the inconvenience of printing the bars over the characteristics, the logarithms are all increased by 10 and are then registered under the name of **tabular logarithmic sines, cosines,...**

The notation used is $L\cos A$, $L\tan\theta$; thus

$$L\sin A = \log\sin A + 10.$$

For instance,

$$L\sin 45^\circ = 10 + \log\sin 45^\circ = 10 + \log\frac{1}{\sqrt{2}}$$

$$= 10 - \frac{1}{2}\log 2 = 9{\cdot}8494850.$$

179. With certain exceptions that need not be here noticed, the rule of proportional parts holds for the natural sines, cosines,... of all angles, and also for their logarithmic sines, cosines,.... In applying this rule it must be remembered that as the angle increases from 0° to 90° the functions sine, tangent, secant increase, while the co-functions cosine, cotangent, cosecant decrease.

Example 1. Find the value of sin 29° 37′ 42″.

From the Tables,

$$\begin{aligned} \sin 29^\circ 38' &= {\cdot}4944476 \\ \sin 29^\circ 37' &= {\cdot}4941948 \\ \text{diff. for } 60'' &= \quad 2528 \end{aligned}$$

$$\therefore \text{ prop}^{\text{l}} \text{ increase for } 42'' = \frac{42}{60} \times 2528 = \quad 1770$$

$$\begin{aligned} \sin 29^\circ 37' &= {\cdot}4941948 \\ \therefore \sin 29^\circ 37' 42'' &= {\cdot}4943718. \end{aligned}$$

Example 2. Find the angle whose cosine is ·7280843.

Let A be the required angle; then from the Tables,

$$\begin{aligned} \cos 43^\circ 16' &= {\cdot}7281716 \\ \cos 43^\circ 17' &= {\cdot}7279722 \\ \text{diff. for } 60'' & \quad 1994 \end{aligned}$$

$$\begin{aligned} \cos 43^\circ 16' &= {\cdot}7281716 \\ \cos A &= {\cdot}7280843 \\ \text{prop}^{\text{l}} \text{ part} & \quad 873 \end{aligned}$$

```
        873
         60
1994 ) 52380 ( 26
       3988
       12500
       11964
```

But $\cos A$ is *less* than cos 43° 16′; hence A must be *greater* than 43° 16′ by $\frac{873}{1994} \times 60''$, that is by 26″ nearly.

Thus the angle is 43° 16′ 26″.

180. In order to illustrate the use of the tabular logarithmic functions we give the following extract from the table of logarithmic sines, cosines,... in Chambers' *Mathematical Tables*.

27 Deg.

′	Sine	Diff.	Cosec.	Secant	D.	Cosine	′
0	9·6570468	2478	10·3429532	10·0501191	644	9·9498809	60
1	9·6572946	2477	10·3427054	10·0501835	644	9·9498165	59
2	9·6575423	2475	10·3424577	10·0502479	645	9·9497521	58
3	9·6577898	2473	10·3422102	10·0503124	646	9·9496876	57
4	9·6580371		10·3419629	10·0503770		9·9496230	56
56	9·6706576	2382	10·3293424	10·0537968	670	9·9462032	4
57	9·6708958	2380	10·3291042	10·0538638	670	9·9461362	3
58	9·6711338	2378	10·3288662	10·0539308	671	9·9460692	2
59	9·6713716	2377	10·3286284	10·0539979	672	9·9460021	1
60	9·6716093		10·3283907	10·0540651		9·9459349	0
′	Cosine	Diff.	Secant	Cosec.	D.	Sine	′

62 Deg.

181. We have quoted here the logarithmic sines, cosecants, secants, and cosines of the angles differing by 1′ between 27° 0′ and 27° 4′, and also between 27° 56′ and 27° 60′. The same extract gives the logarithmic functions of the complements of these angles, namely those between 62° 0′ and 62° 4′, and those between 62° 56′ and 62° 60′.

The column of minutes for 27° is given on the left and increases downwards, the column for 62° is on the right and increases upwards.

The names of the functions printed at the top refer to the angle 27°, the names printed at the foot refer to the angle 62°. Thus

$$L\cos 27°\ 3' = 9{\cdot}9496876, \quad L\operatorname{cosec} 27°\ 58' = 10{\cdot}3288662,$$
$$L\sin 62°\ 2' = 9{\cdot}9460692, \quad L\cos 62°\ 59' = 9{\cdot}6572946.$$

The first *difference column* gives the differences in the logarithms of the sines and cosecants, the second *difference column* gives the differences in the logarithms of the cosines and secants, each difference corresponding to a difference of 1′ in the angle.

Example 1. Find $L \cos 62° 57' 12''$.

From the Tables,
$$L \cos 62° 57' = 9 \cdot 6577898$$
$$L \cos 62° 58' = 9 \cdot 6575423$$
$$\text{diff. for } 60'' \qquad 2475$$

$$\therefore \text{ proportional } \textit{decrease} \text{ for } 12'' = \frac{12}{60} \times 2475 = 495.$$

$$L \cos 62° 57' = 9 \cdot 6577898$$
$$\textit{Subtract} \text{ for } 12'' \qquad 495$$
$$\therefore L \cos 62° 57' 12'' = 9 \cdot 6577403$$

Example 2. Given $L \sec 27° 39' = 10 \cdot 0526648$, diff. for $10'' = 110$, find A when $L \sec A = 10 \cdot 0527253$.

$$L \sec A \qquad = 10 \cdot 0527253$$
$$L \sec 27° 39' = 10 \cdot 0526648$$
$$\text{diff.} \qquad 605$$

$$\therefore \text{ proportional increase} = \frac{605}{110} \times 10'' = 55''.$$

Thus $$A = 27° 39' 55''.$$

EXAMPLES. XV. b.

1. Find $\sin 38° 3' 35''$, having given that
$\sin 38° 4' = \cdot 6165780$, $\sin 38° 3' = \cdot 6163489$.

2. Find $\tan 38° 24' 37 \cdot 5''$, having given that
$\tan 38° 25' = \cdot 7930640$, $\tan 38° 24' = \cdot 7925902$.

3. Find $\operatorname{cosec} 55° 21' 28''$, having given that
$\operatorname{cosec} 55° 22' = 1 \cdot 2153535$, $\operatorname{cosec} 55° 21' = 1 \cdot 2155978$.

4. Find the angle whose secant is $2 \cdot 1809460$, given
$\sec 62° 43' = 2 \cdot 1815435$, $\sec 62° 42' = 2 \cdot 1803139$.

5. Find the angle whose cosine is $\cdot 8600931$, given
$\cos 30° 41' = \cdot 8600007$, $\cos 30° 40' = \cdot 8601491$.

6. Find the angle whose cotangent is $\cdot 8766003$, given
$\cot 48° 46' = \cdot 8764620$, $\cot 48° 45' = \cdot 8769765$.

7. Find $L \sin 44° 17' 33''$, given
$L \sin 44° 18' = 9 \cdot 8441137$, $L \sin 44° 17' = 9 \cdot 8439842$.

8. Find $L \cot 36° \, 26' \, 16''$, given
$L \cot 36° \, 27' = 10·1315840$, $L \cot 36° \, 26' = 10·1318483$.

9. Find $L \cos 55° \, 30' \, 24''$, given
$L \cos 55° \, 31' = 9·7529442$, $L \cos 55° \, 30' = 9·7531280$.

10. Find the angle whose tabular logarithmic sine is 9·8440018, using the data of example 7.

11. Find the angle whose tabular logarithmic cosine is 9·7530075, using the data of example 9.

12. Given $L \tan 24° \, 50' = 9·6653662$, diff. for $1' = 3313$, find $L \tan 24° \, 50' \, 52·5''$.

13. Given $L \operatorname{cosec} 40° \, 5' = 10·1911808$, diff. for $1' = 1502$, find $L \operatorname{cosec} 40° \, 4' \, 17·5''$.

182. Considerable practice in the use of logarithmic Tables will be required before the quickness and accuracy necessary in all practical calculations can be attained. Experience shews that mistakes frequently arise from incorrect quotation from the Tables, and from clumsy arrangement. The student is reminded that care in taking out the logarithms from the Tables is of the first importance, and that in the course of the work he should learn to leave out all needless steps, making his solutions as concise as possible consistent with accuracy.

Example 1. Divide 6·6425693 by ·3873007.

From the Tables,

log 6·6425	= ·8223316		
6	40		
9	5	9	
3	·	20	
log 6·6425693	= ·8223362		
log ·3873007	$= \bar{1}·5880483$		

log ·38730	$= \bar{1}·5880475$	
0		0
7	7	8
log ·3873007	$= \bar{1}·5880483$	

By subtraction, we obtain		1·2342879
From the Tables, log 17·150		= 1·2342641
		238
	9	229
		90
	3	76

Thus the quotient is 17·15093.

Example 2. The hypotenuse of a right-angled triangle is 3·141024 and one side is 2·593167; find the other side.

Let c be the hypotenuse, a the given side, and x the side required; then

$$x^2 = c^2 - a^2 = (c+a)(c-a);$$
$$\therefore\ 2\log x = \log(c+a) + \log(c-a).$$

$$\begin{aligned} c &= 3{\cdot}141024 \\ a &= 2{\cdot}593167 \\ c+a &= 5{\cdot}734191 \\ c-a &= {\cdot}547858 \end{aligned}$$

$$\begin{array}{lrrl}
\text{From the Tables, } \log 5{\cdot}7341 & = & {\cdot}7584653 & \\
9 & & 68 & \\
1 & & & 8 \\
\log {\cdot}54785 & = & \bar{1}{\cdot}7386617 & \\
7 & & 55 & \\
\text{By addition,} & & {\cdot}4971394 &
\end{array}$$

Dividing by 2, we have

$$\begin{array}{lrr}
\log x & = & {\cdot}2485697 \\
\log 1{\cdot}7724 & = & {\cdot}2485617 \\
 & & 80 \\
3 & & 74 \\
 & & 60 \\
2 & & 49
\end{array}$$

Thus the required side is 1·772432.

EXAMPLES. XV. c.

[*In this exercise the logarithms are to be taken from the Tables.*]

1. Multiply 300·2618 by ·0078915194.

2. Find the product of 235·6783 and 357·8438.

3. Find the continued product of

   153·2419, 2·8632503, and ·07583646.

4. Divide 1·0304051 by 27·093524.

5. Divide 357·8364 by ·00318973.

6. Find x from the equation

   $$\cdot 0178345x = 21{\cdot}85632.$$

7. Find the value of

   $$3{\cdot}78956 \times {\cdot}0536872 \div {\cdot}0072916.$$

8. Find the cube of ·83410039.

9. Find the fifth root of 15063·018.

10. Evaluate $\sqrt[5]{384{\cdot}731}$ and $\sqrt[13]{15{\cdot}7324}$.

11. Find the product of the square root of 1034·3963 and the cube root of 353246.

12. Subtract the square of ·7503269 from the square of 1·035627.

13. Find the value of

$$\frac{(34{\cdot}7326)^{\frac{3}{5}} \times \sqrt[6]{2{\cdot}53894}}{\sqrt[5]{4{\cdot}39682}}.$$

Example 3. Find a third proportional to the cube of ·3172564 and the cube root of 23·32873.

Let x be the required third proportional; then

$$({\cdot}3172564)^3 : (23{\cdot}32873)^{\frac{1}{3}} = (23{\cdot}32873)^{\frac{1}{3}} : x;$$

whence
$$x = (23{\cdot}32873)^{\frac{2}{3}} \div ({\cdot}3172564)^3;$$

$$\therefore \log x = \frac{2}{3} \log 23{\cdot}32873 - 3 \log {\cdot}3172564.$$

From the Tables,

$$\begin{array}{lr|l}
\log {\cdot}31725 & = \bar{1}{\cdot}5014016 & \\
6 & 82 & \\
4 & 5 & 5 \\
\hline
 & \bar{1}{\cdot}5014103 & 5 \\
 & 3 & \\
\hline
 & \bar{2}{\cdot}5042311 &
\end{array}
\qquad
\begin{array}{lr|l}
\log 23{\cdot}328 & = 1{\cdot}3678775 & \\
7 & 130 & \\
3 & 5 & 6 \\
\hline
 & 1{\cdot}3678910 & 6 \\
 & 2 & \\
\hline
3 \,\big|\, & 2{\cdot}7357821 & \\
\hline
 & {\cdot}9119274 & \\
 & \bar{2}{\cdot}5042311 &
\end{array}$$

$$\begin{array}{llr}
\text{By subtraction,} \quad \log x & & = 2{\cdot}4076963 \\
\log 255{\cdot}67 & & = 2{\cdot}4076798 \\
\hline
 & & 165 \\
 & 9 & 153 \\
\hline
 & & 120 \\
 & 7 & 119 \\
\hline
\end{array}$$

Thus the third proportional is 255·6797.

14. Find a mean proportional between

$$\cdot 0037258169 \text{ and } \cdot 56301078.$$

15. Find a third proportional to the square of ·43607528 and the square root of ·03751786.

16. Find a fourth proportional to

$$56712\cdot 43,\quad 29\cdot 302564,\quad \cdot 33025107.$$

17. Find the geometric mean between

$$(\cdot 035689)^{\frac{2}{5}} \text{ and } (2\cdot 879432)^{\frac{3}{7}}.$$

18. Find a fourth proportional to

$$\sqrt[3]{32\cdot 7812},\quad \sqrt[5]{357\cdot 814},\quad \sqrt[4]{7836\cdot 43}.$$

19. Find the value of

$$\sin 27^\circ\ 13'\ 12'' \times \cos 46^\circ\ 2'\ 15''.$$

20. Find the value of

$$\cot 97^\circ\ 14'\ 16'' \times \sec 112^\circ\ 13'\ 5''.$$

21. Evaluate

$$\sin 20^\circ\ 13'\ 20'' \times \cot 47^\circ\ 53'\ 15'' \times \sec 42^\circ\ 15'\ 30''.$$

22. Find the value of $ab \sin C$, when

$$a = 324\cdot 1368,\quad b = 417\cdot 2431,\quad C = 113^\circ\ 14'\ 16''.$$

23. If $a : b = \sin A : \sin B$, find a, given

$$b = 378\cdot 25,\quad A = 35^\circ\ 15'\ 33'',\quad B = 119^\circ\ 14'\ 18''.$$

24. Find the smallest values of θ which satisfy the equations

(1) $\tan^3 \theta = \frac{5}{12}$; (2) $3 \sin^2 \theta + 2 \sin \theta = 1$.

25. Find x from the equation

$$x \times \sec 28^\circ\ 17'\ 25'' = \sin 23^\circ\ 18'\ 5'' \times \cot 38^\circ\ 15'\ 13''.$$

26. Find θ from the equation

$$\sin^3 \theta = \cos^2 \alpha \cot \beta,$$

where $\alpha = 32^\circ\ 47'$ and $\beta = 41^\circ\ 19'$.

CHAPTER XVI.

SOLUTION OF TRIANGLES WITH LOGARITHMS.

183. The examples on the solution of triangles in Chap. XIII. furnish a useful exercise on the formulæ connecting the sides and angle of a triangle; but in practical work much of the labour of arithmetical calculation is avoided by the use of logarithms.

We shall now shew how the formulæ of Chap. XIII. may be used or adapted for use in connection with logarithmic Tables.

184. *To find the functions of the half-angles in terms of the sides.*

We have $$2\sin^2\frac{A}{2}=1-\cos A$$

$$=1-\frac{b^2+c^2-a^2}{2bc}$$

$$=\frac{2bc-b^2-c^2+a^2}{2bc}=\frac{a^2-(b^2-2bc+c^2)}{2bc}$$

$$=\frac{a^2-(b-c)^2}{2bc}=\frac{(a+b-c)(a-b+c)}{2bc}.$$

Let $$a+b+c=2s;$$

then $$a+b-c=2s-2c=2(s-c),$$

and $$a-b+c=2s-2b=2(s-b).$$

$$\therefore\quad 2\sin^2\frac{A}{2}=\frac{4(s-c)(s-b)}{2bc}=\frac{2(s-b)(s-c)}{bc};$$

$$\therefore\quad \sin\frac{A}{2}=\sqrt{\frac{(s-b)(s-c)}{bc}}.$$

Again, $2\cos^2\frac{A}{2}=1+\cos A=1+\frac{b^2+c^2-a^2}{2bc}$

$$=\frac{(b+c)^2-a^2}{2bc}=\frac{(b+c+a)(b+c-a)}{2bc};$$

$$\therefore\quad 2\cos^2\frac{A}{2}=\frac{4s(s-a)}{2bc}=\frac{2s(s-a)}{bc};$$

$$\therefore\quad \cos\frac{A}{2}=\sqrt{\frac{s(s-a)}{bc}}.$$

Also $\tan\frac{A}{2}=\sin\frac{A}{2}\div\cos\frac{A}{2}$

$$=\sqrt{\frac{(s-b)(s-c)}{bc}\times\frac{bc}{s(s-a)}};$$

$$\therefore\quad \tan\frac{A}{2}=\sqrt{\frac{(s-b)(s-c)}{s(s-a)}}.$$

185. Similarly it may be proved that

$$\sin\frac{B}{2}=\sqrt{\frac{(s-c)(s-a)}{ca}},\qquad \sin\frac{C}{2}=\sqrt{\frac{(s-a)(s-b)}{ab}};$$

$$\cos\frac{B}{2}=\sqrt{\frac{s(s-b)}{ca}},\qquad \cos\frac{C}{2}=\sqrt{\frac{s(s-c)}{ab}};$$

$$\tan\frac{B}{2}=\sqrt{\frac{(s-c)(s-a)}{s(s-b)}},\qquad \tan\frac{C}{2}=\sqrt{\frac{(s-a)(s-b)}{s(s-c)}}.$$

In each of these formulæ the positive value of the square root must be taken, for each half angle is less than 90°, so that all its functions are positive.

186. *To find* $\sin A$ *in terms of the sides.*

$$\sin A=2\sin\frac{A}{2}\cos\frac{A}{2}$$

$$=2\sqrt{\frac{(s-b)(s-c)}{bc}\times\frac{s(s-a)}{bc}};$$

$$\therefore\quad \sin A=\frac{2}{bc}\sqrt{s(s-a)(s-b)(s-c)}.$$

We may also obtain this formula in another way which is instructive.

We have

$$\sin^2 A = 1 - \cos^2 A = (1 + \cos A)(1 - \cos A)$$

$$= \left(1 + \frac{b^2 + c^2 - a^2}{2bc}\right)\left(1 - \frac{b^2 + c^2 - a^2}{2bc}\right)$$

$$= \frac{(b+c)^2 - a^2}{2bc} \times \frac{a^2 - (b-c)^2}{2bc}$$

$$= \frac{(b+c+a)(b+c-a)(a+b-c)(a-b+c)}{4b^2c^2}$$

$$= \frac{16s(s-a)(s-b)(s-c)}{4b^2c^2};$$

$$\therefore \quad \sin A = \frac{2}{bc}\sqrt{s(s-a)(s-b)(s-c)}.$$

The positive value of the square root must be taken, since the *sine* of an angle of any triangle is always positive.

EXAMPLES. XVI. a.

Prove the following formulæ in any triangle:

1. $b \cos^2 \frac{A}{2} + a \cos^2 \frac{B}{2} = s.$ 2. $s \tan \frac{B}{2} \tan \frac{C}{2} = s - a.$

3. $\frac{\text{vers } A}{\text{vers } B} = \frac{a(a+c-b)}{b(b+c-a)}.$ 4. $b \sin^2 \frac{A}{2} + a \sin^2 \frac{B}{2} = s - c.$

5. $(s-a) \tan \frac{A}{2} = (s-b) \tan \frac{B}{2} = (s-c) \tan \frac{C}{2}.$

6. Find the value of $\tan \frac{B}{2}$, when $a = 10$, $b = 17$, $c = 21$.

7. Find $\cot \frac{C}{2}$, when $a = 13$, $b = 14$, $c = 15$.

8. Prove that

$$\frac{1}{a} \cos^2 \frac{A}{2} + \frac{1}{b} \cos^2 \frac{B}{2} + \frac{1}{c} \cos^2 \frac{C}{2} = \frac{s^2}{abc}.$$

9. Prove that

$$\frac{b-c}{a} \cos^2 \frac{A}{2} + \frac{c-a}{b} \cos^2 \frac{B}{2} + \frac{a-b}{c} \cos^2 \frac{C}{2} = 0.$$

187. *To solve a triangle when the three sides are given.*

From the formula

$$\tan\frac{A}{2}=\sqrt{\frac{(s-b)(s-c)}{s(s-a)}},$$

$$\log\tan\frac{A}{2}=\frac{1}{2}\{\log(s-b)+\log(s-c)-\log s-\log(s-a)\};$$

whence $\frac{A}{2}$ may be obtained by the help of the Tables.

Similarly B can be found from the formula for $\tan\frac{B}{2}$, and then C from the equation $C=180°-A-B$.

In the above solution, we shall require to look out from the tables *four* logarithms only, namely those of s, $s-a$, $s-b$, $s-c$; whereas if we were to solve from the sine or cosine formulæ we should require *six* logarithms; for

$$\cos\frac{A}{2}=\sqrt{\frac{s(s-a)}{bc}} \quad \text{and} \quad \cos\frac{B}{2}=\sqrt{\frac{s(s-b)}{ca}},$$

so that we should have to look out the logarithms of the *six* quantities s, $s-a$, $s-b$, a, b, c.

If therefore *all* the angles have to be found by the use of the tables it is best to solve from the tangent formulæ; but if *one* angle only is required it is immaterial whether the sine, cosine, or tangent formula is used.

In cases where a solution has to be obtained from certain given logarithms, the choice of formulæ must depend on the data.

NOTE. We shall always find the angles to the nearest second, so that, on account of the multiplication by 2, the half-angles should be found to the nearest tenth of a second.

188. In Art. 178 we have mentioned that 10 is added to each of the logarithmic functions before they are registered as *tabular* logarithms; but this device is introduced only as a convenience for the purposes of tabulation, and in practice it will be found that the work is more expeditious if the tabular logarithms are not used. The 10 should be subtracted mentally in copying down the logarithms. Thus we should write

$$\log\sin 64°\ 15'=\bar{1}\cdot9545793, \quad \log\cot 18°\ 35'=\cdot4733850,$$

and in the arrangement of the work care must be taken to keep the mantissæ positive.

Example 1. The sides of a triangle are 35, 49, 63; find the greatest angle; given $\log 2 = \cdot 3010300$, $\log 3 = \cdot 4771213$,

$$L \cos 47^\circ 53' = 9\cdot 8264910, \text{ diff. for } 60'' = 1397.$$

Since the *angles* of a triangle depend only on the *ratios* of the sides and not on their actual magnitudes, we may substitute for the sides any lengths proportional to them. Thus in the present case we may take $a=5$, $b=7$, $c=9$; then C is the greatest angle, and

$$\cos\frac{C}{2} = \sqrt{\frac{s(s-c)}{ab}} = \sqrt{\frac{21}{2} \times \frac{3}{2} \times \frac{1}{5\times 7}} = \sqrt{\frac{9}{20}};$$

$$\therefore \log\cos\frac{C}{2} = \frac{1}{2}(2\log 3 - \log 2 - 1).$$

$$\begin{array}{r} 2\log 3 = \cdot 9542426 \\ 1\cdot 3010300 \\ \hline 2\,)\,\bar{1}\cdot 6532126 \\ \hline \bar{1}\cdot 8266063 \end{array}$$

Thus $\log\cos\frac{C}{2} = \bar{1}\cdot 8266063$

$$\begin{array}{r} \log\cos 47^\circ 53' = \bar{1}\cdot 8264910 \\ \hline \text{diff.} \quad 1153 \end{array}$$

$$\therefore \text{ proportional } \textit{decrease} = \frac{1153}{1397} \times 60'' = 49\cdot 5'';$$

$$\therefore \frac{C}{2} = 47^\circ\, 52'\, 10\cdot 5''.$$

$$\begin{array}{r} 1153 \\ 60 \\ \hline 1397\,)\,69180\,(\,49\cdot 5 \\ 5588 \\ \hline 13300 \\ 12573 \\ \hline 7270 \end{array}$$

Thus the greatest angle is $95^\circ\, 44'\, 21''$.

Example 2. If $a=283$, $b=317$, $c=428$, find all the angles.

$$\tan\frac{A}{2} = \sqrt{\frac{(s-b)(s-c)}{s(s-a)}} = \sqrt{\frac{197\times 86}{514\times 231}};$$

$$\therefore \log\tan\frac{A}{2} = \frac{1}{2}(\log 197 + \log 86 - \log 514 - \log 231).$$

$$\begin{array}{r} 283 \\ 317 \\ 428 \\ \hline 2\,)\,1028 \\ \hline 514 = s \\ 231 = s-a \\ 197 = s-b \\ 86 = s-c \end{array}$$

From the Tables,

$$\begin{array}{r} \log 197 = 2\cdot 2944662 \\ \log 86 = 1\cdot 9344985 \\ \hline 4\cdot 2289647 \\ 5\cdot 0745751 \\ \hline 2\,)\,\bar{1}\cdot 1543896 \end{array} \qquad \begin{array}{r} \log 514 = 2\cdot 7109631 \\ \log 231 = 2\cdot 3636120 \\ \hline 5\cdot 0745751 \end{array}$$

$$\begin{array}{r} \log\tan\frac{A}{2} = \bar{1}\cdot 5771948 \\ \log\tan 20^\circ\, 41' = \bar{1}\cdot 5769585 \\ \hline \text{diff.} \quad 2363 \end{array}$$

But diff. for 60″ is 3822;

$\therefore$ prop^l. increase $=\dfrac{2363}{3822}\times 60'' = 37\cdot 1''$;

$\therefore \dfrac{A}{2} = 20^\circ 41' 37\cdot 1''$ and $A = 41^\circ 23' 14''$.

$$\begin{array}{r} 2363 \\ 60 \\ 3822\,)\,\overline{141780}\,(\,37\cdot 1 \\ 11466 \\ \hline 27120 \\ 26754 \\ \hline 3660 \end{array}$$

Again, $\tan\dfrac{B}{2} = \sqrt{\dfrac{(s-c)(s-a)}{s(s-b)}} = \sqrt{\dfrac{86\times 231}{514\times 197}}$;

$\therefore \log\tan\dfrac{B}{2} = \dfrac{1}{2}(\log 86 + \log 231 - \log 514 - \log 197)$.

$$\begin{array}{rl} \log\ 86 = & 1\cdot 9344985 \\ \log 231 = & 2\cdot 3636120 \\ \hline & 4\cdot 2981105 \\ & 5\cdot 0054293 \\ \hline 2\,) & \overline{1}\cdot 2926812 \\ \hline \end{array} \qquad \begin{array}{rl} \log 514 = & 2\cdot 7109631 \\ \log 197 = & 2\cdot 2944662 \\ \hline & 5\cdot 0054293 \end{array}$$

$$\begin{array}{rr} \log\tan\dfrac{B}{2} = & \overline{1}\cdot 6463406 \\ \log\tan 23^\circ 53' = & \overline{1}\cdot 6461988 \\ \hline \text{diff.} & 1418 \end{array}$$

But diff. for 60″ is 3412;

$\therefore$ prop^l. increase $=\dfrac{1418}{3412}\times 60'' = 24\cdot 9''$.

$\therefore \dfrac{B}{2} = 23^\circ 53' 24\cdot 9''$ and $B = 47^\circ 46' 50''$.

$$\begin{array}{r} 1418 \\ 60 \\ 3412\,)\,\overline{85080}\,(\,24\cdot 9 \\ 6824 \\ \hline 16840 \\ 13648 \\ \hline 31920 \end{array}$$

Thus $A = 41^\circ 23' 14''$, $B = 47^\circ 46' 50''$, $C = 90^\circ 49' 56''$.

EXAMPLES. XVI. b.

1. The sides of a triangle are 5, 8, 11; find the greatest angle; given $\log 7 = \cdot 8450980$,

   $L\sin 56^\circ 47' = 9\cdot 9225205$, $L\sin 56^\circ 48' = 9\cdot 9226032$.

2. If $a=40$, $b=51$, $c=43$, find A; given

   $L\tan 24^\circ 44' 13'' = 9\cdot 6634464$,

   $\log 128 = 2\cdot 1072100$, $\log 603 = 2\cdot 7803173$.

3. The sides a, b, c are as $4:5:6$, find B; given $\log 2$,

   $L\cos 27^\circ 53' = 9\cdot 9464040$, diff. for $1' = 669$.

4. Find the greatest angle of the triangle in which the sides are 5, 6, 7; given $\log 6 = \cdot 7781513$,

$$L \cos 39^\circ\ 14' = 9{\cdot}8890644, \quad \text{diff. for } 1' = 1032.$$

5. If $a=3$, $b=1{\cdot}75$, $c=2{\cdot}75$, find C; given $\log 2$,

$$L \tan 32^\circ\ 18' = 9{\cdot}8008365, \quad \text{diff. for } 1' = 2796.$$

6. If the sides are 24, 22, 14, find the least angle; given

$$L \tan 17^\circ\ 33' = 9{\cdot}500042, \quad \text{diff. for } 1' = 439.$$

7. Find the greatest angle when the sides are 4, 10, 11; given $\log 2$, $\log 3$,

$$L \cos 46^\circ\ 47' = 9{\cdot}8355378, \quad \text{diff. for } 1' = 1345.$$

8. If $a : b : c = 15 : 13 : 14$, find the angles; given $\log 2$, $\log 3$, $\log 7$,

$$L \tan 26^\circ\ 33' = 9{\cdot}6986847, \quad \text{diff. for } 1' = 3159,$$
$$L \tan 29^\circ\ 44' = 9{\cdot}7567587, \quad \text{diff. for } 1' = 2933.$$

9. If $a : b : c = 3 : 4 : 2$, find the angles; given $\log 2$, $\log 3$,

$$L \tan 14^\circ\ 28' = 9{\cdot}4116146, \quad \text{diff. for } 10'' = 870,$$
$$L \tan 52^\circ\ 14' = 10{\cdot}1108395, \quad \text{diff. for } 10'' = 435.$$

189. *To solve a triangle having given two sides and the included angle.*

Let the given parts be b, c, A, and let

$$k = \frac{\sin B}{b} = \frac{\sin C}{c};$$

then

$$\frac{\sin B - \sin C}{\sin B + \sin C} = \frac{kb - kc}{kb + kc} = \frac{b-c}{b+c};$$

$$\therefore \frac{2 \cos \dfrac{B+C}{2} \sin \dfrac{B-C}{2}}{2 \sin \dfrac{B+C}{2} \cos \dfrac{B-C}{2}} = \frac{b-c}{b+c};$$

$$\therefore \frac{\tan \dfrac{B-C}{2}}{\tan \dfrac{B+C}{2}} = \frac{b-c}{b+c};$$

$$\therefore \quad \tan\frac{B-C}{2}=\frac{b-c}{b+c}\tan\frac{B+C}{2}=\frac{b-c}{b+c}\cot\frac{A}{2},$$

since $\dfrac{B+C}{2}=90^\circ-\dfrac{A}{2}$.

$$\therefore \log\tan\frac{B-C}{2}=\log(b-c)-\log(b+c)+\log\cot\frac{A}{2},$$

from which equation we can find $\dfrac{B-C}{2}$.

Also $\dfrac{B+C}{2}=90^\circ-\dfrac{A}{2}$, and is therefore known.

By addition and subtraction we obtain B and C.

From the equation $\qquad a=\dfrac{b\sin A}{\sin B}$,

$$\log a=\log b+\log\sin A-\log\sin B;$$

whence a may be found.

Example 1. If the sides a and b are in the ratio of 7 to 3, and the included angle C is 60°, find A and B; given

$$\log 2=\cdot 3010300, \quad \log 3=\cdot 4771213,$$
$$L\tan 34^\circ\,42'=9\cdot 8403776, \text{ diff. for } 1'=2699.$$

$$\tan\frac{A-B}{2}=\frac{a-b}{a+b}\cot\frac{C}{2}=\frac{7-3}{7+3}\cot 30^\circ=\frac{4}{10}\sqrt{3};$$

$$\therefore \log\tan\frac{A-B}{2}=2\log 2-1+\frac{1}{2}\log 3;$$

$$\therefore \log\tan\frac{A-B}{2}=\bar{1}\cdot 8406207$$

$$\log\tan 34^\circ\,42'=\bar{1}\cdot 8403776$$

$$\text{diff.} \qquad 2431$$

$$\therefore \text{prop}^{l}.\text{ increase}=\frac{2431}{2699}\times 60''=54'';$$

$$\therefore \frac{A-B}{2} \qquad =34^\circ\,42'\,54'';$$

And $\qquad \dfrac{A+B}{2}=90^\circ-\dfrac{C}{2}=60^\circ.$

$$2\log 2=\cdot 6020600$$
$$\tfrac{1}{2}\log 3=\cdot 2385607$$
$$\overline{\cdot 8406207}$$

```
        2431
          60
2699 ) 145860 ( 54
       13495
       -----
        10910
        10796
```

By addition, $\qquad A=94^\circ\,42'\,54''$,

and by subtraction, $\qquad B=25^\circ\,17'\,6''$.

Example 2. If $a=681$, $c=243$, $B=50° 42'$, solve the triangle, by the use of Tables.

$$\tan\frac{A-C}{2}=\frac{a-c}{a+c}\cot\frac{B}{2}=\frac{438}{924}\cot 25° 21';$$

$$\therefore \log\tan\frac{A-C}{2}=\log 438-\log 924 + \log\cot 25° 21'$$

$$\therefore \log\tan\frac{A-C}{2}=\cdot 0002383$$

$$\log\tan 45°=\cdot 0000000$$

$$\text{diff.} \quad 2383$$

$$\begin{aligned}\log 438 &= 2\cdot 6414741\\ \log\cot 25° 21' &= \cdot 3244362\\ &\overline{2\cdot 9659103}\\ \log 924 &= 2\cdot 9656720\\ &\overline{\cdot 0002383}\end{aligned}$$

And diff. for 60″ is 2527;

$$\therefore \text{prop}^{l}.\ \text{increase}=\frac{2383}{2527}\times 60''=57'';$$

$$\therefore \frac{A-C}{2}=45° 0' 57''.$$

$$\begin{array}{r}2383\\60\\2527\,)\,\overline{142980}\,(\,56\cdot 6\\12635\\\overline{16630}\\15162\\\overline{14680}\end{array}$$

Also $\dfrac{A+C}{2}=90°-\dfrac{B}{2}=64° 39'.$

By addition, $A=109° 39' 57''$,

and by subtraction, $C=19° 38' 3''$.

Again, $b=\dfrac{c\sin B}{\sin C}$;

$$\therefore \log b=\log c+\log\sin B-\log\sin C$$
$$=\log 243+\log\sin 50° 42' - \log\sin 19° 38' 3''$$

$$\therefore \log b=2\cdot 7479012$$
$$\log 559\cdot 63 = 2\cdot 7479010$$

$$\therefore b=559\cdot 63.$$

$$\begin{aligned}\log\sin 19° 38' &= \bar{1}\cdot 5263387\\ \frac{3}{60}\times 3540 &= \quad 177\\ \log\sin 19° 38' 3'' &= \bar{1}\cdot 5263564\\ \log 243 &= 2\cdot 3856063\\ \log\sin 50° 42' &= \bar{1}\cdot 8886513\\ &\overline{2\cdot 2742576}\\ \log\sin 19° 38' 3'' &= \bar{1}\cdot 5263564\\ &\overline{2\cdot 7479012}\end{aligned}$$

Thus $A=109° 39' 57''$, $C=19° 38' 3''$, $b=559\cdot 63$.

190. From the formula

$$\tan\frac{B-C}{2}=\frac{b-c}{b+c}\cot\frac{A}{2},$$

it will be seen that if b, c, and $B-C$ are known A can be found; that is, the triangle can be solved when the given parts are two sides and the difference of the angles opposite to them.

EXAMPLES. XVI. c.

1. If $a=9$, $b=6$, $C=60°$, find A and B; given $\log 2$, $\log 3$,
$L\tan 19° \, 6'=9{\cdot}5394287$, $L\tan 19° \, 7'=9{\cdot}5398371$.

2. If $a=1$, $c=9$, $B=65°$, find A and C; given $\log 2$,
$L\cot 32° \, 30'=10{\cdot}1958127$,
$L\tan 51° \, 28'=10{\cdot}0988763$, diff. for $1'=2592$.

3. If $17a=7b$, $C=60°$, find A and B; given $\log 2$, $\log 3$,
$L\tan 35° \, 49'=9{\cdot}8583357$, diff. for $10''=2662$.

4. If $b=27$, $c=23$, $A=44° \, 30'$, find B and C; given $\log 2$,
$L\cot 22° \, 15'=10{\cdot}3881591$,
$L\tan 11° \, 3'=9{\cdot}2906713$, diff. for $1'=6711$.

5. If $c=210$, $a=110$, $B=34° \, 42' \, 30''$, find C and A; given $\log 2$,
$L\cot 17° \, 21' \, 15''=10{\cdot}5051500$.

6. Two sides of a triangle are as 5 : 3 and include an angle of 60° 30′: find the other angles; given $\log 2$,
$L\cot 30° \, 15'=10{\cdot}23420$,
$L\tan 23° \, 13'=9{\cdot}63240$, diff. for $1'=35$.

7. If $a=327$, $c=256$, $B=56° \, 28'$, find A and C; given
$\log 7{\cdot}1={\cdot}8512583$, $\log 5{\cdot}83={\cdot}7656686$,
$L\tan 61° \, 46'=10{\cdot}2700705$,
$L\tan 12° \, 46'=9{\cdot}3552267$, diff. for $1'=5859$.

8. If $b=4c$, $A=65°$, find B and C; given $\log 2$, $\log 3$,
$L\tan 57° \, 30'=10{\cdot}1958127$,
$L\tan 43° \, 18'=9{\cdot}9742133$, diff. for $1'=2531$.

9. If $a=23031$, $b=7677$, $C=30° \, 10' \, 5''$, find A and B; given $\log 2$,
$L\tan 15° \, 5'=9{\cdot}4305727$, diff. for $10''=838$,
$L\cot 61° \, 41'=9{\cdot}7314436$, diff. for $10''=504$.

191. *To solve a triangle having given two angles and a side.*

Let the given parts be denoted by B, C, a; then the third angle A is found from the equation $A = 180° - B - C$,

and
$$b = \frac{a \sin B}{\sin A};$$

$$\therefore \log b = \log a + \log \sin B - \log \sin A;$$

whence b may be found.

Similarly, c may be obtained from the equation
$$\log c = \log a + \log \sin C - \log \sin A.$$

Example. If $b = 1000$, $A = 45°$, $C = 68° 17' 40''$, find the least side, having given

$$\log 2 = \cdot 3010300,\ \log 7 \cdot 6986 = \cdot 8864118,\ \text{diff. for } 1 = 57,$$
$$L \sin 66° 42' = 9 \cdot 9630538,\ \text{diff. for } 1' = 544.$$

$B = 180° - 45° - 68° 17' 40'' = 66° 42' 20''$.

The least side $= a = \dfrac{b \sin A}{\sin B} = \dfrac{1000 \sin 45°}{\sin 66° 42' 20''}$;

$$\therefore \log a = 3 + \log \frac{1}{\sqrt{2}} - \log \sin 66° 42' 20''$$
$$= 3 - \frac{1}{2} \log 2 - \log \sin 66° 42' 20''$$
$$= 3 - \cdot 1135869$$

$$\log \sin 66° 42' = \bar{1} \cdot 9630538$$
$$\frac{20}{60} \times 544 = \quad 181$$
$$\frac{1}{2} \log 2 = \quad \cdot 1505150$$
$$\cdot 1135869$$

$$\therefore \log a = 2 \cdot 8864131$$
$$\log 769 \cdot 86 = 2 \cdot 8864118$$
$$\text{diff.} \quad 13$$

$$\therefore \text{prop}^{l}.\ \text{increase} = \frac{13}{57} = \cdot 22.$$

Thus the least side is 769·8622.

EXAMPLES. XVI. d.

1. If $B = 60° 15'$, $C = 54° 30'$, $a = 100$, find c; given
$L \sin 54° 30' = 9 \cdot 9106860$, $\log 8 \cdot 9646162 = \cdot 9525317$,
$L \sin 65° 15' = 9 \cdot 9581543$.

2. If $A=55°$, $B=65°$, $c=270$, find a; given log 2, log 3,
log 25538 = 4·4071869, $L \sin 55° = 9·9133645$,
log 25539 = 4·4072039.

3. If $A=45° \ 41'$, $C=62° \ 5'$, $b=100$, find c; given
log 9·2788 = ·96749, $L \sin 62° \ 5' = 9·94627$,
$L \sin 72° \ 14' = 9·97878$.

4. If $B=70° \ 30'$, $C=78° \ 10'$, $a=102$, find b and c; given log 2 = ·301, log 1·02 = ·009, log 1·85 = ·267, log 1·92 = ·283,
$L \sin 70° \ 30' = 9·974$, $L \sin 78° \ 10' = 9·990$,
$L \sin 31° \ 20' = 9·716$.

5. If $a=123$, $B=29° \ 17'$, $C=135°$, find c; given log 2,
log 123 = 2·0899051, $L \sin 15° \ 43' = 9·4327777$,
log 3211 = 3·5066403, $D=135$.

6. If $A=44°$, $C=70°$, $b=1006·62$, find a and c; given
$L \sin 44° = 9·8417713$, log 100662 = 5·0028656,
$L \sin 66° = 9·9607302$, log 103543 = 5·0151212,
$L \sin 70° = 9·9729858$, log 7654321 = 6·8839067.

7. If $a=1652$, $B=26° \ 30'$, $C=47° \ 15'$, find b and c;
$L \sin 73° \ 45' = 9·9822938$, log 1·652 = ·2180100,
$L \sin 26° \ 30' = 9·6495274$, log 7·6780 = ·8852481, $D=57$,
$L \sin 47° \ 15' = 9·8658868$, log 1·2636 = ·1016096, $D=344$.

192. *To solve a triangle when two sides and the angle opposite to one of them are given.*

Let a, b, A be given. Then from $\sin B = \frac{b}{a} \sin A$, we have

$$\log \sin B = \log b - \log a + \log \sin A;$$

whence B may be found;

then C is found from the equation $C = 180° - A - B$.

Again, $$c = \frac{a \sin C}{\sin A},$$

$$\therefore \ \log c = \log a + \log \sin C - \log \sin A.$$

If $a<b$, and A is acute the solution is ambiguous and there will be two values of B supplementary to each other, and also two values of C and c. [Art. 147.]

Example. If $b=63$, $c=36$, $C=29°\,23'\,15''$, find B; given

$\log 2=\cdot 3010300$, $\log 7=\cdot 8450980$.

$L\sin 29°\,23'=9\cdot 6907721$, diff. for $1'=2243$,

$L\sin 59°\,10'=9\cdot 9338222$, diff. for $1'=755$.

$$\sin B=\frac{b}{c}\sin C=\frac{63}{36}\sin C$$

$$=\frac{7}{4}\sin 29°\,23'\,15'';$$

$$\therefore\ \log\sin B=\log 7-2\log 2+\log\sin 29°\,23'\,15'';$$

$$\therefore\ \log\sin B=\bar{1}\cdot 9338662$$

$$\log\sin 59°\,10'=\bar{1}\cdot 9338222$$

diff. 440

$$\therefore\ \text{prop}^{l}.\ \text{increase}=\frac{440}{755}\times 60''=35'';$$

$$\therefore\ B=59°\,10'\,35''.$$

Also since $c < b$ there is another value of B supplementary to the above, namely $B=120°\,49'\,25''$.

$\log\sin 29°\,23'=\bar{1}\cdot 6907721$

$\frac{15}{60}\times 2243=$ 561

$\log 7=$ $\cdot 8450980$

$\cdot 5359262$

$2\log 2=$ $\cdot 6020600$

$\bar{1}\cdot 9338662$

```
        440
         60
755 ) 26400 ( 35
      2265
      ----
       3750
       3775
```

EXAMPLES. XVI. e.

1. If $a=145$, $b=178$, $B=41°\ 10'$, find A; given
   $\log 178=2\cdot 2504200$, $L\sin 41°\ 10'=9\cdot 8183919$,
   $\log 145=2\cdot 1613680$, $L\sin 32°\ 25'\ 35''=9\cdot 7293399$.

2. If $A=26°\ 26'$, $b=127$, $a=85$, find B; given
   $\log 1\cdot 27=\cdot 1038037$, $L\sin 26°\ 26'=9\cdot 6485124$,
   $\log 8\cdot 5=\cdot 9294189$, $L\sin 41°\ 41'\ 28''=9\cdot 8228972$.

3. If $c=5$, $b=4$, $C=45°$, find A and B; given $\log 2$,
   $L\sin 34°\ 26'=9\cdot 7525750$, $L\sin 34°\ 27'=9\cdot 7525761$.

4. If $a=1405$, $b=1706$, $A=40°$, find B; given
   $\log 1\cdot 405=\cdot 1476763$, $\log 1\cdot 706=\cdot 2319790$,
   $L\sin 40°=9\cdot 8080675$, $L\sin 51°\ 18'=9\cdot 8923342$,
   diff. for $1'=1012$.

5. If $B=112^\circ 4'$, $b=573$, $c=394$, find A and C; given

$$\log 573=2\cdot7581546, \quad \log 394=2\cdot5954962,$$
$$L\sin 39^\circ 35'=9\cdot8042757, \quad \text{diff. for } 60''=1527,$$
$$L\cos 22^\circ 4'=9\cdot9669614.$$

6. If $b=8\cdot4$, $c=12$, $B=37^\circ 36'$, find A; given

$$\log 7=\cdot8450980, \quad L\sin 37^\circ 36'=9\cdot7854332,$$
$$L\sin 60^\circ 39'=9\cdot9403381, \quad \text{diff. for } 1'=711.$$

7. Supposing the data for the solution of a triangle to be as in the three following cases, point out whether the solution will be ambiguous or not, and find the third side in the obtuse angled triangle in the ambiguous case:

(i) $A=30^\circ$, $a=125$ feet, $c=250$ feet,

(ii) $A=30^\circ$, $a=200$ feet, $c=250$ feet,

(iii) $A=30^\circ$, $a=200$ feet, $c=125$ feet.

Given log 2,

$$\log 6\cdot0389=\cdot7809578, \quad L\sin 38^\circ 41'=9\cdot7958800,$$
$$\log 6\cdot0390=\cdot7809650, \quad L\sin 8^\circ 41'=9\cdot1789001.$$

193. Some formulæ which are not primarily suitable for working with logarithms may be adapted to such work by various artifices.

194. *To adapt the formula $c^2=a^2+b^2$ to logarithmic computation.*

We have $$c^2=a^2\left(1+\frac{b^2}{a^2}\right).$$

Since an angle can always be found whose tangent is equal to a given numerical quantity, we may put $\dfrac{b}{a}=\tan\theta$, and thus obtain

$$c^2=a^2(1+\tan^2\theta)=a^2\sec^2\theta;$$
$$\therefore\quad c=a\sec\theta;$$
$$\therefore\quad \log c=\log a+\log\sec\theta.$$

The angle θ is called a **subsidiary angle** and is found from the equation

$$\log\tan\theta=\log b-\log a.$$

Thus *any expression which can be put into the form of the sum of two squares can be readily adapted to logarithmic work.*

195. *To adapt the formula $c^2=a^2+b^2-2ab\cos C$ to logarithmic computation.*

From the identities

$$\cos C=\cos^2\frac{C}{2}-\sin^2\frac{C}{2}, \text{ and } 1=\cos^2\frac{C}{2}+\sin^2\frac{C}{2},$$

we have

$$c^2=(a^2+b^2)\left(\cos^2\frac{C}{2}+\sin^2\frac{C}{2}\right)-2ab\left(\cos^2\frac{C}{2}-\sin^2\frac{C}{2}\right)$$

$$=(a^2+b^2-2ab)\cos^2\frac{C}{2}+(a^2+b^2+2ab)\sin^2\frac{C}{2}$$

$$=(a-b)^2\cos^2\frac{C}{2}+(a+b)^2\sin^2\frac{C}{2}$$

$$=(a-b)^2\cos^2\frac{C}{2}\left\{1+\left(\frac{a+b}{a-b}\right)^2\tan^2\frac{C}{2}\right\}.$$

Take a subsidiary angle θ, such that

$$\tan\theta=\frac{a+b}{a-b}\tan\frac{C}{2},$$

then $c^2=(a-b)^2\cos^2\frac{C}{2}(1+\tan^2\theta)$

$$=(a-b)^2\cos^2\frac{C}{2}\sec^2\theta;$$

$$\therefore\quad c=(a-b)\cos\frac{C}{2}\sec\theta;$$

$$\therefore\quad \log c=\log(a-b)+\log\cos\frac{C}{2}+\log\sec\theta,$$

where θ is determined from the equation

$$\log\tan\theta=\log(a+b)-\log(a-b)+\log\tan\frac{C}{2}.$$

196. When two sides and the included angle are given, we may solve the triangle by finding the value of the third side first instead of determining the angles first as in Art. 189.

Example. If $a=3$, $c=1$, $B=53°\,7'\,48''$ find b; given

$\log 2=\cdot3010300$, $\log 2\cdot5298=\cdot4030862$, diff. for $1=172$,

$L\cos 26°\,33'\,54''=9\cdot9515452$, $L\tan 26°\,33'\,54''=9\cdot6989700$.

We have $b^2 = c^2 + a^2 - 2ca\cos B$

$$= (a^2+c^2)\left(\cos^2\frac{B}{2} + \sin^2\frac{B}{2}\right) - 2ac\left(\cos^2\frac{B}{2} - \sin^2\frac{B}{2}\right)$$

$$= (a-c)^2\cos^2\frac{B}{2} + (a+c)^2\sin^2\frac{B}{2}$$

$$= (a-c)^2\cos^2\frac{B}{2}\left\{1 + \left(\frac{a+c}{a-c}\right)^2\tan^2\frac{B}{2}\right\}$$

$$= (a-c)^2\cos^2\frac{B}{2}\,(1+\tan^2\theta)\ \ldots\ldots\ldots\ldots(1),$$

where $$\tan\theta = \frac{a+c}{a-c}\tan\frac{B}{2} = 2\tan 26^\circ\,33'\,54'';$$

$$\therefore \log\tan\theta = \log 2 + \log\tan 26^\circ\,33'\,54''$$
$$= \cdot 3010300 + \bar{1}\cdot 6989700$$
$$= 0;$$

whence $\tan\theta = 1$, and $\theta = 45^\circ$.

From (1), $$b = (a-c)\cos\frac{B}{2}\sec\theta$$
$$= 2\sec 45^\circ\cos\frac{B}{2}$$
$$= 2\sqrt{2}\cos 26^\circ\,33'\,54'';$$

$$\therefore \log b = \log 2 + \frac{1}{2}\log 2 + \log\cos 26^\circ\,33'\,54''$$

$\log 2 = \cdot 3010300$
$\frac{1}{2}\log 2 = \cdot 1505150$
$\log\cos 26^\circ\,33'\,54'' = \bar{1}\cdot 9515452$
$\cdot 4030902$

$\therefore \log b = \cdot 4030902$
$\log 2\cdot 5298 = \cdot 4030862$
diff. 40

But diff. for 1 is 172;

$$\therefore \text{proportional increase} = \frac{40}{172} = \frac{10}{43} = \cdot 23.$$

Thus the third side is $2\cdot 529823$.

197. The formula $c^2 = a^2 + b^2 - 2ab\cos C$ may also be adapted to logarithmic computation in two other ways by making use of the identities $\cos C = 2\cos^2\frac{C}{2} - 1$ and $\cos C = 1 - 2\sin^2\frac{C}{2}$.

We shall take the first of these cases, leaving the other as an exercise.

$$c^2=a^2+b^2-2ab\cos C$$
$$=a^2+b^2-2ab\left(2\cos^2\frac{C}{2}-1\right)$$
$$=(a+b)^2-4ab\cos^2\frac{C}{2}$$
$$=(a+b)^2\left\{1-\frac{4ab}{(a+b)^2}\cos^2\frac{C}{2}\right\}.$$

Let $$\frac{4ab}{(a+b)^2}\cos^2\frac{C}{2}=\cos^2\theta,$$

then $$c^2=(a+b)^2(1-\cos^2\theta)=(a+b)^2\sin^2\theta;$$
$$\therefore\ c=(a+b)\sin\theta;$$
$$\therefore\ \log c=\log(a+b)+\log\sin\theta.$$

To determine θ we have the equation

$$\cos\theta=\frac{2\sqrt{ab}}{a+b}\cos\frac{C}{2};$$

$$\therefore\ \log\cos\theta=\log 2+\frac{1}{2}(\log a+\log b)-\log(a+b)+\log\cos\frac{C}{2}.$$

Since $2\sqrt{ab}$ is never greater than $a+b$ and $\cos\frac{C}{2}$ is positive and less than unity, $\cos\theta$ is positive and less than unity, and thus θ is an acute angle.

EXAMPLES. XVI. f.

1. If $a=8$, $b=7$, $c=9$, find the angles; given $\log 2$, $\log 3$,
$L\tan 24°\ 5'=9{\cdot}6502809$, diff. for $60''=3390$,
$L\tan 36°\ 41'=9{\cdot}8721123$, diff. for $60''=2637$.

2. The difference between the angles at the base of a triangle is 24°, and the sides opposite these angles are 175 and 337: find all the angles; given $\log 2$, $\log 3$,
$L\tan 12°=9{\cdot}3274745$, $L\cot 56°\ 6'\ 27''=9{\cdot}8272293$.

3. One of the sides of a right-angled triangle is two-sevenths of the hypotenuse: find the greater of the two acute angles; given

$$\log 2, \log 7, \quad L\sin 14^\circ\ 11' = 9{\cdot}455921, \quad L\sin 14^\circ\ 12' = 9{\cdot}456031.$$

4. Find the greatest side when two of the angles are 78° 14′ and 71° 24′ and the sides joining them is 2183; given

$$\log 2{\cdot}183 = {\cdot}3390537, \quad \log 4{\cdot}2274 = {\cdot}6260733, \quad D = 103,$$
$$L\sin 78^\circ\ 14' = 9{\cdot}9907766, \quad L\sin 30^\circ\ 22' = 9{\cdot}7037486.$$

5. If $b = 2$ ft. 6 in., $c = 2$ ft., $A = 22^\circ\ 20'$, find the other angles; and then shew that the side a is very approximately 1 foot. Given log 2, log 3,

$$L\cot 11^\circ\ 10' = 10{\cdot}70465, \quad L\sin 49^\circ\ 27'\ 34'' = 9{\cdot}88079,$$
$$L\sin 22^\circ\ 20' = 9{\cdot}57977, \quad L\tan 29^\circ\ 22'\ 26'' = 9{\cdot}75041.$$

6. If $a = 1{\cdot}56234$, $b = {\cdot}43766$, $C = 58^\circ\ 42'\ 6''$, find A and B; given $\log 56234 = 4{\cdot}75$,

$$\log\cot 29^\circ\ 21' = {\cdot}250015, \quad \log\cot 29^\circ\ 22' = {\cdot}249715.$$

7. If $a = 9$, $b = 12$, $A = 30^\circ$, find the values of c, having given

$$\begin{aligned}
&\log 12 = 1{\cdot}07918, && L\sin 30^\circ = 9{\cdot}69897,\\
&\log 9 = {\cdot}95424, && L\sin 11^\circ\ 48'\ 39'' = 9{\cdot}31108,\\
&\log 171 = 2{\cdot}23301, && L\sin 41^\circ\ 48'\ 39'' = 9{\cdot}82391,\\
&\log 368 = 2{\cdot}56635, && L\sin 108^\circ\ 11'\ 21'' = 9{\cdot}97774.
\end{aligned}$$

8. The sides of a triangle are 9 and 3, and the difference of the angles opposite to them is 90°: find the angles; having given log 2,

$$L\tan 26^\circ\ 33' = 9{\cdot}6986847, \quad L\tan 26^\circ\ 34' = 9{\cdot}6990006.$$

9. Two sides of a triangle are 1404 and 960 respectively, and an angle opposite to one of them is 32° 15′: find the angle contained by the two sides; having given log 2, log 3,

$$\log 13 = 1{\cdot}1139434, \quad L\operatorname{cosec} 32^\circ\ 15' = 10{\cdot}2727724,$$
$$L\sin 21^\circ\ 23' = 9{\cdot}5621316, \quad L\sin 51^\circ\ 18' = 9{\cdot}8923236.$$

10. If $b : c = 11 : 10$ and $A = 35^\circ\ 25'$, use the formula

$$\tan\frac{1}{2}(B - C) = \tan^2\frac{\phi}{2}\cot\frac{A}{2} \text{ to find } B \text{ and } C;$$

given $\log 1{\cdot}1 = {\cdot}041393$, $L\tan 12^\circ\ 18'\ 36'' = 9{\cdot}338891$,

$$L\cos 24^\circ\ 37'\ 12'' = 9{\cdot}958607, \quad L\cot 17^\circ\ 42'\ 30'' = 10{\cdot}495800,$$
$$L\tan 8^\circ\ 28'\ 56{\cdot}5'' = 9{\cdot}173582.$$

11. If $A=50°$, $b=1071$, $a=873$, find B; given

$$\log 1{\cdot}071 = {\cdot}029789, \quad \log 8{\cdot}73 = {\cdot}941014,$$
$$L\sin 50° = 9{\cdot}884254, \quad L\sin 70° = 9{\cdot}972986,$$
$$L\sin 70°\ 1' = 9{\cdot}973032.$$

12. If $a=3$, $b=1$, $C=53°\ 7'\ 48''$, find c without determining A and B; given $\log 2$,

$$\log 25298 = 4{\cdot}4030862, \quad L\cos 26°\ 33'\ 54'' = 9{\cdot}9515452,$$
$$\log 25299 = 4{\cdot}4031034, \quad L\tan 26°\ 33'\ 54'' = 9{\cdot}6989700.$$

(*In the following Examples the necessary Logarithms must be taken from the Tables.*)

13. Given $a=1000$, $b=840$, $c=1258$, find B.

14. Solve the triangle in which $a=525$, $b=650$, $c=777$.

15. Find the least angle when the sides are proportional to 4, 5, and 6.

16. If $B=90°$, $AC=57{\cdot}321$, $AB=28{\cdot}58$, find A and C.

17. Find the hypotenuse of a right-angled triangle in which the smallest angle is 18° 37′ 29″ and the side opposite to it is 284 feet.

18. The sides of a triangle are 9 and 7 and the angle between them is 60°: find the other angles.

19. How long must a ladder be so that when inclined to the ground at an angle of 72° 15′ it may just reach a window 42·37 feet from the ground?

20. If $a=31{\cdot}95$, $b=21{\cdot}96$, $C=35°$, find A and B.

21. Find B, C, a when $b=25{\cdot}12$, $c=13{\cdot}83$, $A=47°\ 15''$.

22. Find the greatest angle of the triangle whose sides are 1837·2, 2385·6, 2173·84.

23. When $a=21{\cdot}352$, $b=45{\cdot}6843$, $c=37{\cdot}2134$, find A, B, and C.

24. If $b=647{\cdot}324$, $c=850{\cdot}273$, $A=103°\ 12'\ 54''$, find the remaining parts.

25. If $b=23{\cdot}2783$, $A=37°\ 57'$, $B=43°\ 13'$, find the remaining sides.

26. Find a and b when $B=72°\ 43'\ 25''$, $C=47°\ 12'\ 17''$, $c=2484{\cdot}3$.

27. If $AB=4517$, $AC=150$, $A=31° 30''$, find the remaining parts.

28. Find A, B, and b when

$$a=324{\cdot}68,\quad c=421{\cdot}73,\quad C=35° 17' 12''.$$

29. Given $a=321{\cdot}7$, $c=435{\cdot}6$, $A=36° 18' 27''$, find C.

30. If $b=1625$, $c=1665$, $B=52° 19'$, solve the obtuse-angled triangle to which the data belong.

31. If $a=3795$, $B=73° 15' 15''$, $C=42° 18' 30''$, find the other sides.

32. Find the angles of the two triangles which have $b=17$, $c=12$, and $C=43° 12' 12''$.

33. Two sides of a triangle are 2·7402 ft. and ·7401 ft. respectively, and contain an angle 59° 27′ 5″: find the base and altitude of the triangle.

34. The difference between the angles at the base of a triangle is 17° 48′ and the sides subtending these angles are 105·25 ft. and 76·75 ft.: find the angle included by the given sides.

35. From the following data:

(1) $A=43° 15'$, $AB=36{\cdot}5$, $BC=20$,
(2) $A=43° 15'$, $AB=36{\cdot}5$, $BC=30$,
(3) $A=43° 15'$, $AB=36{\cdot}5$, $BC=45$,

point out which solution is impossible and which ambiguous. Find the third side for the triangle the solution of which is neither impossible nor ambiguous.

36. In any triangle prove that $c=(a-b)\sec\theta$, where

$$\tan\theta=\frac{2\sqrt{ab}}{a-b}\sin\frac{C}{2}.$$

If $a=17{\cdot}32$, $b=13{\cdot}47$, $C=47° 13'$, find c without finding A and B.

37. If $\tan\phi=\frac{a+b}{a-b}\tan\frac{C}{2}$, prove that $c=(a-b)\cos\frac{C}{2}\sec\phi$.

If $a=27{\cdot}3$, $b=16{\cdot}8$, $C=45° 12'$, find θ, and thence find c.

CHAPTER XVII.

HEIGHTS AND DISTANCES.

198. Some easy cases of heights and distances depending only on the solution of right-angled triangles have been already dealt with in Chap. VI. The problems in the present chapter are of a more general character, and require for their solution some geometrical skill as well as a ready use of trigonometrical formulæ.

Measurements in one plane.

199. *To find the height and distance of an inaccessible object on a horizontal plane.*

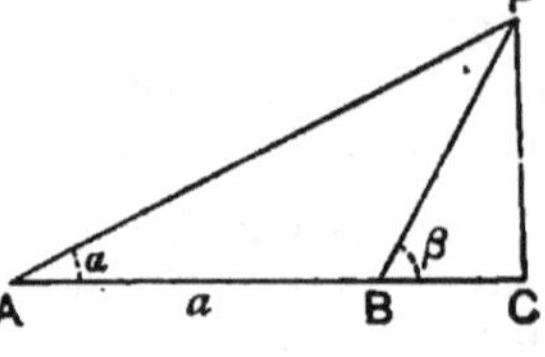

Let A be the position of the observer, CP the object; from P draw PC perpendicular to the horizontal plane; then it is required to find PC and AC.

At A observe the angle of elevation PAC. Measure a base line AB in a direct line from A towards the object, and at B observe the angle of elevation PBC.

Let $\angle PAC = \alpha, \quad \angle PBC = \beta, \quad AB = a.$

From $\triangle PBC$, $\qquad PC = PB \sin \beta.$

From $\triangle PAB$,

$$PB = \frac{AB \sin PAB}{\sin APB} = \frac{a \sin \alpha}{\sin (\beta - \alpha)};$$

$$\therefore \; PC = a \sin \alpha \sin \beta \operatorname{cosec} (\beta - \alpha).$$

Also $\qquad AC = PC \cot \alpha = a \cos \alpha \sin \beta \operatorname{cosec} (\beta - \alpha).$

Each of the above expressions is adapted to logarithmic work; thus if $PC = x$, we have

$$\log x = \log a + \log \sin \alpha + \log \sin \beta + \log \operatorname{cosec} (\beta - \alpha).$$

NOTE. Unless the contrary is stated, it will be supposed that the observer's height is disregarded, and that the angles of elevation are measured from the ground.

Example I. A person walking along a straight road observes that at two consecutive milestones the angles of elevation of a hill in front of him are 30° and 75°: find the height of the hill.

In the adjoining figure,

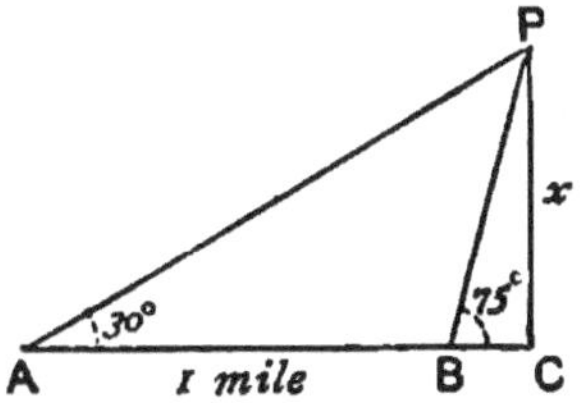

$$\angle PAC = 30°,\ \angle PBC = 75°,\ AB = 1\,\text{mile};$$

$$\angle APB = 75° - 30° = 45°.$$

Let x be the height in yards; then

$$x = PB \sin 75°;$$

but $PB = \dfrac{AB \sin PAB}{\sin APB} = \dfrac{1760 \sin 30°}{\sin 45°}$;

$$\therefore\ x = \frac{1760 \sin 30° \sin 75°}{\sin 45°}$$

$$= 1760 \times \frac{1}{2} \times \sqrt{2} \times \frac{\sqrt{3}+1}{2\sqrt{2}}$$

$$= 440\,(\sqrt{3}+1).$$

If we take $\sqrt{3} = 1{\cdot}732$ and reduce to feet, we find that the height is 3606·24 ft.

EXAMPLES. XVII. a.

1. From the top of a cliff 200 ft. above the sea-level the angles of depression of two boats in the same vertical plane as the observer are 45° and 30°: find their distance apart.

2. A person observes the elevation of a mountain top to be 15°, and after walking a mile directly towards it on level ground the elevation is 75°: find the height of the mountain in feet.

3. From a ship at sea the angle subtended by two forts A and B is 30°. The ship sails 4 miles towards A and the angle is then 48°: prove that the distance of B at the second observation is 6·472 miles.

4. From the top of a tower h ft. high the angles of depression of two objects on the horizontal plane and in a line passing through the foot of the tower are $45° - A$ and $45° + A$. Shew that the distance between them is $2h \tan 2A$.

5. An observer finds that the angular elevation of a tower is A. On advancing a feet towards the tower the elevation is 45° and on advancing b feet nearer the elevation is $90^\circ - A$: find the height of the tower.

6. A person observes that two objects A and B bear due N. and N. 30° W. respectively. On walking a mile in the direction N.W. he finds that the bearings of A and B are N.E. and due E. respectively: find the distance between A and B.

7. A tower stands at the foot of a hill whose inclination to the horizon is 9°; from a point 40 ft. up the hill the tower subtends an angle of 54°: find its height.

8. At a point on a level plane a tower subtends an angle a and a flagstaff c ft. in length at the top of the tower subtends an angle β: shew that the height of the tower is

$$c \sin a \operatorname{cosec} \beta \cos (a+\beta).$$

Example II. The upper three-fourths of a ship's mast subtends at a point on the deck an angle whose tangent is ·6; find the tangent of the angle subtended by the whole mast at the same point.

Let C be the point of observation, and let APB be the mast, AP being the lower fourth of it.

Let $AB=4a$, so that $AP=a$;
also let $AC=b$, $\angle ACB=\theta$, $\angle BCP=\beta$,
so that $\tan \beta = \cdot 6$.

From $\triangle PCA$, $\tan (\theta-\beta)=\dfrac{a}{b}$;

from $\triangle BCA$, $\tan \theta = \dfrac{4a}{b}$;

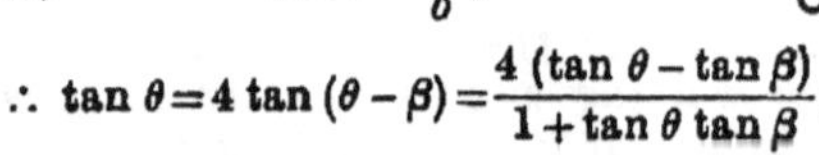

$$\therefore \tan \theta = 4 \tan (\theta - \beta) = \frac{4(\tan \theta - \tan \beta)}{1+\tan \theta \tan \beta};$$

$$\therefore \tan \theta = \frac{4\left(\tan \theta - \frac{3}{5}\right)}{1+\frac{3}{5}\tan \theta} = \frac{4(5\tan \theta - 3)}{5+3\tan \theta}.$$

On reduction, $\tan^2 \theta - 5 \tan \theta + 4 = 0$;

whence $\tan \theta = 1$ or 4.

NOTE. The student should observe that in examples of this class we make use of right-angled triangles in which the horizontal base line forms one side.

Example III. A tower BCD surmounted by a spire DE stands on a horizontal plane. From the extremity A of a horizontal line BA, it is found that BC and DE subtend equal angles. If $BC=9$ ft., $CD=72$ ft., and $DE=36$ ft., find BA.

Let
$$\angle BAC=\angle DAE=\theta,$$
$$\angle DAB=\alpha,\quad AB=x \text{ ft.}$$

Now $BC=9$ ft., $BD=81$ ft., $BE=117$ ft.

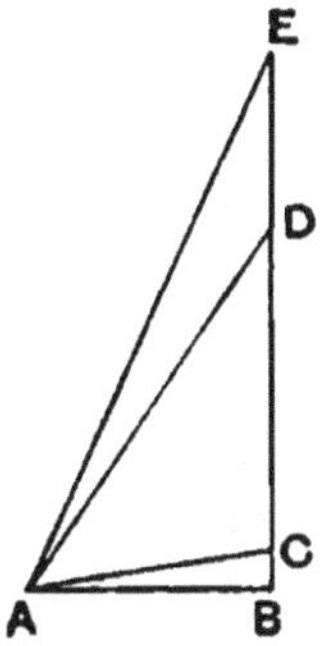

$$\therefore\ \tan(\alpha+\theta)=\frac{BE}{AB}=\frac{117}{x};$$
$$\tan\alpha=\frac{BD}{AB}=\frac{81}{x};$$
$$\tan\theta=\frac{BC}{AB}=\frac{9}{x}.$$

But
$$\tan(\alpha+\theta)=\frac{\tan\alpha+\tan\theta}{1-\tan\alpha\tan\theta};$$

$$\therefore\ \frac{117}{x}=\frac{\frac{81}{x}+\frac{9}{x}}{1-\frac{81}{x}\cdot\frac{9}{x}}=\frac{90}{x}\cdot\frac{x^2}{x^2-81\times 9}.$$

$$117x^2-81\times 9\times 117=90x^2;$$
$$\therefore\ 27x^2=81\times 9\times 117;$$
$$\therefore\ x^2=81\times 39;$$
$$\therefore\ x=9\sqrt{39}.$$

But $\sqrt{39}=6{\cdot}245$ nearly; $\therefore\ x=56{\cdot}205$ nearly.
Thus $AB=56{\cdot}2$ ft. nearly.

9. A flagstaff 20 ft. long standing on a wall 10 ft. high subtends an angle whose tangent is ·5 at a point on the ground: find the tangent of the angle subtended by the wall at this point.

10. A statue standing on the top of a pillar 25 feet high subtends an angle whose tangent is ·125 at a point 60 feet from the foot of the pillar: find the height of the statue.

11. A tower BCD surmounted by a spire DE stands on a horizontal plane. From the extremity A of a horizontal line BA it is found that BC and DE subtend equal angles.

If $\quad BC=9$ ft., $\ CD=280$ ft., and $\ DE=35$ ft.,

prove that $BA=180$ ft. nearly.

12. On the bank of a river there is a column 192 ft. high supporting a statue 24 ft. high. At a point on the opposite bank directly facing the column the statue subtends the same angle as a man 6 ft. high standing at the base of the column: find the breadth of the river.

13. A monument $ABCDE$ stands on level ground. At a point P on the ground the portions AB, AC, AD subtend angles α, β, γ respectively. Supposing that $AB=a$, $AC=b$, $AD=c$, $AP=x$, and $\alpha+\beta+\gamma=180°$, shew that $(a+b+c)x^2=abc$.

Example IV. The altitude of a rock is observed to be 47°; after walking 1000 ft. towards it up a slope inclined at 32° to the horizon the altitude is 77°. Find the vertical height of the rock above the first point of observation, given $\sin 47° = \cdot 731$.

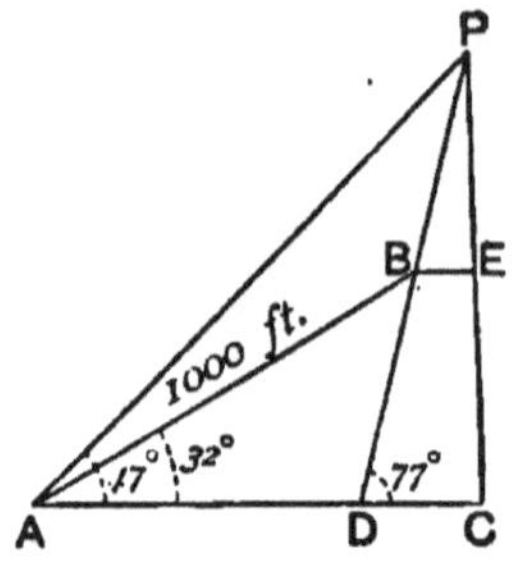

Let P be the top of the rock, A and B the points of observation; then in the figure $\angle PAC=47°$, $\angle BAC=32°$,

$\angle PDC=\angle PBE=77°$, $AB=1000$ ft.

Let x ft. be the height; then

$$x=PA\sin PAC=PA\sin 47°.$$

We have therefore to find PA in terms of AB.

In $\triangle PAB$, $\angle PAB=47°-32°=15°$;

$\angle APB=77°-47°=30°$;

$\therefore \angle ABP=135°$;

$$\therefore PA=\frac{AB\sin ABP}{\sin APB}$$

$$=\frac{1000\sin 135°}{\sin 30°}$$

$$=1000\sqrt{2};$$

$$\therefore x=PA\sin 47°=1000\sqrt{2}\times\cdot 731$$

$$=731\sqrt{2}.$$

If we take $\sqrt{2}=1\cdot 414$, we find that the height is 1034 ft. nearly.

14. From a point on the horizontal plane, the elevation of the top of a hill is 45°. After walking 500 yards towards its summit up a slope inclined at an angle of 15° to the horizon the elevation is 75°: find the height of the hill in feet.

15. From a station B at the base of a mountain its summit A is seen at an elevation of 60°; after walking one mile towards the summit up a plane making an angle of 30° with the horizon to another station C, the angle BCA is observed to be 135°: find the height of the mountain in feet.

16. The elevation of the summit of a hill from a station A is α. After walking c feet towards the summit up a slope inclined at an angle β to the horizon the elevation is γ: shew that the height of the hill is $c \sin\alpha \sin(\gamma-\beta) \operatorname{cosec}(\gamma-\alpha)$ feet.

17. From a point A an observer finds that the elevation of Ben Nevis is 60°; he then walks 800 ft. on a level plane towards the summit and then 800 ft. further up a slope of 30° and finds the elevation to be 75°: shew that the height of Ben Nevis above A is 4478 ft. approximately.

200. In many of the problems which follow, the solution depends upon the knowledge of some geometrical proposition.

Example I. A tower stands on a horizontal plane. From a mound 14 ft. above the plane and at a horizontal distance of 48 ft. from the tower an observer notices a loophole, and finds that the portions of the tower above and below the loophole subtend equal angles. If the height of the loophole is 30 ft., find the height of the tower.

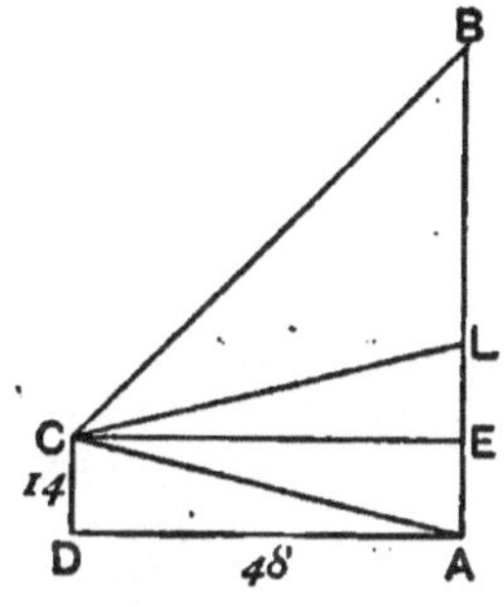

Let AB be the tower, C the point of observation, L the loophole. Draw CD vertical and CE horizontal. Let $AB=x$. We have

$$CD=14,\ AD=EC=48,\ BE=x-14.$$

From $\triangle ADC$, $AC^2=(14)^2+(48)^2=2500$;

$$\therefore AC=50.$$

From $\triangle CEB$, $CB^2=(x-14)^2+(48)^2$

$$=x^2-28x+2500.$$

Now $\angle BCL=\angle ACL$;

hence by Euc. VI. 3, $\dfrac{BC}{AC}=\dfrac{BL}{AL}$;

$$\therefore \frac{\sqrt{x^2-28x+2500}}{50}=\frac{x-30}{30}.$$

By squaring, $9(x^2-28x+2500)=25(x^2-60x+900)$.

On reduction, we obtain $16x^2-1248x=0$; whence $x=78$.

Thus the tower is 78 ft. high.

EXAMPLES. XVII. b.

1. At one side of a road is a flagstaff 25 ft. high fixed on the top of a wall 15 ft. high. On the other side of the road at a point on the ground directly opposite the flagstaff and wall subtend equal angles: find the width of the road.

2. A statue a feet high stands on a column $3a$ feet high. To an observer on a level with the top of the statue, the column and statue subtend equal angles: find the distance of the observer from the top of the statue.

3. A flagstaff a feet high placed on the top of a tower b feet high subtends the same angle as the tower to an observer h feet high standing on the horizontal plane at a distance d feet from the foot of the tower: shew that

$$(a-b)\,d^2=(a+b)\,b^2-2b^2h-(a-b)\,h^2.$$

Example II. A flagstaff is fixed on the top of a wall standing upon a horizontal plane. An observer finds that the angles subtended at a point on this plane by the wall and the flagstaff are α and β. He then walks a distance c directly towards the wall and finds that the flagstaff again subtends an angle β. Find the heights of the wall and flagstaff.

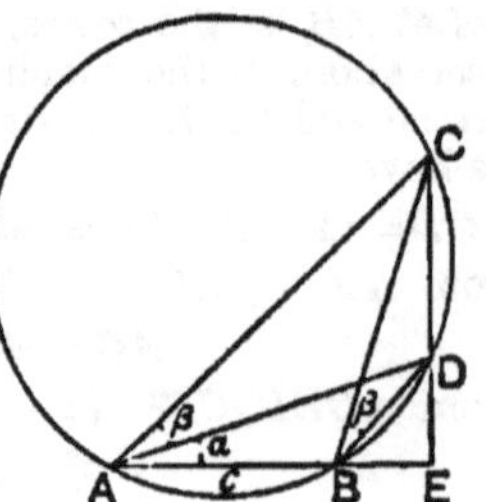

Let ED be the wall, DC the flagstaff, A and B the points of observation.

Then $\angle CAD=\beta=\angle CBD$, so that the four points C, A, B, D are concyclic.

$\therefore\ ABD=$ supp^t. of $\angle ACE$

$$=90^\circ+(\alpha+\beta),\text{ from }\triangle CAE.$$

Hence in $\triangle ADB$,

$$\angle ADB=180^\circ-\alpha-\{90^\circ+(\alpha+\beta)\}$$
$$=90^\circ-(2\alpha+\beta).$$

$$\therefore\ AD=\frac{AB\sin ABD}{\sin ADB}=\frac{c\cos(\alpha+\beta)}{\cos(2\alpha+\beta)}.$$

Hence in $\triangle ADE$,

$$DE=AD\sin\alpha=\frac{c\sin\alpha\cos(\alpha+\beta)}{\cos(2\alpha+\beta)}.$$

And in $\triangle CAD$,

$$CD=\frac{AD\sin CAD}{\sin ACD}=\frac{AD\sin\beta}{\cos(\alpha+\beta)}=\frac{c\sin\beta}{\cos(2\alpha+\beta)}.$$

4. A tower standing on a cliff subtends an angle α at each of two stations in the same horizontal line passing through the base of the cliff and at distances of a feet and b feet from the cliff. Prove that the height of the tower is $(a+b)\tan\alpha$ feet.

5. A column placed on a pedestal 20 feet high subtends an angle of 45° at a point on the ground, and it also subtends an angle of 45° at a point which is 20 feet nearer the pedestal: find the height of the column.

6. A flagstaff on a tower subtends the same angle at each of two places A and B on the ground. The elevations of the top of the flagstaff as seen from A and B are α and β respectively. If $AB=a$, shew that the length of the flagstaff is

$$a\sin(\alpha+\beta-90°)\operatorname{cosec}(\alpha-\beta).$$

Example III. A man walking towards a tower AB on which a flagstaff BC is fixed observes that when he is at a point E, distant c ft. from the tower, the flagstaff subtends its greatest angle. If $\angle BEC=\alpha$, prove that the heights of the tower and flagstaff are $c\tan\left(\frac{\pi}{4}-\frac{\alpha}{2}\right)$ and $2c\tan\alpha$ ft. respectively.

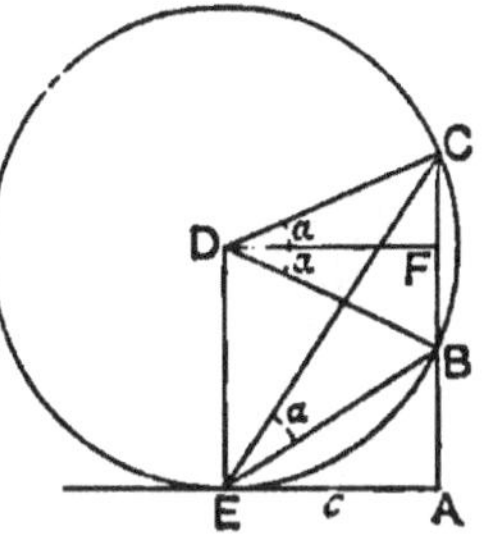

Since E is the point in the horizontal line AE at which BC subtends a maximum angle, it can easily be proved that AE touches the circle passing round the triangle CBE.

[See Hall and Stevens' *Euclid*, p. 242.]

The centre D of this circle lies in the vertical line through E. Draw DF perpendicular to BC, then DF bisects BC and also $\angle CDB$.

By Euc. III. 20,

$$\angle CDB=2\angle CEB=2\alpha;$$
$$\therefore \angle CDF=\angle BDF=\alpha.$$
$$\therefore CB=2CF=2DF\tan\alpha=2c\tan\alpha.$$

Again, $\angle AEB=\angle ECB$ in alternate segment

$$=\frac{1}{2}\angle EDB \text{ at centre}$$
$$=\frac{1}{2}\left(\frac{\pi}{2}-\alpha\right).$$
$$\therefore AB=c\tan AEB=c\tan\left(\frac{\pi}{4}-\frac{\alpha}{2}\right).$$

7. A pillar stands on a pedestal. At a distance of 60 feet from the base of the pedestal the pillar subtends its greatest angle 30°: shew that the length of the pillar is $40\sqrt{3}$ feet, and that the pedestal also subtends 30° at the point of observation.

8. A person walking along a canal observes that two objects are in the same line which is inclined at an angle a to the canal. He walks a distance c further and observes that the objects subtend their greatest angle β: shew that their distance apart is

$$2c \sin a \sin \beta / (\cos a + \cos \beta).$$

9. A tower with a flagstaff stands on a horizontal plane. Shew that the distances from the base at which the flagstaff subtends the same angle and that at which it subtends the greatest possible angle are in geometrical progression.

10. The line joining two stations A and B subtends equal angles at two other stations C and D: prove that

$$AB \sin CBD = CD \sin ADB.$$

11. Two straight lines ABC, DEC meet at C. If

$$\angle DAE = \angle DBE = a, \text{ and } \angle EAB = \beta, \ \angle EBC = \gamma,$$

shew that $$BC = \frac{AB \sin \beta \sin (a+\beta)}{\sin (\gamma - \beta) \sin (a+\beta+\gamma)}.$$

12. Two objects P and Q subtend an angle of 30° at A. Lengths of 20 feet and 10 feet are measured from A at right angles to AP and AQ respectively to points R and S at each of which PQ subtends angles of 30°: find the length of PQ.

13. A ship sailing N.E. is in a line with two beacons which are 5 miles apart, and of which one is due N. of the other. In 3 minutes and also in 21 minutes the beacons are found to subtend a right angle at the ship. Prove that the ship is sailing at the rate of 10 miles an hour, and that the beacons subtend their greatest angle at the ship at the end of $3\sqrt{7}$ minutes.

14. A man walking along a straight road notes when he is in the line of a long straight fence, and observes that 78 yards from this point the fence subtends an angle of 60°, and that 260 yards further on this angle is increased to 120°. When he has walked 260 yards still further, he finds that the fence again subtends an angle of 60°. If a be the angle which the direction of the fence makes with the road, shew that $13 \sin a = 5$. Also shew that the middle point of the fence is 120 yards distant from the road.

Measurements in more than one plane.

201. In Art. 199 the base line AB was measured *directly towards* the object. If this is not possible we may proceed as follows.

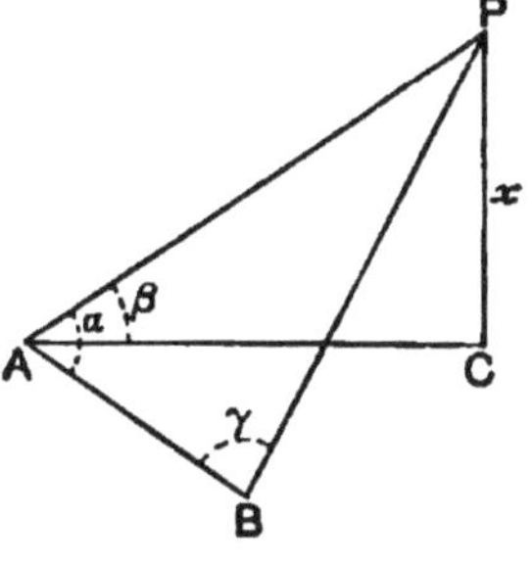

From A measure a base line AB in any convenient direction in the horizontal plane. At A observe the two angles PAB and PAC; and at B observe the angle PBA.

Let $\angle PAB = \alpha$, $\angle PAC = \beta$,

$$\angle PBA = \gamma,$$

$$AB = a, \quad PC = x.$$

From $\Delta\, PAC$,

$$x = PA \sin\beta.$$

From $\Delta\, PAB$,

$$PA = \frac{AB \sin PBA}{\sin APB} = \frac{a \sin\gamma}{\sin(\alpha+\gamma)};$$

$$\therefore\ x = a \sin\beta \sin\gamma \operatorname{cosec}(\alpha+\gamma).$$

202. *To shew how to find the distance between two inaccessible objects.*

Let P and Q be the objects.

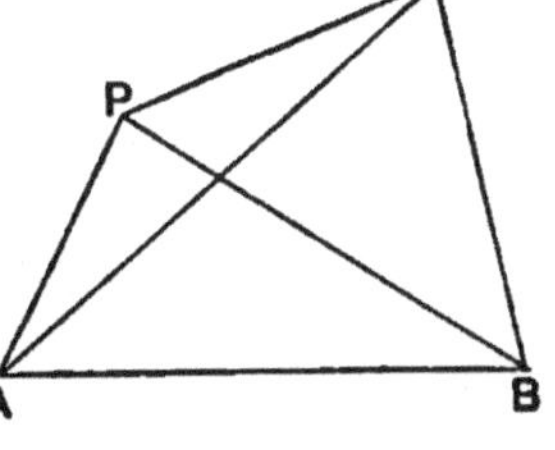

Take any two convenient stations A and B in the same horizontal plane, and measure the distance between them.

At A observe the angles PAQ and QAB. Also if AP, AQ, AB are not in the same plane, measure the angle PAB.

At B observe the angles ABP and ABQ.

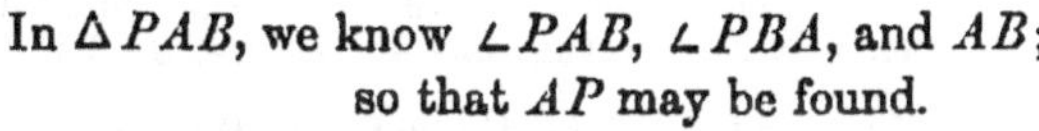

In $\Delta\, PAB$, we know $\angle PAB$, $\angle PBA$, and AB; so that AP may be found.

In $\Delta\, QAB$, we know $\angle QAB$, $\angle QBA$, and AB; so that AQ may be found.

In ΔPAQ, we know AP, AQ, and $\angle PAQ$; so that PQ may be found.

Example 1. The angular elevation of a tower CD at a place A due South of it is 30°, and at a place B due West of A the elevation is 18°. If $AB=a$, shew that the height of the tower is $\dfrac{a}{\sqrt{2+2\sqrt{5}}}$.

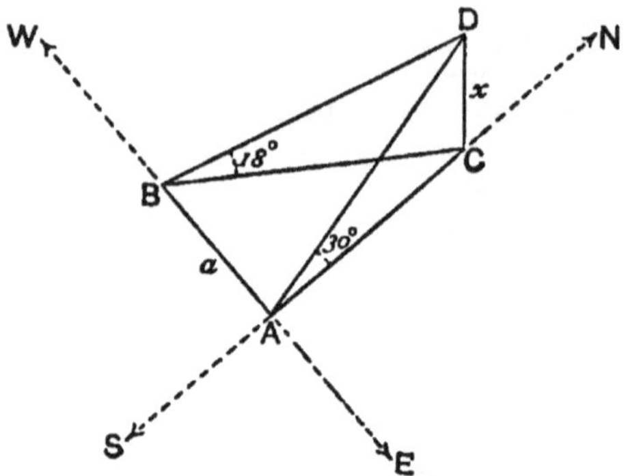

Let $CD=x$.

From the right-angled triangle DCA, $AC=x\cot 30°$.

From the right-angled triangle DCB, $BC=x\cot 18°$.

But $\angle BAC$ is a right angle,

$$\therefore\ BC^2-AC^2=a^2;$$

$$\therefore\ x^2(\cot^2 18° - \cot^2 30°)=a^2;$$

$$\therefore\ x^2(\operatorname{cosec}^2 18° - \operatorname{cosec}^2 30°)=a^2;$$

$$\therefore\ x^2\left\{\left(\frac{4}{\sqrt{5}-1}\right)^2-4\right\}=a^2;$$

$$\therefore\ x^2\{(\sqrt{5}+1)^2-4\}=a^2;$$

$$\therefore\ x^2(2\times 2\sqrt{5})=a^2,$$

which gives the height required.

Example 2. A hill of inclination 1 in 5 faces South. Shew that a road on it which takes a N.E. direction has an inclination 1 in 7.

Let AD running East and West be the ridge of the hill, and let $ABFD$ be a vertical plane through AD. Let C be a point at the foot of the hill, and ABC a section made by a vertical plane running North and South. Draw CG in a N.E. direction in the horizontal plane and let it meet BF in G; draw GH parallel to BA; then if CH is joined it will represent the direction of the road.

Since the inclination of CA is 1 in 5, we may take $AB=a$, and $AC=5a$, so that $BC^2=24a^2$.

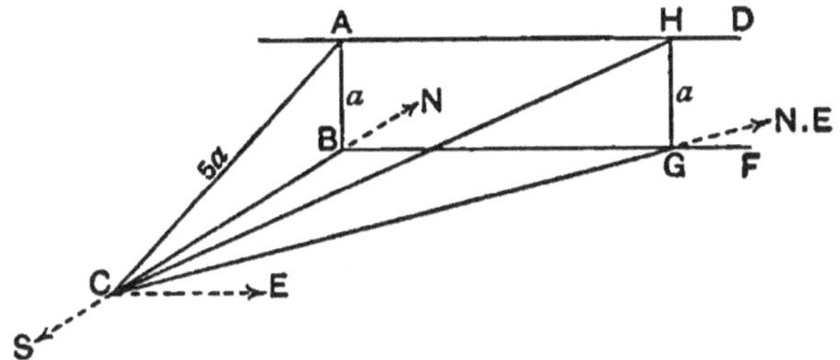

Since CBG is a right-angled isosceles triangle,

$$CG^2=2CB^2=48a^2.$$

Hence in the right-angled triangle CGH,

$$CH^2=48a^2+a^2=49a^2;$$

$$\therefore\ CH=7a=7GH.$$

Thus the slope of the road is 1 in 7.

EXAMPLES. XVII. c.

1. The elevation of a hill at a place P due East of it is 45°, and at a place Q due South of P the elevation is 30°. If the distance from P to Q is 500 yards, find the height of the hill in feet.

2. The elevation of a spire at a point A due West of it is 60°, and at point B due South of A the elevation is 30°. If the spire is 250 feet high, find the distance between A and B.

3. A river flows due North, and a tower stands on its left bank. From a point A up-stream and on the same bank as the tower the elevation of the tower is 60°, and from a point B just opposite on the other bank the elevation is 45°. If the tower is 360 feet high, find the breadth of the river.

4. The elevation of a steeple at a place A due S. of it is 45°, and at a place B due W. of A the elevation is 15°. If $AB=2a$, shew that the height of the steeple is $a(3^{\frac{1}{4}}-3^{-\frac{1}{4}})$.

5. A person due S. of a lighthouse observes that his shadow cast by the light at the top is 24 feet long. On walking 100 yards due E. he finds his shadow to be 30 feet long. Supposing him to be 6 feet high, find the height of the light from the ground.

6. The angles of elevation of a balloon from two stations a mile apart and from a point halfway between them are observed to be 60°, 30°, and 45° respectively. Prove that the height of the balloon is $440\sqrt{6}$ yards.

[*If AD is a median of the triangle ABC,*

then $2AD^2+2BD^2=AB^2+AC^2$.]

7. At each end of a base of length $2a$, the angular elevation of a mountain is θ, and at the middle point of the base the elevation is ϕ. Prove that the height of the mountain is

$$a\sin\theta\sin\phi\sqrt{\operatorname{cosec}(\phi+\theta)\operatorname{cosec}(\phi-\theta)}.$$

8. Two vertical poles, whose heights are a and b, subtend the same angle α at a point in the line joining their feet. If they subtend angles β and γ at any point in the horizontal plane at which the line joining their feet subtends a right angle, prove that

$$(a+b)^2\cot^2\alpha=a^2\cot^2\beta+b^2\cot^2\gamma.$$

9. From the top of a hill a person finds that the angles of depression of three consecutive milestones on a straight level road are α, β, γ. Shew that the height of the hill is

$$5280\sqrt{2}\,/\,\sqrt{\cot^2\alpha-2\cot^2\beta+\cot^2\gamma}\text{ feet.}$$

10. Two chimneys AB and CD are of equal height. A person standing between them in the line AC joining their bases observes the elevation of the one nearer to him to be 60°. After walking 80 feet in a direction at right angles to AC he observes their elevations to be 45° and 30°: find their height and distance apart.

11. Two persons who are 500 yards apart observe the bearing and angular elevation of a balloon at the same instant. One finds the elevation 60° and the bearing S.W., the other finds the elevation 45° and the bearing W. Find the height of the balloon.

12. The side of a hill faces due S. and is inclined to the horizon at an angle α. A straight railway upon it is inclined at an angle β to the horizon: if the bearing of the railway be x degrees E. of N., shew that $\cos x=\cot\alpha\tan\beta$.

EXAMPLES. XVII. d.

[*In the following examples the logarithms are to be taken from the Tables.*]

1. A man in a balloon observes that two churches which he knows to be one mile apart subtend an angle of 11° 25′ 20″ when he is exactly over the middle point between them: find the height of the balloon in miles.

2. There are three points A, B, C in a straight line on a level piece of ground. A vertical pole erected at C has an elevation of 5° 30′ from A and 10° 45′ from B. If AB is 100 yards, find the height of the pole and the distance BC.

3. The angular altitude of a lighthouse seen from a point on the shore is 12° 31′ 46″, and from a point 500 feet nearer the altitude is 26° 33′ 55″: find its height above the sea-level.

4. From a boat the angles of elevation of the highest and lowest points of a flagstaff 30 ft. high on the edge of a cliff are 46° 12′ and 44° 13′: find the height and distance of the cliff.

5. From the top of a hill the angles of depression of two successive milestones on level ground, and in the same vertical plane as the observer, are 5° and 10°. Find the height of the hill in feet and the distance of the nearer milestone in miles.

6. An observer whose eye is 15 feet above the roadway finds that the angle of elevation of the top of a telegraph post is 17° 18′ 35″, and that the angle of depression of the foot of the post is 8° 32′ 15″: find the height of the post and its distance from the observer.

7. Two straight railroads are inclined at an angle of 20° 16′. At the same instant two engines start from the point of intersection, one along each line; one travels at the rate of 20 miles an hour: at what rate must the other travel so that after 3 hours the distance between them shall be 30 miles?

8. An observer finds that from the doorstep of his house the elevation of the top of a spire is $5a$, and that from the roof above the doorstep it is $4a$. If h be the height of the roof above the doorstep, prove that the height of the spire above the doorstep and the horizontal distance of the spire from the house are respectively

$$h \operatorname{cosec} a \cos 4a \sin 5a \text{ and } h \operatorname{cosec} a \cos 4a \cos 5a.$$

If $h = 39$ feet, and $a = 7° \, 17' \, 39''$, calculate the height and the distance.

CHAPTER XVIII.

PROPERTIES OF TRIANGLES AND POLYGONS.

203. *To find the area of a triangle.*

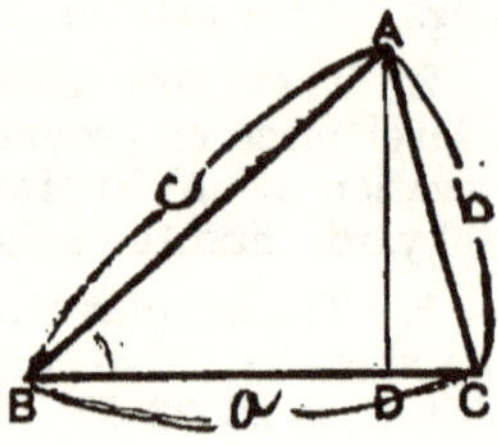

Let Δ denote the area of the triangle ABC. Draw AD perpendicular to BC.

By Euc. I. 41, the area of a triangle is half the area of a rectangle on the same base and of the same altitude.

$$\therefore \quad \Delta = \frac{1}{2}(\text{base} \times \text{altitude})$$

$$= \frac{1}{2} BC \, . \, AD = \frac{1}{2} BC \, . \, AB \sin B$$

$$= \frac{1}{2} ca \sin B.$$

Similarly, it may be proved that

$$\Delta = \frac{1}{2} ab \sin C, \quad \text{and} \quad \Delta = \frac{1}{2} bc \sin A.$$

These three expressions for the area are comprised in the single statement

$$\Delta = \frac{1}{2} (\textit{product of two sides}) \times (\textit{sine of included angle}).$$

Again, $\quad \Delta = \frac{1}{2} bc \sin A = bc \sin \frac{A}{2} \cos \frac{A}{2}$

$$= bc \sqrt{\frac{(s-b)(s-c)}{bc}} \sqrt{\frac{s(s-a)}{bc}}$$

$$= \sqrt{s(s-a)(s-b)(s-c)},$$

which gives the area in terms of the sides.

Again, $\Delta=\frac{1}{2}bc\sin A=\frac{1}{2}\sin A\cdot\frac{a\sin B}{\sin A}\cdot\frac{a\sin C}{\sin A}$

$$=\frac{a^2\sin B\sin C}{2\sin A}$$

$$=\frac{a^2\sin B\sin C}{2\sin(B+C)},$$

which gives the area in terms of one side and the functions of the adjacent angles.

NOTE. Many writers use the symbol S for the area of a triangle, but to avoid confusion between S and s in manuscript work the symbol Δ is preferable.

Example 1. The sides of a triangle are 17, 25, 28: find the lengths of the perpendiculars from the angles upon the opposite sides.

From the formula $\Delta=\frac{1}{2}(\text{base}\times\text{altitude})$,

it is evident that the three perpendiculars are found by dividing 2Δ by the three sides in turn.

Now $\Delta=\sqrt{s(s-a)(s-b)(s-c)}=\sqrt{35\times18\times10\times7}$
$=5\times7\times6=210.$

Thus the perpendiculars are $\frac{420}{17}$, $\frac{420}{25}$, $\frac{420}{28}$, or $\frac{420}{17}$, $\frac{84}{5}$, 15.

Example 2. Two angles of a triangular field are 22·5° and 45°, and the length of the side opposite to the latter is one furlong: find the area.

Let $A=22\frac{1}{2}°$, $B=45°$, then $b=220$ yds., and $C=112\frac{1}{2}°$.

From the formula $\Delta=\frac{b^2\sin A\sin C}{2\sin B}$,

the area in sq. yds. $=\frac{220\times220\times\sin22\frac{1}{2}°\times\sin112\frac{1}{2}°}{2\sin45°}$

$$=\frac{220\times220\times\sin22\frac{1}{2}°\times\cos22\frac{1}{2}°}{2\times2\sin22\frac{1}{2}°\cos22\frac{1}{2}°}$$

$$=110\times110.$$

Expressed in acres, the area $=\frac{110\times110}{4840}=2\frac{1}{2}$.

204. *To find the radius of the circle circumscribing a triangle.*

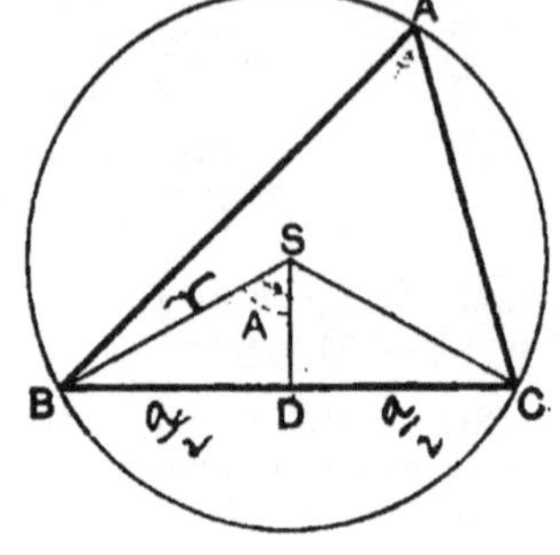

Let S be the centre of the circle circumscribing the triangle ABC, and R its radius.

Bisect $\angle BSC$ by SD, which will also bisect BC at right angles.

Now by Euc. III. 20,

$\angle BSC$ at centre

$$=\text{twice } \angle BAC$$
$$=2A;$$

and $$\frac{a}{2}=BD=BS\sin BSD=R\sin A;$$

$$\therefore\ R=\frac{a}{2\sin A}.$$

Thus $$\frac{a}{\sin A}=\frac{b}{\sin B}=\frac{c}{\sin C}=2R,$$

or $$a=2R\sin A,\quad b=2R\sin B,\quad c=2R\sin C.$$

Example. Shew that $2R^2\sin A\sin B\sin C=\Delta$.

The first side$=\frac{1}{2}\cdot 2R\sin A\cdot 2R\sin B\cdot\sin C$

$$=\frac{1}{2}ab\sin C$$
$$=\Delta.$$

205. From the result of the last article we deduce the following important theorem:

If a chord of length l *subtend an angle* θ *at the circumference of a circle whose radius is* R, *then* $l=2R\sin\theta$.

206. For shortness, the circle circumscribing a triangle may be called the *Circum-circle*, its centre the *Circum-centre*, and its radius the *Circum-radius.*

The circum-radius may be expressed in a form not involving the angles, for

$$R=\frac{a}{2\sin A}=\frac{abc}{2bc\sin A}=\frac{abc}{4\Delta}.$$

207. *To find the radius of the circle inscribed in a triangle.*

Let I be the centre of the circle inscribed in the triangle ABC, and D, E, F the points of contact; then ID, IE, IF are perpendicular to the sides.

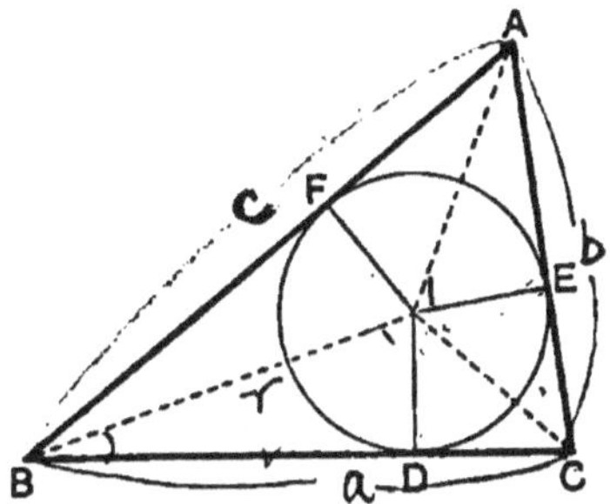

Now Δ = sum of the areas of the triangles BIC, CIA, AIB

$$=\frac{1}{2}ar+\frac{1}{2}br+\frac{1}{2}cr=\frac{1}{2}(a+b+c)r$$

$$=sr;$$

whence
$$r=\frac{\Delta}{s}.$$

208. *To express the radius of the inscribed circle in terms of one side and the functions of the half-angles.*

In the figure of the previous article, we know from Euc. IV. 4 that I is the point of intersection of the lines bisecting the angles, so that

$$\angle IBD=\frac{B}{2},\quad \angle ICD=\frac{C}{2}.$$

$$\therefore\ BD=r\cot\frac{B}{2},\quad CD=r\cot\frac{C}{2}.$$

$$\therefore\ r\left(\cot\frac{B}{2}+\cot\frac{C}{2}\right)=a;$$

$$\therefore\ r\sin\frac{B+C}{2}=a\sin\frac{B}{2}\sin\frac{C}{2};$$

$$\therefore\ r=\frac{a\sin\frac{B}{2}\sin\frac{C}{2}}{\cos\frac{A}{2}}.$$

209. DEFINITION. A circle which touches one side of a triangle and the other two sides produced is said to be an **escribed circle** of the triangle.

Thus the triangle ABC has *three* escribed circles, one touching BC, and AB, AC produced; a second touching CA, and BC, BA produced; a third touching AB, and CA, CB produced.

We shall assume that the student is familiar with the construction of the escribed circles.

[See Hall and Stevens' *Euclid*, p. 255.]

For shortness, we shall call the circle inscribed in a triangle the *In-circle*, its centre the *In-centre*, and its radius the *In-radius*; and similarly the escribed circles may be called the *Ex-circles*, their centres the *Ex-centres*, and their radii the *Ex-radii*.

210. *To find the radius of an escribed circle of a triangle.*

Let I_1 be the centre of the circle touching the side BC and the two sides AB and AC produced. Let D_1, E_1, F_1 be the points of contact; then the lines joining I_1 to these points are perpendicular to the sides.

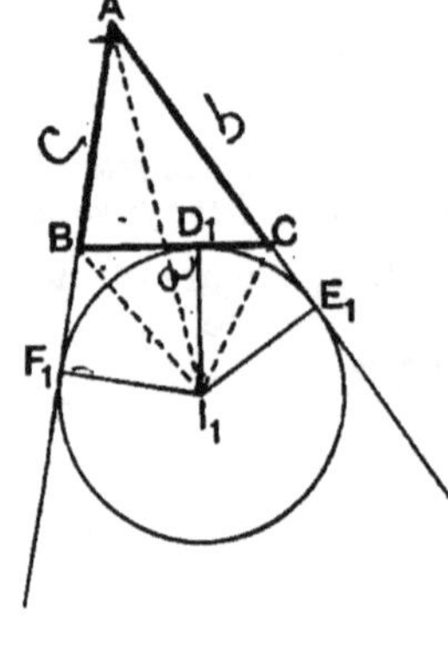

Let r_1 be the radius; then

$$\begin{aligned}\Delta &= \text{area } ABC\\ &= \text{area } ABI_1C - \text{area } BI_1C\\ &= \text{area } BI_1A + \text{area } CI_1A - \text{area } BI_1C\\ &= \frac{1}{2}cr_1 + \frac{1}{2}br_1 - \frac{1}{2}ar_1\\ &= \frac{1}{2}(c+b-a)r_1\\ &= (s-a)r_1;\end{aligned}$$

$$\therefore\ r_1 = \frac{\Delta}{s-a}.$$

Similarly, if r_2, r_3 be the radii of the escribed circles opposite to the angles B and C respectively,

$$r_2 = \frac{\Delta}{s-b}, \quad r_3 = \frac{\Delta}{s-c}.$$

211. *To find the radii of the escribed circles in terms of one side and the functions of the half-angles.*

In the figure of the last article, I_1 is the point of intersection of the lines bisecting the angles B and C externally; so that

$$\angle I_1BD_1 = 90^\circ - \frac{B}{2}, \quad \angle I_1CD_1 = 90^\circ - \frac{C}{2}.$$

$$\therefore \quad BD_1 = r_1 \cot\left(90^\circ - \frac{B}{2}\right) = r_1 \tan\frac{B}{2},$$

$$CD_1 = r_1 \cot\left(90^\circ - \frac{C}{2}\right) = r_1 \tan\frac{C}{2};$$

$$\therefore \quad r_1\left(\tan\frac{B}{2} + \tan\frac{C}{2}\right) = a;$$

$$\therefore \quad r_1 \sin\frac{B+C}{2} = a\cos\frac{B}{2}\cos\frac{C}{2};$$

$$\therefore \quad r_1 = \frac{a\cos\frac{B}{2}\cos\frac{C}{2}}{\cos\frac{A}{2}}.$$

Similarly,

$$r_2 = \frac{b\cos\frac{C}{2}\cos\frac{A}{2}}{\cos\frac{B}{2}}, \quad r_3 = \frac{c\cos\frac{A}{2}\cos\frac{B}{2}}{\cos\frac{C}{2}}.$$

212. By substituting

$$a = 2R\sin A, \quad b = 2R\sin B, \quad c = 2R\sin C,$$

in the formulæ of Art. 208 and Art. 211, we have

$$r = 4R\sin\frac{A}{2}\sin\frac{B}{2}\sin\frac{C}{2},$$

$$r_1 = 4R\sin\frac{A}{2}\cos\frac{B}{2}\cos\frac{C}{2},$$

$$r_2 = 4R\cos\frac{A}{2}\sin\frac{B}{2}\cos\frac{C}{2},$$

$$r_3 = 4R\cos\frac{A}{2}\cos\frac{B}{2}\sin\frac{C}{2}.$$

Example 1. Shew that $\dfrac{r_1 - r}{a} + \dfrac{r_2 - r}{b} = \dfrac{c}{r_3}$.

$$\text{The first side} = \frac{1}{a}\left(\frac{\Delta}{s-a} - \frac{\Delta}{s}\right) + \frac{1}{b}\left(\frac{\Delta}{s-b} - \frac{\Delta}{s}\right)$$

$$= \frac{\Delta}{s(s-a)} + \frac{\Delta}{s(s-b)} = \frac{\Delta(2s-a-b)}{s(s-a)(s-b)}$$

$$= \frac{\Delta c}{s(s-a)(s-b)} = \frac{\Delta c(s-c)}{s(s-a)(s-b)(s-c)}$$

$$= \frac{\Delta c(s-c)}{\Delta^2} = \frac{c(s-c)}{\Delta}$$

$$= \frac{c}{r_3}.$$

Example 2. If $r_1 = r_2 + r_3 + r$, prove that the triangle is right-angled.

By transposition, $r_1 - r = r_2 + r_3$;

$$\therefore 4R \sin\frac{A}{2}\cos\frac{B}{2}\cos\frac{C}{2} - 4R\sin\frac{A}{2}\sin\frac{B}{2}\sin\frac{C}{2}$$

$$= 4R\cos\frac{A}{2}\sin\frac{B}{2}\cos\frac{C}{2} + 4R\cos\frac{A}{2}\cos\frac{B}{2}\sin\frac{C}{2};$$

$$\therefore \sin\frac{A}{2}\left(\cos\frac{B}{2}\cos\frac{C}{2} - \sin\frac{B}{2}\sin\frac{C}{2}\right)$$

$$= \cos\frac{A}{2}\left(\sin\frac{B}{2}\cos\frac{C}{2} + \cos\frac{B}{2}\sin\frac{C}{2}\right);$$

$$\therefore \sin\frac{A}{2}\cos\frac{B+C}{2} = \cos\frac{A}{2}\sin\frac{B+C}{2};$$

$$\therefore \sin^2\frac{A}{2} = \cos^2\frac{A}{2};$$

whence $\dfrac{A}{2} = 45°$, and $A = 90°$.

213. Many important relations connecting a triangle and its circles may be established by elementary geometry.

With the notation of previous articles, since tangents to a circle from the same point are equal, we have $AF=AE,\quad BD=BF,\quad CD=CE;$

$$\therefore\ AF+(BD+CD)=\text{half the sum of the sides};$$

$$\therefore\ AF+a=s.$$

$$\therefore\ AF=s-a=AE.$$

Similarly, $BD=BF=s-b,\quad CD=CE=s-c.$

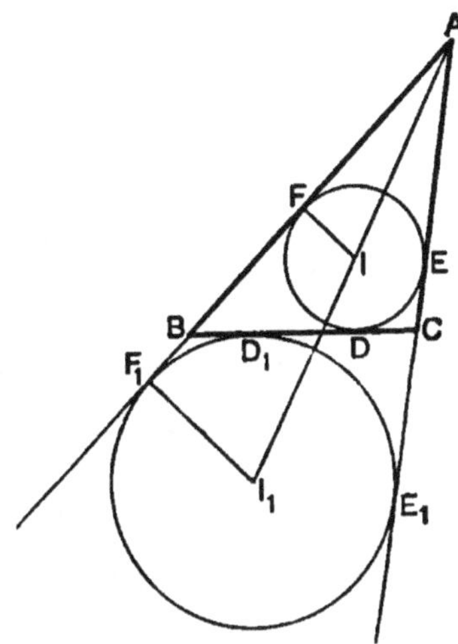

Also $$r=AF\tan\frac{A}{2}=(s-a)\tan\frac{A}{2}.$$

Similarly, $r=(s-b)\tan\frac{B}{2},\quad r=(s-c)\tan\frac{C}{2}.$

Again, $AF_1=AE_1,\quad BF_1=BD_1,\quad CE_1=CD_1;$

$$\therefore\ 2AF_1=AF_1+AE_1=(AB+BD_1)+(AC+CD_1)$$
$$=\text{sum of the sides};$$

$$\therefore\ AF_1=s=AE_1.$$

$$\therefore\ BD_1=BF_1=s-c,\quad CD_1=CE_1=s-b.$$

Also $$r_1=AF_1\tan\frac{A}{2}=s\tan\frac{A}{2}.$$

Similarly, $r_2=s\tan\frac{B}{2},\quad r_3=s\tan\frac{C}{2}.$

EXAMPLES. XVIII. a.

1. Two sides of a triangle are 300 ft. and 120 ft., and the included angle is 150°; find the area.

2. Find the area of the triangle whose sides are 171, 204, 195.

3. Find the sine of the greatest angle of a triangle whose sides are 70, 147, and 119.

4. If the sides of a triangle are 39, 40, 25, find the lengths of the three perpendiculars from the angular points on the opposite sides.

5. One side of a triangle is 30 ft. and the adjacent angles are $22\frac{1}{2}°$ and $112\frac{1}{2}°$, find the area.

6. Find the area of a parallelogram two of whose adjacent sides are 42 and 32 ft., and include an angle of 30°.

7. The area of a rhombus is 648 sq. yds. and one of the angles is 150°: find the length of each side.

8. In a triangle if $a=13$, $b=14$, $c=15$, find r and R.

9. Find r_1, r_2, r_3 in the case of a triangle whose sides are 17, 10, 21.

10. If the area of a triangle is 96, and the radii of the escribed circles are 8, 12, 24, find the sides.

Prove the following formulæ:

11. $\sqrt{rr_1r_2r_3}=\Delta$.

12. $s(s-a)\tan\frac{A}{2}=\Delta$.

13. $rr_1\cot\frac{A}{2}=\Delta$.

14. $4Rrs=abc$.

15. $r_1r_2r_3=rs^2$.

16. $r\cot\frac{B}{2}\cot\frac{C}{2}=r_1$.

17. $Rr(\sin A+\sin B+\sin C)=\Delta$.

18. $r_1r_2+rr_3=ab$.

19. $\cos\frac{A}{2}\sqrt{bc(s-b)(s-c)}=\Delta$.

20. $r_1+r_2=c\cot\frac{C}{2}$.

21. $(r_1-r)(r_2+r_3)=a^2$.

22. $r_1 \cot \frac{A}{2} = r_2 \cot \frac{B}{2} = r_3 \cot \frac{C}{2} = r \cot \frac{A}{2} \cot \frac{B}{2} \cot \frac{C}{2}$.

23. $\frac{1}{r_1} + \frac{1}{r_2} + \frac{1}{r_3} = \frac{1}{r}$.

24. $r_2 r_3 + r_3 r_1 + r_1 r_2 = s^2$.

25. $r_1 + r_2 + r_3 - r = 4R$.

26. $r + r_1 + r_2 - r_3 = 4R \cos C$.

27. $b^2 \sin 2C + c^2 \sin 2B = 4\Delta$.

28. $4R \cos \frac{C}{2} = (a+b) \sec \frac{A-B}{2}$.

29. $a^2 - b^2 = 2Rc \sin (A - B)$.

30. $\frac{a^2 - b^2}{2} \cdot \frac{\sin A \sin B}{\sin (A-B)} = \Delta$.

31. If the perpendiculars from A, B, C to the opposite sides are p_1, p_2, p_3 respectively, prove that

(1) $\frac{1}{p_1} + \frac{1}{p_2} + \frac{1}{p_3} = \frac{1}{r}$; (2) $\frac{1}{p_1} + \frac{1}{p_2} - \frac{1}{p_3} = \frac{1}{r_3}$.

Prove the following identities:

32. $(r_1 - r)(r_2 - r)(r_3 - r) = 4Rr^2$.

33. $\left(\frac{1}{r} - \frac{1}{r_1}\right)\left(\frac{1}{r} - \frac{1}{r_2}\right)\left(\frac{1}{r} - \frac{1}{r_3}\right) = \frac{4R}{r^2 s^2}$.

34. $4\Delta (\cot A + \cot B + \cot C) = a^2 + b^2 + c^2$.

35. $\frac{b-c}{r_1} + \frac{c-a}{r_2} + \frac{a-b}{r_3} = 0$.

36. $a^2 b^2 c^2 (\sin 2A + \sin 2B + \sin 2C) = 32\Delta^3$.

37. $a \cos A + b \cos B + c \cos C = 4R \sin A \sin B \sin C$.

38. $a \cot A + b \cot B + c \cot C = 2(R + r)$.

39. $(b+c) \tan \frac{A}{2} + (c+a) \tan \frac{B}{2} + (a+b) \tan \frac{C}{2}$
$= 4R (\cos A + \cos B + \cos C)$.

40. $r (\sin A + \sin B + \sin C) = 2R \sin A \sin B \sin C$.

41. $\cos^2 \frac{A}{2} + \cos^2 \frac{B}{2} + \cos^2 \frac{C}{2} = 2 + \frac{r}{2R}$.

Inscribed and circumscribed Polygons.

214. *To find the perimeter and area of a regular polygon of* n *sides inscribed in a circle.*

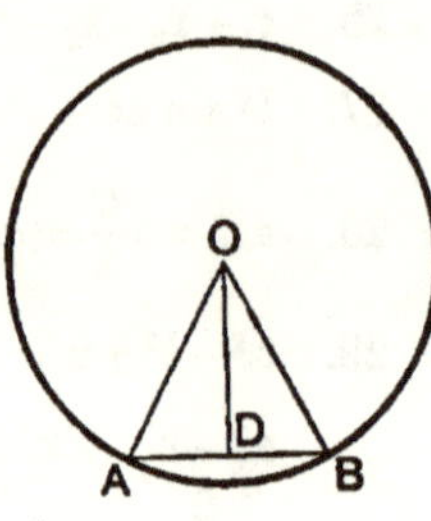

Let r be the radius of the circle, and AB a side of the polygon.

Join OA, OB, and draw OD bisecting $\angle AOB$; then AB is bisected at right angles in D.

And $$\angle AOB = \frac{1}{n}\text{ (four right angles)}$$
$$= \frac{2\pi}{n}.$$

Perimeter of polygon $= nAB = 2nAD = 2nOA \sin AOD$
$$= 2nr \sin \frac{\pi}{n}.$$

Area of polygon $= n$ (area of triangle AOB)
$$= \frac{1}{2} nr^2 \sin \frac{2\pi}{n}.$$

215. *To find the perimeter and area of a regular polygon of* n *sides circumscribed about a given circle.*

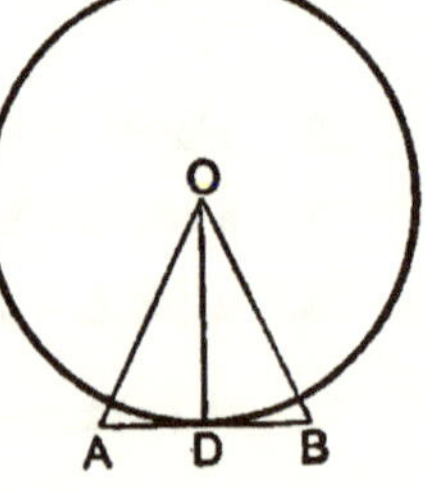

Let r be the radius of the circle, and AB a side of the polygon. Let AB touch the circle at D. Join OA, OB, OD; then OD bisects AB at right angles, and also bisects $\angle AOB$.

Perimeter of polygon
$$= nAB = 2nAD = 2nOD \tan AOD$$
$$= 2nr \tan \frac{\pi}{n}.$$

Area of polygon $= n$ (area of triangle AOB)
$$= nOD \, . \, AD$$
$$= nr^2 \tan \frac{\pi}{n}.$$

216. There is no need to burden the memory with the formulæ of the last two articles, as in any particular instance they are very readily obtained.

Example 1. The side of a regular dodecagon is 2 ft., find the radius of the circumscribed circle.

Let r be the required radius. In the adjoining figure we have

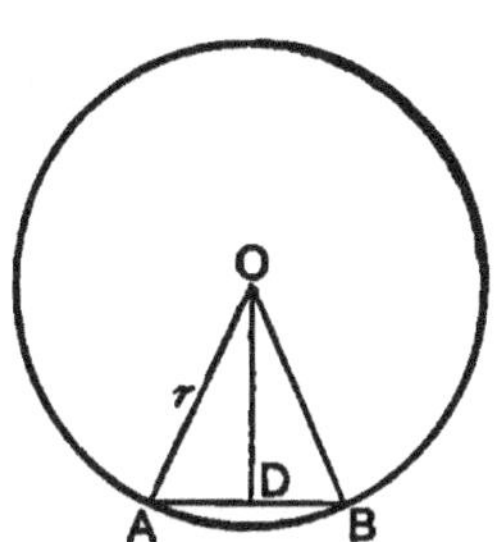

$$AB=2, \quad \angle AOB=\frac{2\pi}{12}.$$

$$AB=2AD=2r\sin\frac{\pi}{12};$$

$$\therefore\ 2r\sin 15^\circ=2;$$

$$\therefore\ r=\frac{1}{\sin 15^\circ}=\frac{2\sqrt{2}}{\sqrt{3}-1}=\sqrt{2}\,(\sqrt{3}+1).$$

Thus the radius is $\sqrt{6}+\sqrt{2}$ feet.

Example 2. A regular pentagon and a regular decagon have the same perimeter, prove that their areas are as 2 to $\sqrt{5}$.

Let AB be one of the n sides of a regular polygon, O the centre of the circumscribed circle, OD perpendicular to AB.

Then if $AB=a$,

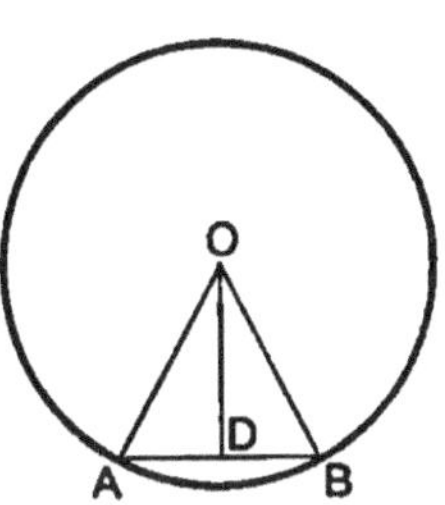

$$\text{area of polygon}=nAD\,.\,OD$$

$$=nAD\,.\,AD\cot\frac{\pi}{n}$$

$$=\frac{na^2}{4}\cot\frac{\pi}{n}.$$

Denote the perimeter of the pentagon and decagon by $10c$. Then each side of the pentagon is $2c$, and its area is $5c^2\cot\frac{\pi}{5}$.

Each side of the decagon is c, and its area is $\frac{5}{2}c^2\cot\frac{\pi}{10}$.

$$\therefore\ \frac{\text{Area of pentagon}}{\text{Area of decagon}}=\frac{2\cot 36^\circ}{\cot 18^\circ}=\frac{2\cos 36^\circ\sin 18^\circ}{\sin 36^\circ\cos 18^\circ}=\frac{2\cos 36^\circ}{2\cos^2 18^\circ}$$

$$=\frac{2\cos 36^\circ}{1+\cos 36^\circ}=\frac{2(\sqrt{5}+1)}{4}\div\left(1+\frac{\sqrt{5}+1}{4}\right)$$

$$=\frac{2(\sqrt{5}+1)}{5+\sqrt{5}}=\frac{2}{\sqrt{5}}.$$

217. *To find the area of a circle.*

Let r be the radius of the circle, and let a regular polygon of n sides be described about it. Then from the adjoining figure, we have

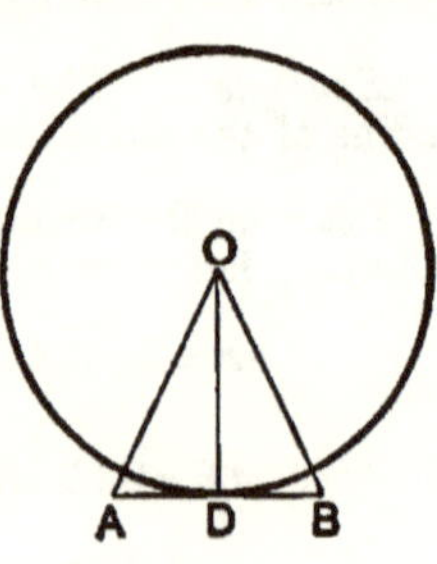

$$\text{area of polygon} = n\,(\text{area of triangle } AOB)$$
$$= n\left(\frac{1}{2}AB\,.\,OD\right)$$
$$= \frac{1}{2}OD\,.\,nAB$$
$$= \frac{r}{2} \times \text{perimeter of polygon.}$$

By increasing the number of sides without limit, the area and the perimeter of the polygon may be made to differ as little as we please from the area and the circumference of the circle. Hence

$$\text{area of a circle} = \frac{r}{2} \times \text{circumference}$$
$$= \frac{r}{2} \times 2\pi r \quad \text{[Art. 59.]}$$
$$= \pi r^2.$$

218. *To find the area of the sector of a circle.*

Let θ be the circular measure of the angle of the sector; then by Euc. VI. 33,

$$\frac{\text{area of sector}}{\text{area of circle}} = \frac{\theta}{2\pi};$$

$$\therefore \text{ area of sector} = \frac{\theta}{2\pi} \times \pi r^2 = \frac{1}{2}r^2\theta.$$

EXAMPLES. XVIII. b.

[*In this Exercise take* $\pi = \frac{22}{7}$.]

1. Find the area of a regular decagon inscribed in a circle whose radius is 3 feet; given $\sin 36° = {\cdot}588$.

2. Find the perimeter and area of a regular quindecagon described about a circle whose diameter is 3 yards; given

$$\tan 12° = \cdot 213.$$

3. Shew that the areas of the inscribed and circumscribed circles of a regular hexagon are in the ratio of 3 to 4.

4. Find the area of a circle inscribed in a regular pentagon whose area is 250 sq. ft.; given $\cot 36° = 1\cdot 376$.

5. Find the perimeter of a regular octagon inscribed in a circle whose area is 1386 sq. inches; given $\sin 22° \, 30' = \cdot 382$.

6. Find the perimeter of a regular pentagon described about a circle whose area is 616 sq. ft.; given $\tan 36° = \cdot 727$.

7. Find the diameter of the circle circumscribing a regular quindecagon, whose inscribed circle has an area of 2464 sq. ft.; given $\sec 12° = 1\cdot 022$.

8. Find the area of a regular dodecagon inscribed in a circle whose regular inscribed pentagon has an area of 50 sq. ft.

9. A regular pentagon and a regular decagon have the same area, prove that the ratio of their perimeters is $\sqrt{2} : \sqrt[4]{5}$.

10. Two regular polygons of n sides and $2n$ sides have the same perimeter; shew that the ratio of their areas is

$$2 \cos \frac{\pi}{n} : 1 + \cos \frac{\pi}{n}.$$

11. If $2a$ be the side of a regular polygon of n sides, R and r the radii of the circumscribed and inscribed circles, prove that

$$R + r = a \cot \frac{\pi}{2n}.$$

12. Prove that the square of the side of a regular pentagon inscribed in a circle is equal to the sum of the squares of the sides of a regular hexagon and decagon inscribed in the same circle.

13. With reference to a given circle, A_1 and B_1 are the areas of the inscribed and circumscribed regular polygons of n sides, A_2 and B_2 are corresponding quantities for regular polygons of $2n$ sides: prove that

(1) A_2 is a geometric mean between A_1 and B_1;

(2) B_2 is a harmonic mean between A_2 and B_1.

The Ex-central Triangle.

*219. Let ABC be a triangle, I_1, I_2, I_3 its ex-centres ; then $I_1I_2I_3$ is called the **Ex-central triangle** of ABC.

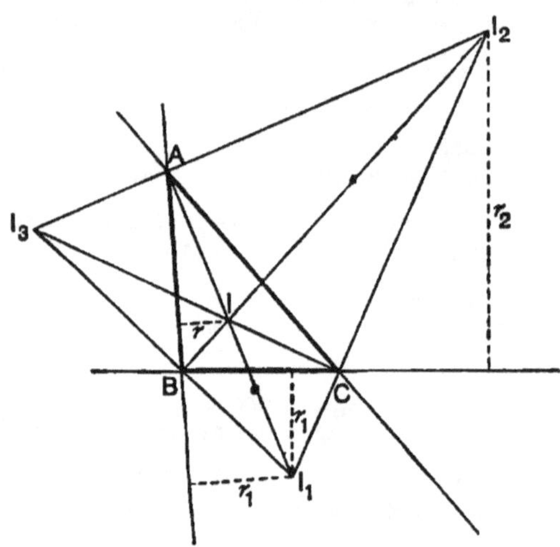

Let I be the in-centre; then from the construction for finding the positions of the in-centre and ex-centres, it follows that :

(i) The points I, I_1 lie on the line bisecting the angle BAC; the points I, I_2 lie on the line bisecting the angle ABC; the points I, I_3 lie on the line bisecting the angle ACB.

(ii) The points I_2, I_3 lie on the line bisecting the angle BAC externally ; the points I_3, I_1 lie on the line bisecting the angle ABC externally ; the points I_1, I_2 lie on the line bisecting the angle ACB externally.

(iii) The line AI_1 is perpendicular to I_2I_3 ; the line BI_2 is perpendicular to I_3I_1 ; the line CI_3 is perpendicular to I_1I_2. Thus the triangle ABC is the *Pedal triangle* of its ex-central triangle $I_1I_2I_3$. [See Art. 223.]

(iv) The angles IBI_1 and ICI_1 are right angles ; hence the points B, I, C, I_1 are concyclic. Similarly, the points C, I, A, I_2, and the points A, I, B, I_3 are concyclic.

(v) The lines AI_1, BI_2, CI_3 meet at the in-centre I, which is therefore the *Orthocentre* of the ex-central triangle $I_1I_2I_3$.

(vi) Each of the four points I, I_1, I_2, I_3 is the orthocentre of the triangle formed by joining the other three points.

***220.** *To find the distances between the in-centre and ex-centres.*

With the figure of the last article,

$$II_1 = AI_1 - AI = r_1 \operatorname{cosec}\frac{A}{2} - r \operatorname{cosec}\frac{A}{2} = (r_1 - r)\operatorname{cosec}\frac{A}{2}$$

$$= 4R\left(\sin\frac{A}{2}\cos\frac{B}{2}\cos\frac{C}{2} - \sin\frac{A}{2}\sin\frac{B}{2}\sin\frac{C}{2}\right)\operatorname{cosec}\frac{A}{2}$$

$$= 4R\cos\frac{B+C}{2} = 4R\sin\frac{A}{2}.$$

Thus the distances are

$$4R\sin\frac{A}{2},\quad 4R\sin\frac{B}{2},\quad 4R\sin\frac{C}{2}.$$

***221.** *To find the sides and angles of the ex-central triangle.*

With the figure of Art. 219,

$$\angle BI_1C = \angle BI_1I + \angle CI_1I$$

$$= \angle BCI + \angle CBI \qquad \text{[Euc. III. 21]}$$

$$= \frac{C}{2} + \frac{B}{2} = 90^\circ - \frac{A}{2}.$$

Thus the angles are

$$90^\circ - \frac{A}{2},\quad 90^\circ - \frac{B}{2},\quad 90^\circ - \frac{C}{2}.$$

Again,

$$I_1I_2 = I_1C + I_2C = r_1 \operatorname{cosec}\left(90^\circ - \frac{C}{2}\right) + r_2 \operatorname{cosec}\left(90^\circ - \frac{C}{2}\right)$$

$$= 4R\left(\sin\frac{A}{2}\cos\frac{B}{2}\cos\frac{C}{2} + \cos\frac{A}{2}\sin\frac{B}{2}\cos\frac{C}{2}\right)\sec\frac{C}{2}$$

$$= 4R\sin\frac{A+B}{2} = 4R\cos\frac{C}{2}.$$

Thus the sides are

$$4R\cos\frac{A}{2},\quad 4R\cos\frac{B}{2},\quad 4R\cos\frac{C}{2}.$$

***222.** *To find the area and circum-radius of the ex-central triangle.*

The area$=\frac{1}{2}$ (product of two sides) × (sine of included angle)

$$=\frac{1}{2}\times 4R\cos\frac{B}{2}\times 4R\cos\frac{C}{2}\times\sin\left(90^\circ-\frac{A}{2}\right)$$

$$=8R^2\cos\frac{A}{2}\cos\frac{B}{2}\cos\frac{C}{2}.$$

$$\text{The circum-radius}=\frac{I_2I_3}{2\sin I_2I_1I_3}=\frac{4R\cos\frac{A}{2}}{2\sin\left(90^\circ-\frac{A}{2}\right)}=2R.$$

The Pedal Triangle.

***223.** Let G, H, K be the feet of the perpendiculars from the angular points on the opposite sides of the triangle ABC; then GHK is called the **Pedal triangle** of ABC.

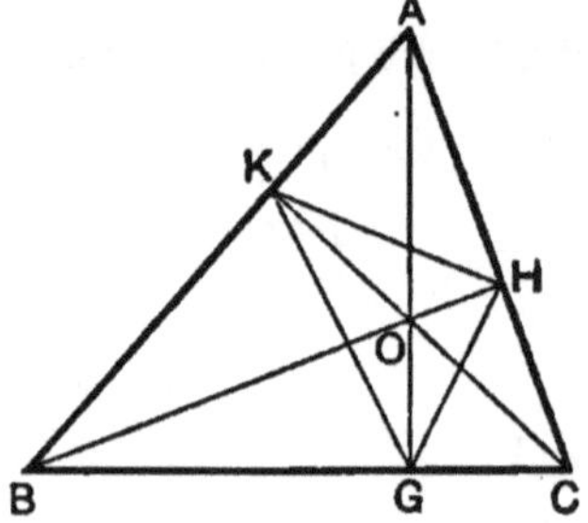

The three perpendiculars AG, BH, CK meet in a point O which is called the **Orthocentre** of the triangle ABC.

***224.** *To find the sides and angles of the pedal triangle.*

In the figure of the last article, the points K, O, G, B are concyclic;

$$\therefore\ \angle OGK=\angle OBK=90^\circ-A.$$

Also the points H, O, G, C are concyclic;

$$\therefore\ \angle OGH=\angle OCH=90^\circ-A;$$

$$\therefore\ \angle KGH=180^\circ-2A.$$

Thus the angles of the pedal triangle are

$$180^\circ-2A,\quad 180^\circ-2B,\quad 180^\circ-2C.$$

Again, $\angle AKH=180^\circ-\angle BKH=\angle BCH$,

since the points B, K, H, C are concyclic;

$$\therefore\ \angle AKH=C.$$

$$\therefore \frac{HK}{AH} = \frac{\sin A}{\sin AKH} = \frac{\sin A}{\sin C} = \frac{a}{c};$$

$$\therefore HK = \frac{a}{c} . AH = \frac{a}{c} . c \cos A = a \cos A.$$

Thus the sides of the pedal triangle are

$$a \cos A, \quad b \cos B, \quad c \cos C.$$

In terms of R, the equivalent forms become

$$R \sin 2A, \quad R \sin 2B, \quad R \sin 2C.$$

If the angle ACB of the given triangle is obtuse, the expressions $180° - 2C$ and $c \cos C$ are both negative, and the values we have obtained require some modification. We leave the student to shew that in this case the angles are $2A, 2B, 2C - 180°$, and the sides $a \cos A, \quad b \cos B, \quad -c \cos C$.

***225.** *To find the area and circum-radius of the pedal triangle.*

The area $= \frac{1}{2}$ (product of two sides) $\times$ (sine of included angle)

$$= \frac{1}{2} R \sin 2B . R \sin 2C . \sin (180 - 2A)$$

$$= \frac{1}{2} R^2 \sin 2A \sin 2B \sin 2C.$$

The circum-radius $= \dfrac{HK}{2 \sin HGK} = \dfrac{R \sin 2A}{2 \sin (180° - 2A)} = \dfrac{R}{2}$.

NOTE. The circum-circle of the pedal triangle is the nine points circle of the triangle ABC. Thus the radius of the nine points circle of the triangle ABC is $\frac{R}{2}$. [See Hall and Stevens' *Euclid*, p. 281.]

***226.** In Art. 224, we have proved that OG, OH, OK bisect the angles HGK, KHG, GKH respectively, so that O is the in-centre of the triangle GHK. Thus the orthocentre of a triangle is the in-centre of the pedal triangle.

Again, the line CGB which is at right angles to OG bisects $\angle HGK$ externally. Similarly the lines AHC and BKA bisect $\angle KHG$ and $\angle GKH$ externally, so that ABC is the ex-central triangle of its pedal triangle GHK.

*227. In Art. 219, we have seen that ABC is the pedal triangle of its ex-central triangle $I_1I_2I_3$. Certain theorems depending on this connection are more evident from the adjoining figure, in which the fact that ABC is the pedal triangle of $I_1I_2I_3$ is brought more prominently into view. For instance, the circum-circle of the triangle ABC is the nine points circle of the triangle $I_1I_2I_3$, and passes through the middle points of II_1, II_2, II_3 and of I_2I_3, I_3I_1, I_1I_2.

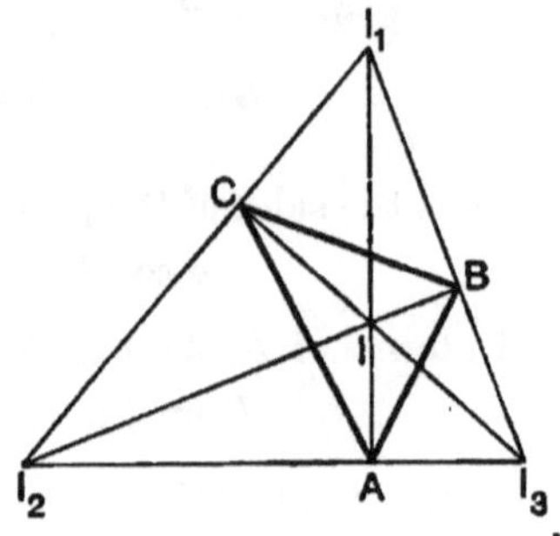

*228. *To find the distance between the in-centre and circum-centre.*

Let S be the circum-centre and I the in-centre. Produce AI to meet the circum-circle in H; join CH and CI.

Draw IE perpendicular to AC. Produce HS to meet the circumference in L, and join CL. Then

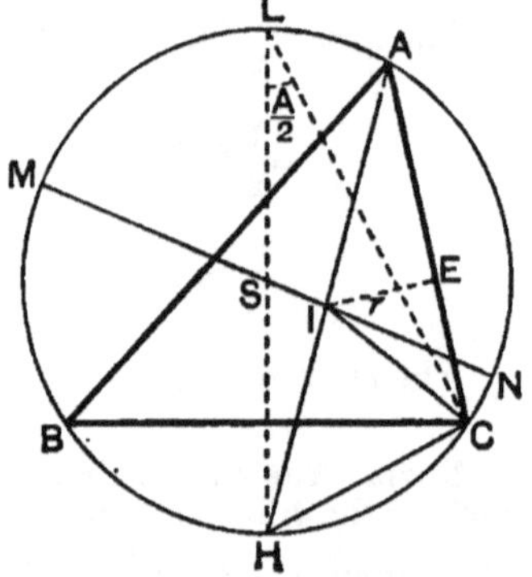

$$\angle HIC = \angle IAC + \angle ICA$$
$$= \frac{A}{2} + \frac{C}{2};$$
$$\angle HCI = \angle ICB + \angle BCH$$
$$= \frac{C}{2} + \angle BAH$$
$$= \frac{C}{2} + \frac{A}{2};$$

$$\therefore \angle HCI = \angle HIC;$$

$$\therefore HI = HC = 2R \sin\frac{A}{2}.$$

Also $$AI = IE \operatorname{cosec}\frac{A}{2} = r \operatorname{cosec}\frac{A}{2};$$

$$\therefore AI \, . \, IH = 2Rr.$$

Produce SI to meet the circumference in M and N.

By Euc. III. 35,

$$AI \,.\, IH = MI \,.\, IN = (R + SI)(R - SI);$$

$$\therefore \; 2Rr = R^2 - SI^2;$$

that is, $$SI^2 = R^2 - 2Rr.$$

***229.** *To find the distance of an ex-centre from the circum-centre.*

Let S be the circum-centre, and I the in-centre; then AI produced passes through the ex-centre I_1.

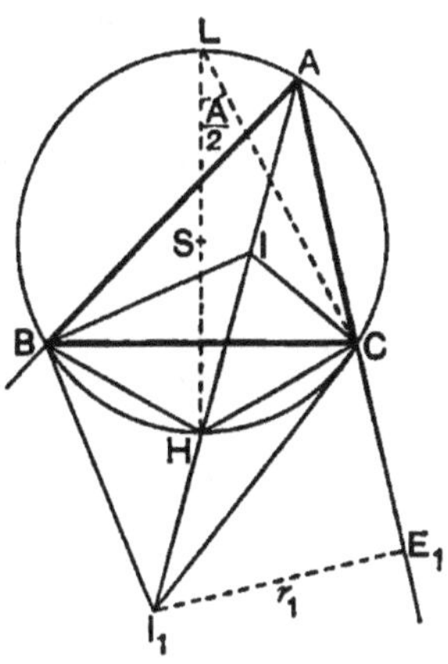

Let AI_1 meet the circum-circle in H; join CI, BI, CH, BH, CI_1, BI_1. Draw I_1E_1 perpendicular to AC.

Produce HS to meet the circumference in L, and join CL.

The angles IBI_1 and ICI_1 are right angles; hence the circle on II_1 as diameter passes through B and C.

The chords BH and CH of the circum-circle subtend equal angles at A, and are therefore equal.

But from the last article, $HC = HI$;

$$\therefore \; HB = HC = HI;$$

hence H is the centre of the circle round IBI_1C.

$$\therefore \; HI_1 = HC = 2R \sin \frac{A}{2}.$$

Now $$SI_1^2 - R^2 = \text{square of tangent from } I_1$$

$$= I_1H \,.\, I_1A$$

$$= 2R \sin \frac{A}{2} \,.\, r_1 \operatorname{cosec} \frac{A}{2}$$

$$= 2Rr_1.$$

$$\therefore \; SI_1^2 = R^2 + 2Rr_1.$$

***230.** *To find the distance of the orthocentre from the circumcentre.*

With the usual notation, we have

$$SO^2 = SA^2 + AO^2 - 2SA \, . \, AO \cos SAO.$$

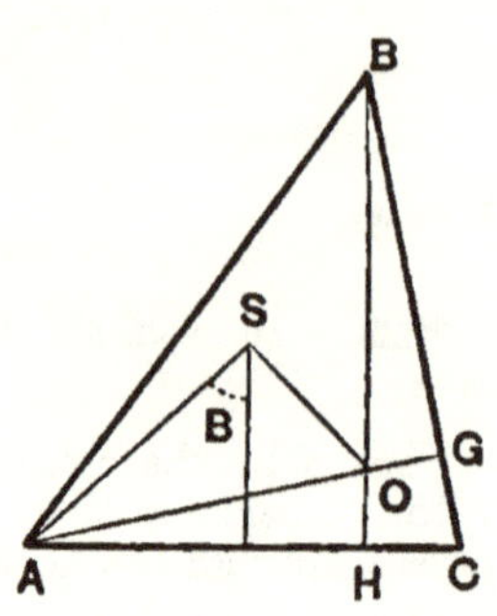

Now $AS = R$;

$$\begin{aligned} AO &= AH \operatorname{cosec} C \\ &= c \cos A \operatorname{cosec} C \\ &= 2R \sin C \cos A \operatorname{cosec} C \\ &= 2R \cos A; \end{aligned}$$

$$\begin{aligned} \angle SAO &= \angle SAC - \angle OAC \\ &= (90^\circ - B) - (90^\circ - C) \\ &= C - B. \end{aligned}$$

$$\begin{aligned} \therefore \; SO^2 &= R^2 + 4R^2 \cos^2 A - 4R^2 \cos A \cos (C - B) \\ &= R^2 - 4R^2 \cos A \{\cos (B + C) + \cos (C - B)\} \\ &= R^2 - 8R^2 \cos A \cos B \cos C. \end{aligned}$$

The student may apply a similar method to establish the results of the last two articles.

*EXAMPLES. XVIII. c.

1\. Shew that the distance of the in-centre from A is

$$4R \sin \frac{B}{2} \sin \frac{C}{2}.$$

2\. Shew that the distances of the ex-centre I_1 from the angular points A, B, C are

$$4R \cos \frac{B}{2} \cos \frac{C}{2}, \quad 4R \sin \frac{A}{2} \cos \frac{C}{2}, \quad 4R \sin \frac{A}{2} \cos \frac{B}{2}.$$

3\. Prove that the area of the ex-central triangle is equal to

$$(1) \;\; 2Rs; \qquad (2) \;\; \frac{1}{2} \Delta \operatorname{cosec} \frac{A}{2} \operatorname{cosec} \frac{B}{2} \operatorname{cosec} \frac{C}{2}.$$

4\. Shew that

$$r \, . \, II_1 \, . \, II_2 \, . \, II_3 = 4R \, . \, IA \, . \, IB \, . \, IC.$$

5\. Shew that the perimeter and in-radius of the pedal triangle are respectively

$$4R \sin A \sin B \sin C \quad \text{and} \quad 2R \cos A \cos B \cos C.$$

6. If g, h, k denote the sides of the pedal triangle, prove that

(1) $\frac{g}{a^2}+\frac{h}{b^2}+\frac{k}{c^2}=\frac{a^2+b^2+c^2}{2abc}$;

(2) $\frac{(b^2-c^2)g}{a^2}+\frac{(c^2-a^2)h}{b^2}+\frac{(a^2-b^2)k}{c^2}=0.$

7. Prove that the ex-radii of the pedal triangle are
$2R\sin A\cos B\cos C$, $2R\cos A\sin B\cos C$, $2R\cos A\cos B\sin C$.

8. Prove that any formula which connects the sides and angles of a triangle holds if we replace

(1) a, b, c by $a\cos A$, $b\cos B$, $c\cos C$,
and A, B, C by $180°-2A$, $180°-2B$, $180°-2C$;

(2) a, b, c by $a \operatorname{cosec}\frac{A}{2}$, $b\operatorname{cosec}\frac{B}{2}$, $c\operatorname{cosec}\frac{C}{2}$,

and A, B, C by $90°-\frac{A}{2}$, $90°-\frac{B}{2}$, $90°-\frac{C}{2}$.

9. Prove that the radius of the circum-circle is never less than the diameter of the in-circle.

10. If $R=2r$, shew that the triangle is equilateral.

11. Prove that

$$SI^2+SI_1^2+SI_2^2+SI_3^2=12R^2.$$

12. Prove that

(1) $a.AI^2+b.BI^2+c.CI^2=abc$;

(2) $a.AI_1^2-b.BI_1^2-c.CI_1^2=abc.$

13. If GHK be the pedal triangle, and O the orthocentre, prove that

(1) $\frac{OG}{AG}+\frac{OH}{BH}+\frac{OK}{CK}=1$;

(2) $\frac{OG}{OG+a\cot A}+\frac{OH}{OH+b\cot B}+\frac{OK}{OK+c\cot C}=1.$

14. If GHK be the pedal triangle, shew that the sum of the circum-radii of the triangles AHK, BKG, CGH is equal to $R+r$.

15. If $A_1B_1C_1$ is the ex-central triangle of ABC, and $A_2B_2C_2$ the ex-central triangle of $A_1B_1C_1$, and $A_3B_3C_3$ the ex-central triangle of $A_2B_2C_2$, and so on: find the angles of the triangle $A_nB_nC_n$, and prove that when n is indefinitely increased the triangle becomes equilateral.

16. Prove that

(1) $OS^2 = 9R^2 - a^2 - b^2 - c^2$;

(2) $OI^2 = 2r^2 - 4R^2 \cos A \cos B \cos C$;

(3) $OI_1^2 = 2r_1^2 - 4R^2 \cos A \cos B \cos C$.

17. If f, g, h denote the distances of the circum-centre of the pedal triangle from the angular points of the original triangle, shew that

$$4(f^2+g^2+h^2) = 11R^2 + 8R^2 \cos A \cos B \cos C.$$

Quadrilaterals.

***231.** *To prove that the area of a quadrilateral is equal to*

$$\frac{1}{2}\,(\textit{product of the diagonals}) \times (\textit{sine of included angle}).$$

Let the diagonals AC, BD intersect at P, and let $\angle DPA = \alpha$, and let S denote the area of the quadrilateral.

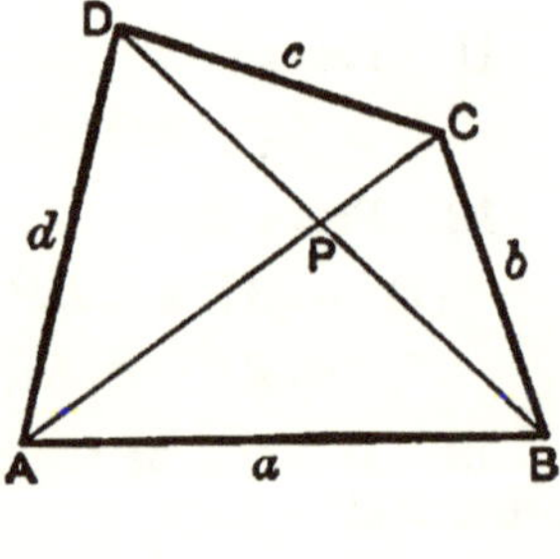

$$\Delta DAC = \Delta APD + \Delta CPD$$

$$= \frac{1}{2} DP \,.\, AP \sin \alpha + \frac{1}{2} DP \,.\, PC \sin(\pi - \alpha)$$

$$= \frac{1}{2} DP(AP + PC) \sin \alpha$$

$$= \frac{1}{2} DP \,.\, AC \sin \alpha.$$

Similarly $\qquad \Delta ABC = \frac{1}{2} BP \,.\, AC \sin \alpha.$

$$\therefore \quad S = \frac{1}{2}(DP + BP)\, AC \sin \alpha$$

$$= \frac{1}{2} DB \,.\, AC \sin \alpha.$$

***232.** *To find the area of a quadrilateral in terms of the sides and the sum of two opposite angles.*

Let $ABCD$ be the quadrilateral, and let a, b, c, d be the lengths of its sides, S its area.

By equating the two values of BD^2 found from the triangles BAD, BCD, we have

$$a^2+d^2-2ad\cos A=b^2+c^2-2bc\cos C;$$

$$\therefore\ a^2+d^2-b^2-c^2=2ad\cos A-2bc\cos C\ \text{.........(1)}.$$

Also $\quad S=$ sum of areas of triangles BAD, BCD

$$=\frac{1}{2}ad\sin A+\frac{1}{2}bc\sin C;$$

$$\therefore\ 4S=2ad\sin A+2bc\sin C\ \text{..............(2)}.$$

Square (2) and add to the square of (1);

$$\therefore\ 16S^2+(a^2+d^2-b^2-c^2)^2=4a^2d^2+4b^2c^2-8abcd\cos(A+C).$$

Let $A+C=2\alpha$; then

$$\cos(A+C)=\cos 2\alpha=2\cos^2\alpha-1;$$

$$\therefore\ 16S^2=4(ad+bc)^2-(a^2+d^2-b^2-c^2)^2-16abcd\cos^2\alpha.$$

But the first two terms on the right

$$=(2ad+2bc+a^2+d^2-b^2-c^2)(2ad+2bc-a^2-d^2+b^2+c^2)$$

$$=\{(a+d)^2-(b-c)^2\}\{(b+c)^2-(a-d)^2\}$$

$$=(a+d+b-c)(a+d-b+c)(b+c+a-d)(b+c-a+d)$$

$$=(2\sigma-2c)(2\sigma-2b)(2\sigma-2d)(2\sigma-2a),$$

where $a+b+c+d=2\sigma$,

$$=16(\sigma-a)(\sigma-b)(\sigma-c)(\sigma-d).$$

Thus $\quad S^2=(\sigma-a)(\sigma-b)(\sigma-c)(\sigma-d)-abcd\cos^2\alpha$,
where 2σ denotes the sum of the sides, 2α the sum of either pair of opposite angles.

***233.** In the case of a *cyclic quadrilateral*, $A+C=180°$, so that $\alpha=90°$; hence

$$S=\sqrt{(\sigma-a)(\sigma-b)(\sigma-c)(\sigma-d)}.$$

This formula may be obtained directly as in the last article by making use of the condition $A+C=180°$ during the course of the work. In this case $\cos C=-\cos A$, and $\sin C=\sin A$, so that the expressions (1) and (2) become

$$a^2+d^2-b^2-c^2=2(ad+bc)\cos A,$$

and $$4S=2(ad+bc)\sin A;$$

whence by eliminating A we obtain

$$16S^2+(a^2+d^2-b^2-c^2)^2=4(ad+bc)^2.$$

***234.** *To find the diagonals and the circum-radius of a cyclic quadrilateral.*

If $ABCD$ is a cyclic quadrilateral, we have just proved that

$$2(ad+bc)\cos A=a^2+d^2-b^2-c^2.$$

Now $$BD^2=a^2+d^2-2ad\cos A$$

$$=a^2+d^2-\frac{ad(a^2+d^2-b^2-c^2)}{ad+bc}$$

$$=\frac{bc(a^2+d^2)+ad(b^2+c^2)}{ad+bc}$$

$$=\frac{(ab+cd)(ac+bd)}{ad+bc}.$$

Similarly, we may prove that

$$AC^2=\frac{(ad+bc)(ac+bd)}{ab+cd}.$$

Thus $$AC.BD=ac+bd,$$ [Compare Euc. VI. D.]

and $$\frac{AC}{BD}=\frac{ad+bc}{ab+cd}.$$

The circle passing round the quadrilateral circumscribes the triangle ABD; hence

$$\text{the circum-radius}=\frac{BD}{2\sin A}$$

$$=\frac{(ad+bc)BD}{2(ad+bc)\sin A}=\frac{(ad+bc)BD}{4S}$$

$$=\frac{1}{4S}\sqrt{(ab+cd)(ac+bd)(ad+bc)}.$$

Example. A quadrilateral $ABCD$ is such that one circle can be inscribed in it and another circle circumscribed about it; shew that $\tan^2 \frac{A}{2} = \frac{bc}{ad}$.

If a circle can be inscribed in a quadrilateral, the sum of one pair of the opposite sides is equal to that of the other pair;

$$\therefore\ a+c=b+d.$$

Since the quadrilateral is cyclic,

$$\cos A = \frac{a^2+d^2-b^2-c^2}{2(ad+bc)}. \qquad \text{[Art. 233.]}$$

But $a-d=b-c$, so that $a^2-2ad+d^2=b^2-2bc+c^2$;

$$\therefore\ a^2+d^2-b^2-c^2=2(ad-bc);$$

$$\therefore\ \cos A = \frac{ad-bc}{ad+bc};$$

$$\therefore\ \tan^2 \frac{A}{2} = \frac{1-\cos A}{1+\cos A} = \frac{bc}{ad}.$$

*EXAMPLES. XVIII. d.

1. If a circle can be inscribed in a quadrilateral, shew that its radius is S/σ where S is the area and 2σ the sum of the sides of the quadrilateral.

2. If the sides of a cyclic quadrilateral be 3, 3, 4, 4, shew that a circle can be inscribed in it, and find the radii of the inscribed and circumscribed circles.

3. If the sides of a cyclic quadrilateral be 1, 2, 4, 3, shew that the cosine of the angle between the two greatest sides is $\frac{5}{7}$, and that the radius of the inscribed circle is ·98 nearly.

4. The sides of a cyclic quadrilateral are 60, 25, 52, 39: shew that two of the angles are right angles, and find the diagonals and the area.

5. The sides of a quadrilateral are 4, 5, 8, 9, and one diagonal is 9: find the area.

6. If a circle can be inscribed in a cyclic quadrilateral, shew that the area of the quadrilateral is $\sqrt{abcd}$, and that the radius of the circle is

$$2\sqrt{abcd}/(a+b+c+d).$$

7. If the sides of a quadrilateral are given, shew that the area is a maximum when the quadrilateral can be inscribed in a circle.

8. If the sides of a quadrilateral are 23, 29, 37, 41 inches, prove that the maximum area is 7 sq. ft.

9. If $ABCD$ is a cyclic quadrilateral, prove that

$$\tan^2\frac{B}{2}=\frac{(\sigma-a)(\sigma-b)}{(\sigma-c)(\sigma-d)}.$$

10. If f, g denote the diagonals of a quadrilateral and β the angle between them, prove that

$$2fg\cos\beta=(a^2+c^2)\sim(b^2+d^2).$$

11. If β is the angle between the diagonals of any quadrilateral, prove that the area is

$$\frac{1}{4}\{(a^2+c^2)\sim(b^2+d^2)\}\tan\beta.$$

12. Prove that the area of a quadrilateral in which a circle can be inscribed is

$$\sqrt{abcd}\sin\frac{A+C}{2}.$$

13. If a circle can be inscribed in a quadrilateral whose diagonals are f and g, prove that

$$4S^2=f^2g^2-(ac-bd)^2.$$

14. If β is the angle between the diagonals of a cyclic quadrilateral, prove that

(1) $(ac+bd)\sin\beta=(ad+bc)\sin A$;

(2) $\cos\beta=\dfrac{(a^2+c^2)\sim(b^2+d^2)}{2(ac+bd)}$;

(3) $\tan^2\dfrac{\beta}{2}=\dfrac{(\sigma-b)(\sigma-d)}{(\sigma-a)(\sigma-c)}$ or $\dfrac{(\sigma-a)(\sigma-c)}{(\sigma-b)(\sigma-d)}$.

15. If f, g are the diagonals of a quadrilateral, shew that

$$S=\frac{1}{4}\sqrt{4f^2g^2-(a^2+c^2-b^2-d^2)^2}.$$

16. In a cyclic quadrilateral, prove that the product of the segments of a diagonal is

$$abcd\,(ac+bd)/(ab+cd)(ad+bc).$$

235. The following exercise consists of miscellaneous questions involving the properties of triangles.

EXAMPLES. XVIII. e.

1. If the sides of a triangle are 242, 1212, 1450 yards, shew that the area is 6 acres.

2. One of the sides of a triangle is 200 yards and the adjacent angles are 22·5° and 67·5°: find the area.

3. If $r_1=2r_2=2r_3$, shew that $3a=4b$.

4. If a, b, c are in A. P., shew that r_1, r_2, r_3 are in H. P.

5. Find the area of a triangle whose sides are

$$\frac{y}{z}+\frac{z}{x},\quad \frac{z}{x}+\frac{x}{y},\quad \frac{x}{y}+\frac{y}{z}.$$

6. If $\sin A : \sin C=\sin(A-B) : \sin(B-C)$, shew that a^2, b^2, c^2 are in A. P.

Prove that

7. $$\frac{a\sin A+b\sin B+c\sin C}{4\cos\frac{A}{2}\cos\frac{B}{2}\cos\frac{C}{2}}=\frac{a^2+b^2+c^2}{2s}.$$

8. $$\left(\frac{a^2}{\sin A}+\frac{b^2}{\sin B}+\frac{c^2}{\sin C}\right)\sin\frac{A}{2}\sin\frac{B}{2}\sin\frac{C}{2}=\Delta.$$

9. $$(r_2+r_3)(r_3+r_1)(r_1+r_2)=4R(r_2r_3+r_3r_1+r_1r_2).$$

10. $$\tan\frac{A}{2}+\tan\frac{B}{2}+\tan\frac{C}{2}=\frac{r_1+r_2+r_3}{(r_2r_3+r_3r_1+r_1r_2)^{\frac{1}{2}}}.$$

11. $$bc\cot\frac{A}{2}+ca\cot\frac{B}{2}+ab\cot\frac{C}{2}=4Rs^2\left(\frac{1}{a}+\frac{1}{b}+\frac{1}{c}-\frac{3}{s}\right).$$

12. $$\left(\frac{1}{r}+\frac{1}{r_1}+\frac{1}{r_2}+\frac{1}{r_3}\right)^2=\frac{4}{r}\left(\frac{1}{r_1}+\frac{1}{r_2}+\frac{1}{r_3}\right).$$

13. The perimeter of a right-angled triangle is 70, and the in-radius is 6: find the sides.

14. If f, g, h are the perpendiculars from the circum-centre on the sides, prove that

$$\frac{a}{f}+\frac{b}{g}+\frac{c}{h}=\frac{abc}{4fgh}.$$

15. An equilateral triangle and a regular hexagon have the same perimeter: shew that the areas of their inscribed circles are as 4 to 9.

*16. Shew that the perimeter of the pedal triangle is equal to

$$abc/2R^2.$$

*17. Shew that the area of the ex-central triangle is equal to

$$abc\,(a+b+c)/4\Delta.$$

18. In the ambiguous case, if A, a, b are the given parts, and c_1, c_2 the two values of the third side, shew that the distance between the circum-centres of the two triangles is $\frac{c_1 \sim c_2}{2 \sin A}$.

*19. If β be the angle between the diagonals of a cyclic quadrilateral, shew that

$$\sin\beta = \frac{2S}{ac+bd}.$$

*20. Shew that

$$r^3 . II_1 . II_2 . II_3 = IA^2 . IB^2 . IC^2.$$

*21. Shew that the sum of the squares of the sides of the ex-central triangle is equal to $8R(4R+r)$.

*22. If circles can be inscribed in and circumscribed about a quadrilateral, and if β be the angle between the diagonals, shew that

$$\cos\beta = (ac \sim bd)/(ac+bd).$$

23. If l, m, n are the lengths of the medians of a triangle, prove that

(1) $4(l^2+m^2+n^2) = 3(a^2+b^2+c^2)$;

(2) $(b^2-c^2)\,l^2 + (c^2-a^2)\,m^2 + (a^2-b^2)\,n^2 = 0$;

(3) $16(l^4+m^4+n^4) = 9(a^4+b^4+c^4)$.

24. Shew that the radii of the escribed circles are the roots of the equation

$$x^3 - (4R+r)\,x^2 + s^2x - s^2r = 0.$$

25. If Δ_1, Δ_2, Δ_3 be the areas of the triangles cut off by tangents to the in-circle parallel to the sides of a triangle, prove that

$$\frac{\Delta_1}{(s-a)^2} = \frac{\Delta_2}{(s-b)^2} = \frac{\Delta_3}{(s-c)^2} = \frac{\Delta}{s^2}.$$

*26. The triangle LMN is formed by joining the points of contact of the in-circle; shew that it is similar to the ex-central triangle, and that their areas are as r^2 to $4R^2$.

27. In the triangle PQR formed by drawing tangents at A, B, C to the circum-circle, prove that the angles and sides are

$$180^\circ - 2A, \quad 180^\circ - 2B, \quad 180^\circ - 2C;$$

and
$$\frac{a}{2\cos B\cos C}, \quad \frac{b}{2\cos C\cos A}, \quad \frac{c}{2\cos A\cos B}.$$

28. If p, q, r be the lengths of the bisectors of the angles of a triangle, prove that

(1) $\dfrac{1}{p}\cos\dfrac{A}{2} + \dfrac{1}{q}\cos\dfrac{B}{2} + \dfrac{1}{r}\cos\dfrac{C}{2} = \dfrac{1}{a} + \dfrac{1}{b} + \dfrac{1}{c}$;

(2) $\dfrac{pqr}{4\Delta} = \dfrac{abc\,(a+b+c)}{(b+c)(c+a)(a+b)}$.

29. If the perpendiculars AG, BH, CK are produced to meet the circum-circle in L, M, N, prove that

(1) area of triangle $LMN = 8\Delta\cos A\cos B\cos C$;

(2) $AL\sin A + BM\sin B + CN\sin C = 8R\sin A\sin B\sin C$.

30. If r_a, r_b, r_c be the radii of the circles inscribed between the in-circle and the sides containing the angles A, B, C respectively, shew that

(1) $r_a = r\tan^2\dfrac{\pi - A}{4}$; (2) $\sqrt{r_b r_c} + \sqrt{r_c r_a} + \sqrt{r_a r_b} = r$.

*31. Lines drawn through the angular points of a triangle ABC parallel to the sides of the pedal triangle form a triangle XYZ: shew that the perimeter and area of XYZ are respectively

$$2R\tan A\tan B\tan C \quad \text{and} \quad R^2\tan A\tan B\tan C.$$

*32. A straight line cuts three concentric circles in A, B, C and passes at a distance p from their centre: shew that the area of the triangle formed by the tangents at A, B, C is

$$\frac{BC \,.\, CA \,.\, AB}{2p}.$$

MISCELLANEOUS EXAMPLES. F.

1. If $\alpha+\beta+\gamma+\delta=180°$, shew that

$$\cos\alpha\cos\beta+\cos\gamma\cos\delta=\sin\alpha\sin\beta+\sin\gamma\sin\delta.$$

2. Prove that

$$\cos(15°-A)\sec 15°-\sin(15°-A)\operatorname{cosec} 15°=4\sin A.$$

3. Shew that in a triangle

$$\cot A+\sin A\operatorname{cosec} B\operatorname{cosec} C$$

retains the same value if any two of the angles A, B, C are interchanged.

4. If $a=2$, $b=\sqrt{8}$, $A=30°$, solve the triangle.

5. Shew that

(1) $\cot 18°=\sqrt{5}\cot 36°$;

(2) $16\sin 36°\sin 72°\sin 108°\sin 144°=5$.

6. Find the number of ciphers before the first significant digit in $(\cdot 0396)^{50}$, given

$$\log 2=\cdot 30103,\quad \log 3=\cdot 47712,\quad \log 11=1\cdot 04139.$$

7. An observer finds that the angle subtended by the line joining two points A and B on the horizontal plane is 30°. On walking 50 yards directly towards A the angle increases to 75°: find his distance from B at each observation.

8. Prove that $\cos^2\alpha+\cos^2\beta+\cos^2\gamma+\cos^2(\alpha+\beta+\gamma)$

$$=2+2\cos(\beta+\gamma)\cos(\gamma+\alpha)\cos(\alpha+\beta).$$

9. Shew that

(1) $\tan 40°+\cot 40°=2\sec 10°$;

(2) $\tan 70°+\tan 20°=2\operatorname{cosec} 40°$.

10. Prove that

(1) $2\sin 4\alpha-\sin 10\alpha+\sin 2\alpha=16\sin\alpha\cos\alpha\cos 2\alpha\sin^2 3\alpha$;

(2) $\sin\frac{2\pi}{7}+\sin\frac{4\pi}{7}-\sin\frac{6\pi}{7}=4\sin\frac{\pi}{7}\sin\frac{3\pi}{7}\sin\frac{5\pi}{7}$.

11. If $B=30°$, $b=3\sqrt{2}-\sqrt{6}$, $c=6-2\sqrt{3}$, solve the triangle.

12. From a ship which is sailing N.E., the bearing of a rock is N.N.W. After the ship has sailed 10 miles the rock bears due W.: find the distance of the ship from the rock at each observation.

13. Shew that in any triangle

$$\frac{b^2-c^2}{\cos B+\cos C}+\frac{c^2-a^2}{\cos C+\cos A}+\frac{a^2-b^2}{\cos A+\cos B}=0.$$

14. If $\cos(\theta-a)$, $\cos\theta$, $\cos(\theta+a)$ are in harmonical progression, shew that

$$\cos\theta=\sqrt{2}\cos\frac{a}{2}.$$

15. If $\sin\beta$ be the geometric mean between $\sin a$ and $\cos a$, prove that $$\cos 2\beta=2\cos^2\left(\frac{\pi}{4}+a\right).$$

16. Shew that the distances of the orthocentre from the sides are $2R\cos B\cos C$, $2R\cos C\cos A$, $2R\cos A\cos B$.

17. If $$\cos\theta=\frac{\cos u-e}{1-e\cos u},$$
prove that $$\tan\frac{\theta}{2}=\sqrt{\frac{1+e}{1-e}}\tan\frac{u}{2}.$$

18. If the sides of a right-angled triangle are

$$2(1+\sin\theta)+\cos\theta \quad\text{and}\quad 2(1+\cos\theta)+\sin\theta,$$

prove that the hypotenuse is

$$3+2(\cos\theta+\sin\theta).$$

*19. Prove that the distances of the in-centre of the ex-central triangle $I_1I_2I_3$ from its ex-centres are

$$8R\sin\frac{B+C}{4},\quad 8R\sin\frac{C+A}{4},\quad 8R\sin\frac{A+B}{4}.$$

*20. Prove that the distances between the ex-centres of the ex-central triangle $I_1I_2I_3$ are

$$8R\cos\frac{B+C}{4},\quad 8R\cos\frac{C+A}{4},\quad 8R\cos\frac{A+B}{4}.$$

21. If

$$(1+\cos\alpha)(1+\cos\beta)(1+\cos\gamma)=(1-\cos\alpha)(1-\cos\beta)(1-\cos\gamma),$$

shew that each expression is equal to $\pm\sin\alpha\sin\beta\sin\gamma$.

22. If the sum of four angles is 180°, shew that the sum of the products of their sines taken two together is equal to the sum of the products of their cosines taken two together.

*23. In a triangle, shew that

(1) $II_1 \,.\, II_2 \,.\, II_3 = 16R^2r$; (2) $II_1^2 + I_2I_3^2 = 16R^2$.

24. Find the angles of a triangle whose sides are proportional to

$$(1) \quad \cos\frac{A}{2}, \quad \cos\frac{B}{2}, \quad \cos\frac{C}{2};$$

$$(2) \quad \sin 2A, \quad \sin 2B, \quad \sin 2C.$$

25. Prove that the expression

$$\sin^2(\theta+\alpha)+\sin^2(\theta+\beta)-2\cos(\alpha-\beta)\sin(\theta+\alpha)\sin(\theta+\beta)$$

is independent of θ.

*26. If a, b, c, d are the sides of a quadrilateral described about a circle, prove that

$$ab\sin^2\frac{A}{2}=cd\sin^2\frac{C}{2}.$$

27. Tangents parallel to the three sides are drawn to the in-circle. If p, q, r be the lengths of the parts of the tangents within the triangle, prove that $\frac{p}{a}+\frac{q}{b}+\frac{r}{c}=1$.

[*The Tables will be required for Examples* 28 *and* 29.]

28. From the top of a cliff 1566 ft. in height a train, which is travelling at a uniform speed in a straight line to a tunnel immediately below the observer, is seen to pass two consecutive stations at an interval of 3 minutes. The angles of depression of the two stations are 13° 14′ 12″ and 56° 24′ 36″ respectively; how fast is the train travelling?

29. A harbour lies in a direction 46° 8′ 8·6″ South of West from a fort, and at a distance of 27·23 miles from it. A ship sets out from the harbour at noon and sails due East at 10 miles an hour; when will the ship be 20 miles from the fort?

CHAPTER XIX.

GENERAL VALUES AND INVERSE FUNCTIONS.

236. The equation $\sin\theta = \frac{1}{2}$ is satisfied by $\theta = \frac{\pi}{6}$, and by $\theta = \pi - \frac{\pi}{6}$, and all angles coterminal with these will have the same sine. This example shews that there are an infinite number of angles whose sine is equal to a given quantity. Similar remarks apply to the other functions.

We proceed to shew how to express by a single formula all angles which have a given sine, cosine, or tangent.

237. From the results proved in Chap. IX., it is easily seen that in going once through the four quadrants, there are two and only two positions of the boundary line which give angles with the same sine, cosine, or tangent.

Thus if $\sin\alpha$ has a given value, the positions of the radius vector are OP and OP' bounding the angles α and $\pi - \alpha$. [Art. 92.]

If $\cos\alpha$ has a given value, the positions of the radius vector are OP and OP' bounding the angles α and $2\pi - \alpha$. [Art. 105.]

If $\tan\alpha$ has a given value, the positions of the radius vector are OP and OP' bounding the angles α and $\pi + \alpha$. [Art. 97.]

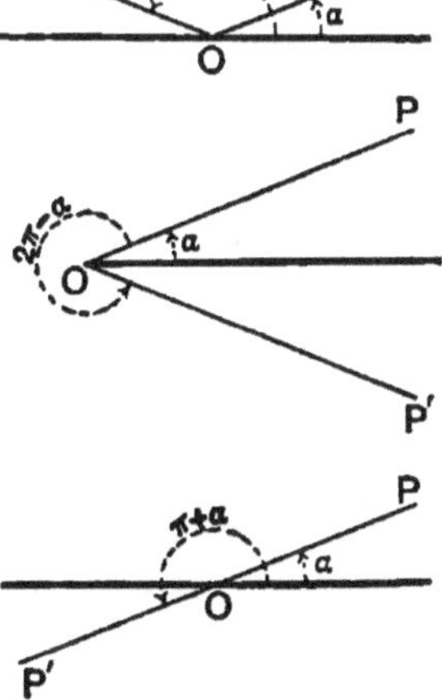

238. *To find a formula for all the angles which have a given sine.*

Let a be the smallest positive angle which has a given sine. Draw OP and OP' bounding the angles a and $\pi - a$; then the required angles are those coterminal with OP and OP'.

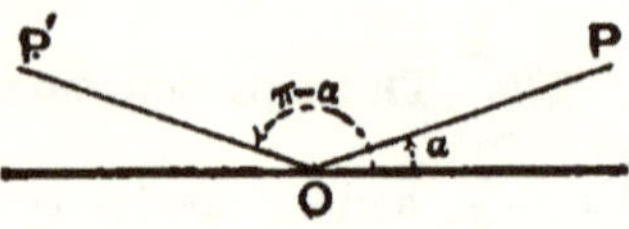

The positive angles are

$$2p\pi + a \text{ and } 2p\pi + (\pi - a),$$

where p is zero, or any positive integer.

The negative angles are

$$-(\pi + a) \text{ and } -(2\pi - a),$$

and those which may be obtained from them by the addition of any negative multiple of 2π; that is, angles denoted by

$$2q\pi - (\pi + a) \text{ and } 2q\pi - (2\pi - a),$$

where q is zero, or any negative integer.

These angles may be grouped as follows:

$$\left.\begin{matrix} 2p\pi + a, \\ (2q-2)\pi + a, \end{matrix}\right\} \text{ and } \left\{\begin{matrix} (2p+1)\pi - a, \\ (2q-1)\pi - a, \end{matrix}\right.$$

and it will be noticed that even multiples of π are followed by $+a$, and odd multiples of π by $-a$.

Thus all angles equi-sinal with a are included in the formula

$$n\pi + (-1)^n a,$$

where n is zero, or any integer positive or negative.

This is also the formula for all angles which have the same cosecant as a.

Example 1. Write down the general solution of $\sin\theta = \dfrac{\sqrt{3}}{2}$.

The least value of θ which satisfies the equation is $\dfrac{\pi}{3}$; therefore the general solution is $n\pi + (-1)^n \dfrac{\pi}{3}$.

Example 2. Find the general solution of $\sin^2\theta = \sin^2 a$.

This equation gives either $\sin\theta = +\sin a$..........................(1),

or $\sin\theta = -\sin a = \sin(-a)$............(2).

From (1), $\theta = n\pi + (-1)^n \alpha$;
and from (2), $\theta = n\pi + (-1)^n (-\alpha)$.

Both values are included in the formula $\theta = n\pi \pm \alpha$.

239. *To find a formula for all the angles which have a given cosine.*

Let α be the smallest positive angle which has a given cosine. Draw OP and OP' bounding the angles α and $2\pi - \alpha$; then the required angles are those coterminal with OP and OP'.

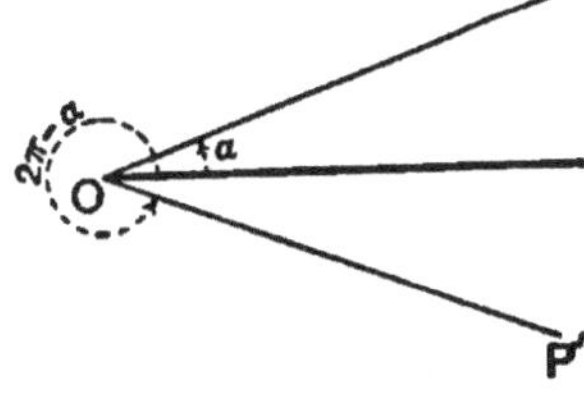

The positive angles are

$2p\pi + \alpha$ and $2p\pi + (2\pi - \alpha)$,

where p is zero, or any positive integer.

The negative angles are

$$-\alpha \text{ and } -(2\pi - \alpha),$$

and those which may be obtained from them by the addition of any negative multiple of 2π; that is, angles denoted by

$$2q\pi - \alpha \text{ and } 2q\pi - (2\pi - \alpha),$$

where q is zero, or any negative integer.

The angles may be grouped as follows:

$$\left.\begin{matrix} 2p\pi + \alpha, \\ 2q\pi - \alpha, \end{matrix}\right\} \text{ and } \left\{\begin{matrix} (2p+2)\pi - \alpha, \\ (2q-2)\pi + \alpha, \end{matrix}\right.$$

and it will be noticed that the multiples of π are always even, but may be followed by $+\alpha$ or by $-\alpha$.

Thus all angles equi-cosinal with α are included in the formula

$$2n\pi \pm \alpha,$$

where n is zero, or any integer positive or negative.

This is also the formula for all angles which have the same secant as α.

Example 1. Find the general solution of $\cos\theta = -\dfrac{1}{2}$.

The least value of θ is $\pi - \dfrac{\pi}{3}$, or $\dfrac{2\pi}{3}$; hence the general solution is $2n\pi \pm \dfrac{2\pi}{3}$.

240. *To find a formula for all the angles which have a given tangent.*

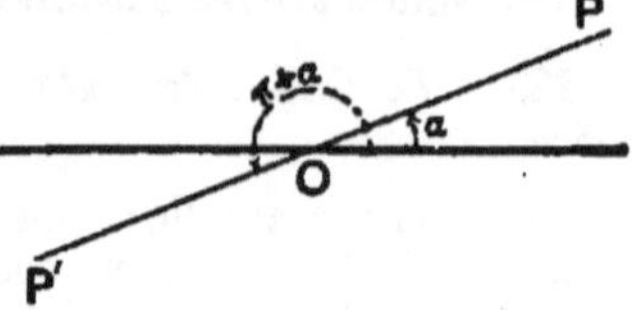

Let α be the smallest positive angle which has a given tangent. Draw OP and OP' bounding the angles α and $\pi+\alpha$; then the required angles are those coterminal with OP and OP'.

The positive angles are

$$2p\pi+\alpha \text{ and } 2p\pi+(\pi+\alpha),$$

where p is zero, or any positive integer.

The negative angles are

$$-(\pi-\alpha) \text{ and } -(2\pi-\alpha),$$

and those which may be obtained from them by the addition of any negative multiple of 2π; that is, angles denoted by

$$2q\pi-(\pi-\alpha) \text{ and } 2q\pi-(2\pi-\alpha),$$

where q is zero, or any negative integer.

The angles may be grouped as follows :

$$\left.\begin{matrix} 2p\pi+\alpha, \\ (2q-2)\pi+\alpha, \end{matrix}\right\} \text{ and } \left\{\begin{matrix} (2p+1)\pi+\alpha, \\ (2q-1)\pi+\alpha, \end{matrix}\right.$$

and it will be noticed that whether the multiple of π is even or odd, it is always followed by $+\alpha$. Thus all angles equi-tangential with α are included in the formula

$$\theta=n\pi+\alpha.$$

This is also the formula for all the angles which have the same cotangent as α.

Example. Solve the equation $\cot 4\theta=\cot\theta$.

The general solution is $4\theta=n\pi+\theta$;

whence $3\theta=n\pi$, or $\theta=\dfrac{n\pi}{3}$.

241. *All angles which are both equi-sinal and equi-cosinal with α are included in the formula $2n\pi+\alpha$.*

All angles equi-cosinal with α are included in the formula $2n\pi\pm\alpha$; so that the multiple of π is even. But in the formula $n\pi+(-1)^n\alpha$, which includes all angles equi-sinal with α, when the multiple of π is even, α must be preceded by the $+$ sign. Thus the formula is $2n\pi+\alpha$.

242. In the solution of equations, the general value of the angle should always be given.

Example. Solve the equation $\cos 9\theta = \cos 5\theta - \cos \theta$.

By transposition, $(\cos 9\theta + \cos \theta) - \cos 5\theta = 0$;

$$\therefore\ 2\cos 5\theta \cos 4\theta - \cos 5\theta = 0;$$

$$\therefore\ \cos 5\theta\,(2\cos 4\theta - 1) = 0;$$

$\therefore$ either $\cos 5\theta = 0$, or $2\cos 4\theta - 1 = 0$.

From the first equation, $5\theta = 2n\pi \pm \frac{\pi}{2}$, or $\theta = \frac{(4n \pm 1)\pi}{10}$;

and from the second, $4\theta = 2n\pi \pm \frac{\pi}{3}$, or $\theta = \frac{(6n \pm 1)\pi}{12}$.

EXAMPLES. XIX. a.

Find the general solution of the equations:

1. $\sin \theta = \frac{1}{2}$.
2. $\sin \theta = \frac{1}{\sqrt{2}}$.
3. $\cos \theta = \frac{1}{2}$.
4. $\tan \theta = \sqrt{3}$.
5. $\cot \theta = -\sqrt{3}$.
6. $\sec \theta = -\sqrt{2}$.
7. $\cos^2 \theta = \frac{1}{2}$.
8. $\tan^2 \theta = \frac{1}{3}$.
9. $\operatorname{cosec}^2 \theta = \frac{4}{3}$.
10. $\cos \theta = \cos \alpha$.
11. $\tan^2 \theta = \tan^2 \alpha$.
12. $\sec^2 \theta = \sec^2 \alpha$.
13. $\tan 2\theta = \tan \theta$.
14. $\operatorname{cosec} 3\theta = \operatorname{cosec} 3\alpha$.
15. $\cos 3\theta = \cos 2\theta$.
16. $\sin 5\theta + \sin \theta = \sin 3\theta$.
17. $\cos \theta - \cos 7\theta = \sin 4\theta$.
18. $\sin 4\theta - \sin 3\theta + \sin 2\theta - \sin \theta = 0$.
19. $\cos \theta + \cos 3\theta + \cos 5\theta + \cos 7\theta = 0$.
20. $\sin 5\theta \cos \theta = \sin 6\theta \cos 2\theta$.
21. $\sin 11\theta \sin 4\theta + \sin 5\theta \sin 2\theta = 0$.
22. $\sqrt{2} \cos 3\theta - \cos \theta = \cos 5\theta$.
23. $\sin 7\theta - \sqrt{3} \cos 4\theta = \sin \theta$.
24. $1 + \cos \theta = 2 \sin^2 \theta$.
25. $\tan^2 \theta + \sec \theta = 1$.
26. $\cot^2 \theta - 1 = \operatorname{cosec} \theta$.
27. $\cot \theta - \tan \theta = 2$.
28. If $2\cos \theta = -1$ and $2 \sin \theta = \sqrt{3}$, find θ.
29. If $\sec \theta = \sqrt{2}$ and $\tan \theta = -1$, find θ.

243. In the following examples, the solution is simplified by the use of some particular artifice.

Example 1. Solve the equation $\cos m\theta = \sin n\theta$.

Here $$\cos m\theta = \cos\left(\frac{\pi}{2} - n\theta\right);$$

$$\therefore\ m\theta = 2k\pi \pm \left(\frac{\pi}{2} - n\theta\right),$$

where k is zero, or any integer.

By transposition, we obtain

$$(m+n)\,\theta = \left(2k + \frac{1}{2}\right)\pi, \text{ or } (m-n)\,\theta = \left(2k - \frac{1}{2}\right)\pi.$$

This equation may also be solved through the medium of the sine. For we have

$$\sin\left(\frac{\pi}{2} - m\theta\right) = \sin n\theta;$$

$$\therefore\ \frac{\pi}{2} - m\theta = p\pi + (-1)^p n\theta,$$

where p is zero or any integer;

$$\therefore\ \{m + (-1)^p n\}\,\theta = \left(\frac{1}{2} - p\right)\pi.$$

NOTE. The general solution can frequently be obtained in several ways. The various forms which the result takes are merely different modes of expressing the same series of angles.

Example 2. Solve $\sqrt{3}\cos\theta + \sin\theta = 1$.

Multiply every term by $\frac{1}{2}$, then

$$\frac{\sqrt{3}}{2}\cos\theta + \frac{1}{2}\sin\theta = \frac{1}{2},$$

$$\therefore\ \cos\frac{\pi}{6}\cos\theta + \sin\frac{\pi}{6}\sin\theta = \frac{1}{2};$$

$$\therefore\ \cos\left(\theta - \frac{\pi}{6}\right) = \frac{1}{2};$$

$$\therefore\ \theta - \frac{\pi}{6} = 2n\pi \pm \frac{\pi}{3};$$

$$\therefore\ \theta = 2n\pi + \frac{\pi}{2} \text{ or } 2n\pi - \frac{\pi}{6}.$$

NOTE. In examples of this type, it is a common mistake to square the equation; but this process is objectionable, because it introduces solutions which do not belong to the given equation. Thus in the present instance,

$$\sqrt{3}\cos\theta = 1 - \sin\theta;$$

by squaring, $$3\cos^2\theta = (1-\sin\theta)^2.$$

But the solutions of this equation include the solutions of

$$-\sqrt{3}\cos\theta = 1 - \sin\theta,$$

as well as those of the given equation.

Example 3. Solve $\cos 2\theta = \cos\theta + \sin\theta$.

From this equation, $\cos^2\theta - \sin^2\theta = \cos\theta + \sin\theta$;

$$\therefore (\cos\theta + \sin\theta)(\cos\theta - \sin\theta) = \cos\theta + \sin\theta;$$

$\therefore$ either $\cos\theta + \sin\theta = 0$ ....(1),

or $\cos\theta - \sin\theta = 1$ ....(2).

From (1), $\tan\theta = -1$,

$$\therefore \theta = n\pi - \frac{\pi}{4}.$$

From (2), $$\frac{1}{\sqrt{2}}\cos\theta - \frac{1}{\sqrt{2}}\sin\theta = \frac{1}{\sqrt{2}};$$

$$\therefore \cos\theta\cos\frac{\pi}{4} - \sin\theta\sin\frac{\pi}{4} = \frac{1}{\sqrt{2}}.$$

$$\therefore \cos\left(\theta + \frac{\pi}{4}\right) = \frac{1}{\sqrt{2}};$$

$$\therefore \theta + \frac{\pi}{4} = 2n\pi \pm \frac{\pi}{4};$$

$$\therefore \theta = 2n\pi \text{ or } 2n\pi - \frac{\pi}{2}.$$

EXAMPLES. XIX. b.

Find the general solution of the equations:

1. $\tan p\theta = \cot q\theta$.
2. $\sin m\theta + \cos n\theta = 0$.
3. $\cos\theta - \sqrt{3}\sin\theta = 1$.
4. $\sin\theta - \sqrt{3}\cos\theta = 1$.
5. $\cos\theta = \sqrt{3}(1 - \sin\theta)$.
6. $\sin\theta + \sqrt{3}\cos\theta = \sqrt{2}$.

Find the general solution of the equations:

7. $\cos\theta - \sin\theta = \frac{1}{\sqrt{2}}$.

8. $\cos\theta + \sin\theta + \sqrt{2} = 0$.

9. $\operatorname{cosec}\theta + \cot\theta = \sqrt{3}$.

10. $\cot\theta - \cot 2\theta = 2$.

11. $2\sin\theta\sin 3\theta = 1$.

12. $\sin 3\theta = 8\sin^3\theta$.

13. $\tan\theta + \tan 3\theta = 2\tan 2\theta$.

14. $\cos\theta - \sin\theta = \cos 2\theta$.

15. $\operatorname{cosec}\theta + \sec\theta = 2\sqrt{2}$.

16. $\sec\theta - \operatorname{cosec}\theta = 2\sqrt{2}$.

17. $\sec 4\theta - \sec 2\theta = 2$.

18. $\cos 3\theta + 8\cos^3\theta = 0$.

19. $1 + \sqrt{3}\tan^2\theta = (1 + \sqrt{3})\tan\theta$.

20. $\tan^3\theta + \cot^3\theta = 8\operatorname{cosec}^3 2\theta + 12$.

21. $\sin\theta = \sqrt{2}\sin\phi, \qquad \sqrt{3}\cos\theta = \sqrt{2}\cos\phi$.

22. $\operatorname{cosec}\theta = \sqrt{3}\operatorname{cosec}\phi, \qquad \cot\theta = 3\cot\phi$.

23. $\sec\phi = \sqrt{2}\sec\theta, \qquad \cot\theta = \sqrt{3}\cot\phi$.

24. Explain why the same two series of angles are given by the equations

$$\theta + \frac{\pi}{4} = n\pi + (-1)^n\frac{\pi}{6} \quad \text{and} \quad \theta - \frac{\pi}{4} = 2n\pi \pm \frac{\pi}{3}.$$

25. Shew that the formulæ

$$\left(2n + \frac{1}{4}\right)\pi \pm a \quad \text{and} \quad \left(n - \frac{1}{4}\right)\pi + (-1)^n\left(\frac{\pi}{2} - a\right)$$

comprise the same angles, and illustrate by a figure.

Inverse Circular Functions.

244. If $\sin\theta = s$, we know that θ may be *any* angle whose sine is s. It is often convenient to express this statement *inversely* by writing $\theta = \sin^{-1}s$.

In this *inverse notation* θ stands alone on one side of the equation, and may be regarded as an angle whose value is only known through the medium of its sine. Similarly, $\tan^{-1}\sqrt{3}$ indicates in a concise form any one of the angles whose tangent is $\sqrt{3}$. But all these angles are given by the formula $n\pi + \frac{\pi}{3}$. Thus

$$\theta = \tan^{-1}\sqrt{3} \quad \text{and} \quad \theta = n\pi + \frac{\pi}{3}$$

are equivalent statements expressed in different forms.

245. Expressions of the form $\cos^{-1}x$, $\sin^{-1}a$, $\tan^{-1}b$ are called **Inverse Circular Functions.**

It must be remembered that these expressions denote angles, and that -1 *is not an algebraical index*; that is,

$$\sin^{-1}x \textit{ is not the same as } (\sin x)^{-1} \textit{ or } \frac{1}{\sin x}.$$

246. From Art. 244, we see that an inverse function has an infinite number of values.

If f denote any one of the circular functions, and $f^{-1}(x)=A$, the **principal value** of $f^{-1}(x)$ is the smallest numerical value of A. Thus the principal values of

$$\cos^{-1}\frac{1}{2}, \quad \sin^{-1}\left(-\frac{1}{2}\right), \quad \cos^{-1}\left(-\frac{1}{\sqrt{2}}\right), \quad \tan^{-1}(-1)$$

are $\quad 60°, \quad -30°, \quad 135°, \quad -45°.$

Hence if x be positive, the principal values of $\sin^{-1}x$, $\cos^{-1}x$, $\tan^{-1}x$ all lie between 0 and 90°.

If x be negative, the principal values of $\sin^{-1}x$ and $\tan^{-1}x$ lie between 0 and $-90°$, and the principal value of $\cos^{-1}x$ lies between 90° and 180°.

In numerical instances we shall usually suppose that the *principal value* is selected.

247. If $\sin\theta=x$, we have $\cos\theta=\sqrt{1-x^2}$.

Expressed in the inverse notation, these equations become

$$\theta=\sin^{-1}x, \quad \theta=\cos^{-1}\sqrt{1-x^2}.$$

In each of these two statements, θ has an infinite number of values; but, as the formulæ for the general values of the sine and cosine are not identical, we cannot assert that the equation

$$\sin^{-1}x=\cos^{-1}\sqrt{1-x^2}$$

is identically true. This will be seen more clearly from a numerical instance. If $x=\frac{1}{2}$, then $\sqrt{1-x^2}=\frac{\sqrt{3}}{2}$.

Here $\sin^{-1}x$ may be any one of the angles

$$30°, \; 150°, \; 390°, \; 510°, \ldots;$$

and $\cos^{-1}\sqrt{1-x^2}$ may be any one of the angles

$$30°, \; 330°, \; 390°, \; 690°, \ldots.$$

248. From the relations established in the previous chapters, we may deduce corresponding relations connecting the inverse functions. Thus in the identity

$$\cos 2\theta = \frac{1-\tan^2\theta}{1+\tan^2\theta},$$

let $\tan\theta = a$, so that $\theta = \tan^{-1} a$; then

$$\cos(2\tan^{-1} a) = \frac{1-a^2}{1+a^2};$$

$$\therefore\ 2\tan^{-1} a = \cos^{-1}\frac{1-a^2}{1+a^2}.$$

Similarly, the formula

$$\cos 3\theta = 4\cos^3\theta - 3\cos\theta$$

when expressed in the inverse notation becomes

$$3\cos^{-1} a = \cos^{-1}(4a^3 - 3a).$$

249. *To prove that*

$$\tan^{-1} x + \tan^{-1} y = \tan^{-1}\frac{x+y}{1-xy}.$$

Let $\tan^{-1} x = \alpha$, so that $\tan\alpha = x$;

and $\tan^{-1} y = \beta$, so that $\tan\beta = y$.

We require $\alpha+\beta$ in the form of an inverse tangent.

Now
$$\tan(\alpha+\beta) = \frac{\tan\alpha + \tan\beta}{1-\tan\alpha\tan\beta}$$

$$= \frac{x+y}{1-xy};$$

$$\therefore\ \alpha+\beta = \tan^{-1}\frac{x+y}{1-xy};$$

that is,
$$\tan^{-1} x + \tan^{-1} y = \tan^{-1}\frac{x+y}{1-xy}.$$

By putting $y = x$, we obtain

$$2\tan^{-1} x = \tan^{-1}\frac{2x}{1-x^2}.$$

NOTE. It is useful to remember that

$$\tan(\tan^{-1} x + \tan^{-1} y) = \frac{x+y}{1-xy}.$$

Example 1. Prove that

$$\tan^{-1}5-\tan^{-1}3+\tan^{-1}\frac{7}{9}=n\pi+\frac{\pi}{4}.$$

$$\text{The first side}=\tan^{-1}\frac{5-3}{1+15}+\tan^{-1}\frac{7}{9}$$

$$=\tan^{-1}\frac{1}{8}+\tan^{-1}\frac{7}{9}$$

$$=\tan^{-1}\frac{\frac{1}{8}+\frac{7}{9}}{1-\frac{7}{72}}=\tan^{-1}1$$

$$=n\pi+\frac{\pi}{4}.$$

NOTE. The value of n cannot be assigned until we have selected some particular values for the angles $\tan^{-1}5$, $\tan^{-1}3$, $\tan^{-1}\frac{7}{9}$. If we choose the *principal values*, then $n=0$.

Example 2. Prove that

$$\sin^{-1}\frac{4}{5}+\cos^{-1}\frac{12}{13}+\sin^{-1}\frac{16}{65}=\frac{\pi}{2}.$$

We may write this identity in the form

$$\sin^{-1}\frac{4}{5}+\cos^{-1}\frac{12}{13}=\frac{\pi}{2}-\sin^{-1}\frac{16}{65}=\cos^{-1}\frac{16}{65}.$$

Let $\alpha=\sin^{-1}\frac{4}{5}$, so that $\sin\alpha=\frac{4}{5}$;

and $\beta=\cos^{-1}\frac{12}{13}$, so that $\cos\beta=\frac{12}{13}$.

We have to express $\alpha+\beta$ as an inverse cosine.

Now $\cos(\alpha+\beta)=\cos\alpha\cos\beta-\sin\alpha\sin\beta$; whence by reading off the values of the functions from the figures in the margin, we have

$$\cos(\alpha+\beta)=\frac{3}{5}\cdot\frac{12}{13}-\frac{4}{5}\cdot\frac{5}{13}$$

$$=\frac{16}{65};$$

$$\therefore\ \alpha+\beta=\cos^{-1}\frac{16}{65}.$$

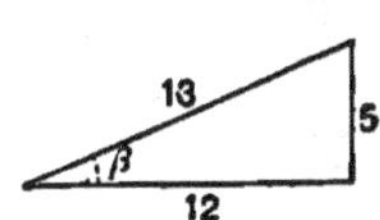

It is sometimes convenient to work entirely in terms of the tangent or cotangent.

Example 3. Prove that

$$2\cot^{-1}7+\cos^{-1}\frac{3}{5}=\operatorname{cosec}^{-1}\frac{125}{117}.$$

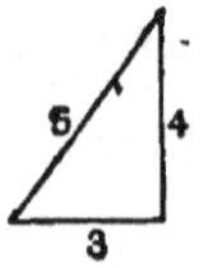

The first side $=\cot^{-1}\dfrac{7^2-1}{2\times 7}+\cot^{-1}\dfrac{3}{4}$

$$=\cot^{-1}\frac{24}{7}+\cot^{-1}\frac{3}{4}$$

$$=\cot^{-1}\frac{\frac{24}{7}\times\frac{3}{4}-1}{\frac{24}{7}+\frac{3}{4}}$$

$$=\cot^{-1}\frac{44}{117}=\operatorname{cosec}^{-1}\frac{125}{117}.$$

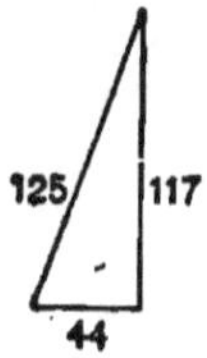

5 4 3

125 117 44

EXAMPLES. XIX. c.

Prove the following statements :

1. $\sin^{-1}\dfrac{12}{13}=\cot^{-1}\dfrac{5}{12}.$

2. $\operatorname{cosec}^{-1}\dfrac{17}{8}=\tan^{-1}\dfrac{8}{15}.$

3. $\sec(\tan^{-1}x)=\sqrt{1+x^2}.$

4. $2\tan^{-1}\dfrac{1}{3}=\tan^{-1}\dfrac{3}{4}.$

5. $\tan^{-1}\dfrac{4}{3}-\tan^{-1}1=\tan^{-1}\dfrac{1}{7}.$

6. $\tan^{-1}\dfrac{2}{11}+\cot^{-1}\dfrac{24}{7}=\tan^{-1}\dfrac{1}{2}.$

7. $\cot^{-1}\dfrac{4}{3}-\cot^{-1}\dfrac{15}{8}=\cot^{-1}\dfrac{84}{13}.$

8. $2\tan^{-1}\dfrac{1}{5}+\tan^{-1}\dfrac{1}{4}=\tan^{-1}\dfrac{32}{43}.$

9. $\tan^{-1}\dfrac{1}{2}+\tan^{-1}\dfrac{1}{3}=\tan^{-1}\dfrac{5}{6}+\tan^{-1}\dfrac{1}{11}.$

10. $\tan^{-1}\dfrac{1}{7}+\tan^{-1}\dfrac{1}{8}+\tan^{-1}\dfrac{1}{18}=\cot^{-1}3.$

11. $\tan^{-1}\frac{3}{5}+\sin^{-1}\frac{3}{5}=\tan^{-1}\frac{27}{11}$.

12. $2\cot^{-1}\frac{5}{4}=\tan^{-1}\frac{40}{9}$. **13.** $2\tan^{-1}\frac{8}{15}=\sin^{-1}\frac{240}{289}$.

14. $\sin(2\sin^{-1}x)=2x\sqrt{1-x^2}$.

15. $\cos^{-1}x=2\sin^{-1}\sqrt{\frac{1-x}{2}}$.

16. $2\tan^{-1}\sqrt{\frac{x}{a}}=\cos^{-1}\frac{a-x}{a+x}$.

17. $2\tan^{-1}\frac{1}{8}+\tan^{-1}\frac{1}{7}+2\tan^{-1}\frac{1}{5}=\frac{\pi}{4}$.

18. $\sin^{-1}a-\cos^{-1}b=\cos^{-1}\{b\sqrt{1-a^2}+a\sqrt{1-b^2}\}$.

19. $\sin^{-1}\frac{4}{5}+\cos^{-1}\frac{2}{\sqrt{5}}=\cot^{-1}\frac{2}{11}$.

20. $\cos^{-1}\frac{63}{65}+2\tan^{-1}\frac{1}{5}=\sin^{-1}\frac{3}{5}$.

21. $\tan^{-1}m+\tan^{-1}n=\cos^{-1}\frac{1-mn}{\sqrt{(1+m^2)(1+n^2)}}$.

22. $\cos^{-1}\frac{20}{29}-\tan^{-1}\frac{16}{63}=\cos^{-1}\frac{1596}{1885}$.

23. $\cos^{-1}\sqrt{\frac{2}{3}}-\cos^{-1}\frac{\sqrt{6}+1}{2\sqrt{3}}=\frac{\pi}{6}$.

24. $\tan(2\tan^{-1}x)=2\tan(\tan^{-1}x+\tan^{-1}x^3)$.

25. $\tan^{-1}a=\tan^{-1}\frac{a-b}{1+ab}+\tan^{-1}\frac{b-c}{1+bc}+\tan^{-1}c$.

26. If $\tan^{-1}x+\tan^{-1}y+\tan^{-1}z=\pi$, prove that

$$x+y+z=xyz.$$

27. If $u=\cot^{-1}\sqrt{\cos a}-\tan^{-1}\sqrt{\cos a}$, prove that

$$\sin u=\tan^2\frac{a}{2}.$$

250. We shall now shew how to solve equations expressed in the inverse notation.

Example 1. Solve $\tan^{-1}2x+\tan^{-1}3x=n\pi+\frac{3\pi}{4}$.

We have
$$\tan^{-1}\frac{2x+3x}{1-6x^2}=n\pi+\frac{3\pi}{4};$$
$$\therefore \frac{2x+3x}{1-6x^2}=\tan\left(n\pi+\frac{3\pi}{4}\right)=-1;$$
$$\therefore 6x^2-5x-1=0, \text{ or } (6x+1)(x-1)=0;$$
$$\therefore x=1, \text{ or } -\frac{1}{6}.$$

Example 2. Solve $\sin^{-1}x+\sin^{-1}(1-x)=\cos^{-1}x$.

By transposition, $\sin^{-1}(1-x)=\cos^{-1}x-\sin^{-1}x$.

Let $\cos^{-1}x=\alpha$, and $\sin^{-1}x=\beta$; then
$$\sin^{-1}(1-x)=\alpha-\beta;$$
$$\therefore 1-x=\sin(\alpha-\beta)=\sin\alpha\cos\beta-\cos\alpha\sin\beta.$$

But $\cos\alpha=x$, and therefore $\sin\alpha=\sqrt{1-x^2}$;

also $\sin\beta=x$, and therefore $\cos\beta=\sqrt{1-x^2}$;
$$\therefore 1-x=(1-x^2)-x^2=1-2x^2;$$
$$\therefore 2x^2-x=0;$$
whence
$$x=0, \text{ or } \frac{1}{2}.$$

EXAMPLES. XIX. d.

Solve the equations:

1. $\sin^{-1}x=\cos^{-1}x$. 2. $\tan^{-1}x=\cot^{-1}x$.

3. $\tan^{-1}(x+1)-\tan^{-1}(x-1)=\cot^{-1}2$.

4. $\cot^{-1}x+\cot^{-1}2x=\frac{3\pi}{4}$.

5. $\sin^{-1}x-\cos^{-1}x=\sin^{-1}(3x-2)$.

6. $\cos^{-1}x-\sin^{-1}x=\cos^{-1}x\sqrt{3}$.

7. $\tan^{-1}\frac{x-1}{x-2}+\tan^{-1}\frac{x+1}{x+2}=\frac{\pi}{4}$.

8. $2\cot^{-1}2+\cos^{-1}\frac{3}{5}=\operatorname{cosec}^{-1}x$.

9. $\tan^{-1}x+\tan^{-1}(1-x)=2\tan^{-1}\sqrt{x-x^2}$.

10. $\cos^{-1}\frac{1-a^2}{1+a^2}-\cos^{-1}\frac{1-b^2}{1+b^2}=2\tan^{-1}x$.

11. $\sin^{-1}\frac{2a}{1+a^2}+\tan^{-1}\frac{2x}{1-x^2}=\cos^{-1}\frac{1-b^2}{1+b^2}$.

12. $\cot^{-1}\frac{x^2-1}{2x}+\tan^{-1}\frac{2x}{x^2-1}+\frac{4\pi}{3}=0$.

13. Shew that we can express

$$\sin^{-1}\frac{2ab}{a^2+b^2}+\sin^{-1}\frac{2cd}{c^2+d^2} \text{ in the form } \sin^{-1}\frac{2xy}{x^2+y^2}$$

where x and y are rational functions of a, b, c, d.

14. If $\sin[2\cos^{-1}\{\cot(2\tan^{-1}x)\}]=0$, find x.

15. If $2\tan^{-1}(\cos\theta)=\tan^{-1}(2\operatorname{cosec}\theta)$, find θ.

16. If $\sin(\pi\cos\theta)=\cos(\pi\sin\theta)$, shew that

$$2\theta=\pm\sin^{-1}\frac{3}{4}.$$

17. If $\sin(\pi\cot\theta)=\cos(\pi\tan\theta)$, and n is any integer, shew that either $\cot 2\theta$ or $\operatorname{cosec} 2\theta$ is of the form $\frac{4n+1}{4}$.

18. If $\tan(\pi\cot\theta)=\cot(\pi\tan\theta)$, and n is any integer, shew that

$$\tan\theta=\frac{2n+1}{4}\pm\frac{\sqrt{4n^2+4n-15}}{4}.$$

19. Find all the positive integral solutions of

$$\tan^{-1}x+\cot^{-1}y=\tan^{-1}3.$$

MISCELLANEOUS EXAMPLES. G.

1. If the sines of the angles of a triangle are in the ratio of 4 : 5 : 6, shew that the cosines are in the ratio of 12 : 9 : 2.

2. Solve the equations:

(1) $2\cos^3\theta+\sin^2\theta-1=0$; (2) $\sec^3\theta-2\tan^2\theta=2$.

3. If $\tan\beta=2\sin\alpha\sin\gamma\,\mathrm{cosec}\,(\alpha+\gamma)$, prove that $\cot\alpha$, $\cot\beta$, $\cot\gamma$ are in arithmetical progression.

4. In a triangle shew that

$$4r(r_1+r_2+r_3)=2(bc+ca+ab)-(a^2+b^2+c^2).$$

5. Prove that

(1) $\tan^{-1}\dfrac{1}{3}-\tan^{-1}\dfrac{1}{5}+\tan^{-1}\dfrac{1}{7}=\tan^{-1}\dfrac{3}{11}$;

(2) $\sin^{-1}\dfrac{3}{5}+\sin^{-1}\dfrac{8}{17}+\sin^{-1}\dfrac{36}{85}=\dfrac{\pi}{2}$.

6. Find the greatest angle of the triangle whose sides are 185, 222, 259; given $\log 6=\cdot 7781513$,

$$L\cos 39^\circ\, 14'=9\cdot 8890644, \text{ diff. for } 1'=1032.$$

7. If $\tan(\alpha+\theta)=n\tan(\alpha-\theta)$, prove that $\dfrac{\sin 2\theta}{\sin 2\alpha}=\dfrac{n-1}{n+1}$.

8. If in a triangle $8R^2=a^2+b^2+c^2$, prove that one of the angles is a right angle.

9. The area of a regular polygon of n sides inscribed in a circle is three-fourths of the area of the circumscribed regular polygon with the same number of sides: find n.

10. $ABCD$ is a straight sea-wall. From B the straight lines drawn to two boats are each inclined at 45° to the direction of the wall, and from C the angles of inclination are 15° and 75°. If $BC=400$ yards, find the distance between the boats, and the distance of each from the sea-wall.

CHAPTER XX.

FUNCTIONS OF SUBMULTIPLE ANGLES.

251. Trigonometrical ratios of $22\frac{1}{2}°$ or $\frac{\pi}{8}$.

From the identity

$$2\sin^2 22\tfrac{1}{2}° = 1 - \cos 45°,$$

we have
$$4\sin^2 22\tfrac{1}{2}° = 2 - 2\cos 45° = 2 - \sqrt{2};$$

$$\therefore\ 2\sin 22\tfrac{1}{2}° = \sqrt{2 - \sqrt{2}} \quad \text{..................(1)}.$$

In like manner from

$$2\cos^2 22\tfrac{1}{2}° = 1 + \cos 45°,$$

we obtain
$$2\cos 22\tfrac{1}{2}° = \sqrt{2 + \sqrt{2}} \quad \text{..................(2)}.$$

In each of these cases the positive sign must be taken before the radical, since $22\frac{1}{2}°$ is an acute angle.

Again,
$$\tan 22\tfrac{1}{2}° = \frac{1 - \cos 45°}{\sin 45°} = \operatorname{cosec} 45° - \cot 45°;$$

$$\therefore\ \tan 22\tfrac{1}{2}° = \sqrt{2} - 1.$$

252. We have seen that $2\cos\frac{\pi}{8} = \sqrt{2 + \sqrt{2}}$;

but
$$4\cos^2\frac{\pi}{16} = 2 + 2\cos\frac{\pi}{8};$$

$$\therefore\ 4\cos^2\frac{\pi}{16} = 2 + \sqrt{2 + \sqrt{2}};$$

$$\therefore\ 2\cos\frac{\pi}{16} = \sqrt{2 + \sqrt{2 + \sqrt{2}}}.$$

Similarly,
$$2\cos\frac{\pi}{32} = \sqrt{2 + \sqrt{2 + \sqrt{2 + \sqrt{2}}}};$$

and so on.

253. Suppose that $\cos A = \frac{1}{2}$ and that it is required to find $\sin \frac{A}{2}$.

$$\sin \frac{A}{2} = \sqrt{\frac{1-\cos A}{2}} = \sqrt{\frac{1}{2}\left(1-\frac{1}{2}\right)} = \sqrt{\frac{1}{4}} = \pm\frac{1}{2}.$$

This case differs from those of the two previous articles in that the datum is less precise. All we know of the angle A is contained in the statement that its cosine is equal to $\frac{1}{2}$, and without some further knowledge respecting A we cannot remove the ambiguity of sign in the value found for $\sin \frac{A}{2}$.

We now proceed to a more general discussion.

254. *Given* $\cos A$ *to find* $\sin \frac{A}{2}$ *and* $\cos \frac{A}{2}$ *and to explain the presence of the two values in each case.*

From the identities

$$2 \sin^2 \frac{A}{2} = 1 - \cos A, \quad \text{and} \quad 2 \cos^2 \frac{A}{2} = 1 + \cos A,$$

we have

$$\sin \frac{A}{2} = \pm\sqrt{\frac{1-\cos A}{2}}, \quad \text{and} \quad \cos \frac{A}{2} = \pm\sqrt{\frac{1+\cos A}{2}}.$$

Thus corresponding to *one* value of $\cos A$, there are *two* values for $\sin \frac{A}{2}$, and *two* values for $\cos \frac{A}{2}$.

The presence of these two values may be explained as follows. If $\cos A$ is given and nothing further is stated about the angle A, all we know is that A belongs to a certain group of *equicosinal angles.* Let a be the smallest positive angle belonging to this group, then $A = 2n\pi \pm a$. Thus in finding $\sin \frac{A}{2}$ and $\cos \frac{A}{2}$ we are really finding the values of

$$\sin \frac{1}{2}(2n\pi \pm a) \quad \text{and} \quad \cos \frac{1}{2}(2n\pi \pm a).$$

Now $$\sin\frac{1}{2}(2n\pi \pm a) = \sin\left(n\pi \pm \frac{a}{2}\right)$$

$$= \sin n\pi \cos\frac{a}{2} \pm \cos n\pi \sin\frac{a}{2}$$

$$= \pm \sin\frac{a}{2},$$

for $\sin n\pi = 0$ and $\cos n\pi = \pm 1$.

Again, $$\cos\frac{1}{2}(2n\pi \pm a) = \cos n\pi \cos\frac{a}{2} \mp \sin n\pi \sin\frac{a}{2}$$

$$= \pm \cos\frac{a}{2}.$$

Thus there are two values for $\sin\frac{A}{2}$ and two values for $\cos\frac{A}{2}$ when $\cos A$ is given and nothing further is known respecting A.

255. Geometrical Illustration. Let a be the smallest positive angle which has the same cosine as A; then

$$A = 2n\pi \pm a,$$

and we have to find the sine and cosine of $\frac{A}{2}$, that is of

$$n\pi \pm \frac{a}{2}.$$

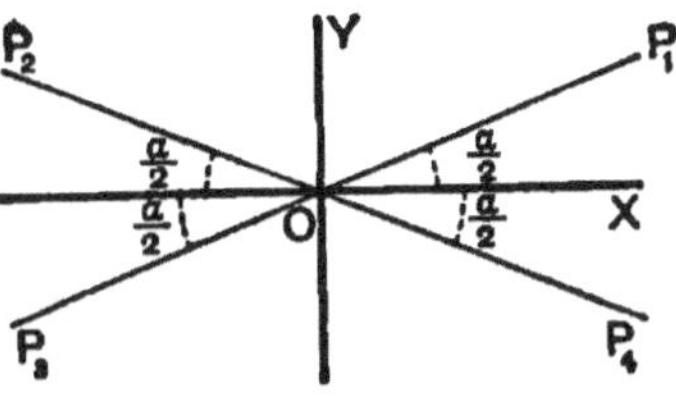

Each of the angles denoted by this formula is bounded by one of the lines OP_1, OP_2, OP_3, OP_4. Now

$$\sin XOP_2 = \sin\frac{a}{2}, \quad \sin XOP_3 = -\sin\frac{a}{2}, \quad \sin XOP_4 = -\sin\frac{a}{2},$$

$$\cos XOP_2 = -\cos\frac{a}{2}, \quad \cos XOP_3 = -\cos\frac{a}{2}, \quad \cos XOP_4 = \cos\frac{a}{2}.$$

Thus the values of $\sin\frac{A}{2}$ are $\pm\sin\frac{a}{2}$, and the values of $\cos\frac{A}{2}$ are $\pm\cos\frac{a}{2}$.

256. If $\cos A$ is given, and A lies between certain known limits, the ambiguities of sign in the formulæ of Art. 254 may be removed.

Example. If $\cos A = -\frac{7}{25}$, and A lies between 450° and 540°, find $\sin\frac{A}{2}$ and $\cos\frac{A}{2}$.

$$\sin\frac{A}{2} = \sqrt{\frac{1-\cos A}{2}} = \sqrt{\frac{1}{2}\left(1+\frac{7}{25}\right)} = \sqrt{\frac{16}{25}} = \pm\frac{4}{5};$$

$$\cos\frac{A}{2} = \sqrt{\frac{1+\cos A}{2}} = \sqrt{\frac{1}{2}\left(1-\frac{7}{25}\right)} = \sqrt{\frac{9}{25}} = \pm\frac{3}{5}.$$

Now $\frac{A}{2}$ lies between 225° and 270°, so that $\sin\frac{A}{2}$ and $\cos\frac{A}{2}$ are both negative;

$$\therefore \sin\frac{A}{2} = -\frac{4}{5}, \text{ and } \cos\frac{A}{2} = -\frac{3}{5}.$$

257. *To find* $\sin\frac{A}{2}$ *and* $\cos\frac{A}{2}$ *in terms of* $\sin A$ *and to explain the presence of four values in each case.*

We have $$\sin^2\frac{A}{2} + \cos^2\frac{A}{2} = 1,$$

and $$2\sin\frac{A}{2}\cos\frac{A}{2} = \sin A.$$

By addition, $$\left(\sin\frac{A}{2} + \cos\frac{A}{2}\right)^2 = 1 + \sin A;$$

by subtraction, $$\left(\sin\frac{A}{2} - \cos\frac{A}{2}\right)^2 = 1 - \sin A.$$

$$\therefore \sin\frac{A}{2} + \cos\frac{A}{2} = \pm\sqrt{1+\sin A} \quad \text{............(1)},$$

and $$\sin\frac{A}{2} - \cos\frac{A}{2} = \pm\sqrt{1-\sin A} \quad \text{............(2)}.$$

By addition and subtraction, we obtain $\sin\frac{A}{2}$ and $\cos\frac{A}{2}$; and since there is a double sign before each radical, there are *four* values for $\sin\frac{A}{2}$, and *four* values for $\cos\frac{A}{2}$ corresponding to *one* value of $\sin A$.

The presence of these four values may be explained as follows.

If $\sin A$ is given and nothing else is stated about the angle A all we know is that A belongs to a certain group of *equi-sinal angles.* Let a be the smallest positive angle belonging to this group, then $A = n\pi + (-1)^n a$. Thus in finding $\sin\frac{A}{2}$ and $\cos\frac{A}{2}$ we are really finding

$$\sin\frac{1}{2}\{n\pi + (-1)^n a\}, \text{ and } \cos\frac{1}{2}\{n\pi + (-1)^n a\}.$$

First suppose n even and equal to $2m$; then

$$\sin\frac{1}{2}\{n\pi + (-1)^n a\} = \sin\left(m\pi + \frac{a}{2}\right)$$

$$= \sin m\pi \cos\frac{a}{2} + \cos m\pi \sin\frac{a}{2}$$

$$= \pm\sin\frac{a}{2},$$

since $\sin m\pi = 0$, and $\cos m\pi = \pm 1$.

Next suppose n odd and equal to $2m+1$; then

$$\sin\frac{1}{2}\{n\pi + (-1)^n a\} = \sin\left(m\pi + \frac{\pi}{2} - \frac{a}{2}\right)$$

$$= \sin m\pi \cos\left(\frac{\pi}{2} - \frac{a}{2}\right) + \cos m\pi \sin\left(\frac{\pi}{2} - \frac{a}{2}\right)$$

$$= \pm\sin\left(\frac{\pi}{2} - \frac{a}{2}\right).$$

Thus we have *four* values for $\sin\frac{A}{2}$ when $\sin A$ is given and nothing further is known respecting A.

In like manner it may be shewn that

$\cos\frac{A}{2}$ has the *four* values $\pm\cos\frac{a}{2}$, $\quad \pm\cos\left(\frac{\pi}{2} - \frac{a}{2}\right)$.

258. Geometrical Illustration. Let α be the smallest positive angle which has the same sine as A; then

$$A = n\pi + (-1)^n \alpha,$$

and we have to find the sine and cosine of $\frac{A}{2}$, that is of

$$\frac{1}{2}\{n\pi + (-1)^n \alpha\}.$$

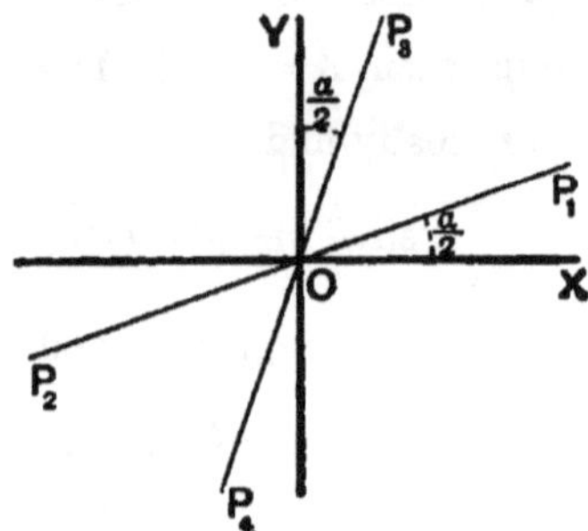

If n is even and equal to $2m$, this expression becomes $m\pi + \frac{\alpha}{2}$.

If n is odd and equal to $2m+1$, the expression becomes

$$m\pi + \left(\frac{\pi}{2} - \frac{\alpha}{2}\right).$$

The angles denoted by the formula $m\pi + \frac{\alpha}{2}$ are bounded by one of the lines OP_1 or OP_2; and those denoted by the formula $m\pi + \left(\frac{\pi}{2} - \frac{\alpha}{2}\right)$ are bounded by one of the lines OP_3 or OP_4.

Now $\quad \sin XOP_1 = \sin \frac{\alpha}{2}$;

$$\sin XOP_2 = -\sin XOP_1 = -\sin \frac{\alpha}{2};$$

$$\sin XOP_3 = \sin \left(\frac{\pi}{2} - \frac{\alpha}{2}\right);$$

$$\sin XOP_4 = -\sin XOP_3 = -\sin \left(\frac{\pi}{2} - \frac{\alpha}{2}\right).$$

Thus the values of $\sin \frac{A}{2}$ are

$$\pm \sin \frac{\alpha}{2} \quad \text{and} \quad \pm \sin \left(\frac{\pi}{2} - \frac{\alpha}{2}\right).$$

Similarly the values of $\cos \frac{A}{2}$ are $\pm \cos \frac{\alpha}{2}$ and $\pm \cos \left(\frac{\pi}{2} - \frac{\alpha}{2}\right)$.

259. If in addition to the value of $\sin A$ we know that A lies between certain limits, the ambiguities of sign in equations (1) and (2) of Art. 257 may be removed.

Example 1. Find $\sin\frac{A}{2}$ and $\cos\frac{A}{2}$ in terms of $\sin A$ when A lies between 450° and 630°.

In this case $\frac{A}{2}$ lies between 225° and 315°. From the adjoining figure it is evident that between these limits $\sin\frac{A}{2}$ is greater than $\cos\frac{A}{2}$ and is negative.

$$\therefore\ \sin\frac{A}{2}+\cos\frac{A}{2}=-\sqrt{1+\sin A},$$

and
$$\sin\frac{A}{2}-\cos\frac{A}{2}=-\sqrt{1-\sin A}.$$

$$\therefore\ 2\sin\frac{A}{2}=-\sqrt{1+\sin A}-\sqrt{1-\sin A},$$

and
$$2\cos\frac{A}{2}=-\sqrt{1+\sin A}+\sqrt{1-\sin A}.$$

Example 2. Determine the limits between which A must lie in order that

$$2\cos A=-\sqrt{1+\sin 2A}-\sqrt{1-\sin 2A}.$$

The given relation is obtained by combining

$$\sin A+\cos A=-\sqrt{1+\sin 2A} \quad\text{................(1)},$$

and
$$\sin A-\cos A=+\sqrt{1-\sin 2A} \quad\text{.................(2)}.$$

From (1), we see that of $\sin A$ and $\cos A$ *the numerically greater is negative.*

From (2), we see that the cosine is the greater.

Hence we have to choose limits between which $\cos A$ is numerically greater than $\sin A$ and is negative. From the figure we see that A lies between $2n\pi+\frac{3\pi}{4}$ and $2n\pi+\frac{5\pi}{4}$.

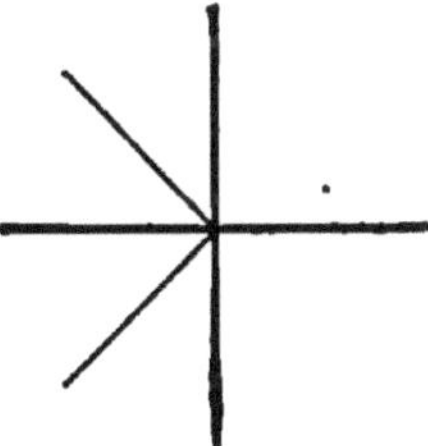

Example 3. Trace the changes of $\cos\theta - \sin\theta$ in sign and magnitude as θ increases from 0 to 2π.

$$\cos\theta - \sin\theta = \sqrt{2}\left(\frac{1}{\sqrt{2}}\cos\theta - \frac{1}{\sqrt{2}}\sin\theta\right)$$

$$= \sqrt{2}\left(\cos\theta\cos\frac{\pi}{4} - \sin\theta\sin\frac{\pi}{4}\right)$$

$$= \sqrt{2}\cos\left(\theta + \frac{\pi}{4}\right).$$

As θ increases from 0 to $\frac{\pi}{4}$, the expression is positive and decreases from 1 to 0.

As θ increases from $\frac{\pi}{4}$ to $\frac{3\pi}{4}$, the expression is negative and increases numerically from 0 to $-\sqrt{2}$.

As θ increases from $\frac{3\pi}{4}$ to $\frac{5\pi}{4}$, the expression is negative and decreases numerically from $-\sqrt{2}$ to 0.

As θ increases from $\frac{5\pi}{4}$ to $\frac{7\pi}{4}$, the expression is positive and increases from 0 to $\sqrt{2}$.

As θ increases from $\frac{7\pi}{4}$ to 2π, the expression is positive and decreases from $\sqrt{2}$ to 1.

260. *To find the sine and cosine of* 9°.

Since $\cos 9° > \sin 9°$ and is positive, we have

$$\sin 9° + \cos 9° = +\sqrt{1 + \sin 18°},$$

and

$$\sin 9° - \cos 9° = -\sqrt{1 - \sin 18°}.$$

$$\therefore\ \sin 9° + \cos 9° = +\sqrt{1 + \frac{\sqrt{5}-1}{4}} = +\frac{1}{2}\sqrt{3+\sqrt{5}},$$

and

$$\sin 9° - \cos 9° = -\sqrt{1 - \frac{\sqrt{5}-1}{4}} = -\frac{1}{2}\sqrt{5-\sqrt{5}}.$$

$$\therefore\ \sin 9° = \frac{1}{4}\left\{\sqrt{3+\sqrt{5}} - \sqrt{5-\sqrt{5}}\right\},$$

and

$$\cos 9° = \frac{1}{4}\left\{\sqrt{3+\sqrt{5}} + \sqrt{5+\sqrt{5}}\right\}.$$

EXAMPLES. XX. a.

1. When A lies between $-270°$ and $-360°$, prove that

$$\sin\frac{A}{2} = -\sqrt{\frac{1-\cos A}{2}}.$$

2. If $\cos A = \frac{119}{169}$, find $\sin\frac{A}{2}$ and $\cos\frac{A}{2}$ when A lies between $270°$ and $360°$.

3. If $\cos A = -\frac{161}{289}$, find $\sin\frac{A}{2}$ and $\cos\frac{A}{2}$ when A lies between $540°$ and $630°$.

4. Find $\sin\frac{A}{2}$ and $\cos\frac{A}{2}$ in terms of $\sin A$ when A lies between $270°$ and $450°$.

5. Find $\sin\frac{A}{2}$ and $\cos\frac{A}{2}$ in terms of $\sin A$ when $\frac{A}{2}$ lies between $225°$ and $315°$.

6. Find $\sin\frac{A}{2}$ and $\cos\frac{A}{2}$ in terms of $\sin A$ when A lies between $-450°$ and $-630°$.

7. If $\sin A = \frac{24}{25}$, find $\sin\frac{A}{2}$ and $\cos\frac{A}{2}$ when A lies between $90°$ and $180°$.

8. If $\sin A = -\frac{240}{289}$, find $\sin\frac{A}{2}$ and $\cos\frac{A}{2}$ when A lies between $270°$ and $360°$.

9. Determine the limits between which A must lie in order that

(1) $2\sin A = \sqrt{1+\sin 2A} - \sqrt{1-\sin 2A}$;

(2) $2\cos A = -\sqrt{1+\sin 2A} + \sqrt{1-\sin 2A}$;

(3) $2\sin A = -\sqrt{1+\sin 2A} + \sqrt{1-\sin 2A}$.

10. If $A=240°$, is the following statement correct?

$$2\sin\frac{A}{2}=\sqrt{1+\sin A}-\sqrt{1-\sin A}.$$

If not, how must it be modified?

11. Prove that

(1) $\tan 7\frac{1}{2}°=\sqrt{6}-\sqrt{3}+\sqrt{2}-2$;

(2) $\cot 142\frac{1}{2}°=\sqrt{2}+\sqrt{3}-2-\sqrt{6}$.

12. Shew that $\sin 9°$ lies between ·156 and ·157.

13. Prove that

(1) $2\sin 11°\,15'=\sqrt{2-\sqrt{2+\sqrt{2}}}$;

(2) $\tan 11°\,15'=\sqrt{4+2\sqrt{2}}-(\sqrt{2}+1)$.

14. When θ varies from 0 to 2π trace the changes in sign and magnitude of

(1) $\cos\theta+\sin\theta$; (2) $\sin\theta-\sqrt{3}\cos\theta$.

15. When θ varies from 0 to π, trace the changes in sign and magnitude of

(1) $\dfrac{\tan\theta+\cot\theta}{\tan\theta-\cot\theta}$; (2) $\dfrac{2\sin\theta-\sin 2\theta}{2\sin\theta+\sin 2\theta}$.

261. *To find* $\tan\frac{A}{2}$ *when* $\tan A$ *is given and to explain the presence of the two values.*

Denote $\tan A$ by t; then

$$t=\tan A=\frac{2\tan\frac{A}{2}}{1-\tan^2\frac{A}{2}};$$

$$\therefore\ t\tan^2\frac{A}{2}+2\tan\frac{A}{2}-t=0;$$

$$\therefore\ \tan\frac{A}{2}=\frac{-2\pm\sqrt{4+4t^2}}{2t}=\frac{-1\pm\sqrt{1+t^2}}{t}.$$

The presence of these two values may be explained as follows.

If α be the smallest positive angle which has the given tangent, then $A = n\pi + \alpha$, and we are really finding the value of

$$\tan \frac{1}{2}(n\pi + \alpha).$$

(1) Let n be even and equal to $2m$; then

$$\tan \frac{1}{2}(n\pi + \alpha) = \tan\left(m\pi + \frac{\alpha}{2}\right) = \tan \frac{\alpha}{2}.$$

(2) Let n be odd and equal to $2m+1$; then

$$\tan \frac{1}{2}(n\pi + \alpha) = \tan\left(m\pi + \frac{\pi}{2} + \frac{\alpha}{2}\right) = \tan\left(\frac{\pi}{2} + \frac{\alpha}{2}\right).$$

Thus $\tan \frac{A}{2}$ has the two values $\tan \frac{\alpha}{2}$ and $\tan\left(\frac{\pi}{2} + \frac{\alpha}{2}\right)$.

Example 1. If $A = 170°$, prove that $\tan \frac{A}{2} = \frac{-1 - \sqrt{1 + \tan^2 A}}{\tan A}$.

Here $\frac{A}{2}$ is an acute angle, so that $\tan \frac{A}{2}$ must be positive. Hence in the formula $\frac{-1 \pm \sqrt{1 + \tan^2 A}}{\tan A}$ the numerator must have the same sign as the denominator. But when $A = 170°$, $\tan A$ is negative, and therefore we must choose the sign which will make the numerator negative; thus $\tan \frac{A}{2} = \frac{-1 - \sqrt{1 + \tan^2 A}}{\tan A}$.

Example 2. Given $\cos A = \cdot 6$, find $\tan \frac{A}{2}$, and explain the double answer.

$$\tan^2 \frac{A}{2} = \frac{1 - \cos A}{1 + \cos A} = \frac{\cdot 4}{1 \cdot 6} = \frac{1}{4};$$

$$\therefore \tan \frac{A}{2} = \pm \frac{1}{2}.$$

Here all we know of the angle A is that it must be one of a group of equi-cosinal angles. Let α be the smallest positive angle of this group; then $A = 2n\pi \pm \alpha$.

$$\therefore \tan \frac{A}{2} = \tan\left(n\pi \pm \frac{\alpha}{2}\right) = \tan\left(\pm \frac{\alpha}{2}\right) = \pm \tan \frac{\alpha}{2}.$$

Thus we have two values differing only in sign.

262. When any one of the functions of an acute angle A is given, we may in some cases conveniently obtain the functions of $\frac{A}{2}$, as in the following example.

Example. Given $\cos A = \frac{b}{a}$, to find the functions of $\frac{A}{2}$.

Make a right-angled triangle PQR in which the hypotenuse $PQ = a$, and base $QR = b$; then

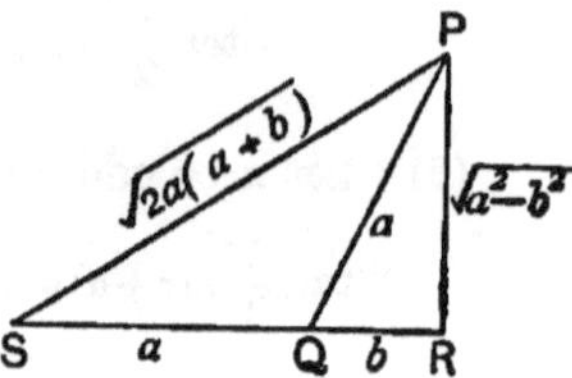

$$\cos PQR = \frac{QR}{PQ} = \frac{b}{a} = \cos A;$$

$$\therefore \angle PQR = A.$$

Produce RQ to S making $QS = QP$;

$$\therefore \angle PSQ = \angle SPQ = \frac{1}{2} \angle PQR = \frac{A}{2}.$$

Now $SR = a + b$, and $PR = \sqrt{a^2 - b^2}$,

$$\therefore PS^2 = (a+b)^2 + (a^2 - b^2) = 2a^2 + 2ab;$$

$$\therefore PS = \sqrt{2a(a+b)}.$$

The functions of $\frac{A}{2}$ may now be written down in terms of the sides of the triangle PRS.

263. From Art. 125, we have

$$\cos A = 4\cos^3 \frac{A}{3} - 3\cos\frac{A}{3}.$$

Thus it appears that if $\cos A$ be given we have a *cubic* equation to find $\cos\frac{A}{3}$; so that $\cos\frac{A}{3}$ has *three* values.

Similarly, from the equation

$$\sin A = 3\sin\frac{A}{3} - 4\sin^3\frac{A}{3}$$

it appears that corresponding to *one* value of $\sin A$ there are *three* values of $\sin\frac{A}{3}$.

It will be a useful exercise to prove these two statements analytically as in Arts. 254 and 257. In the next article we shall give a geometrical explanation for the case of the cosine.

264. *Given* $\cos A$ *to find* $\cos\frac{A}{3}$, *and to explain the presence of the three values.*

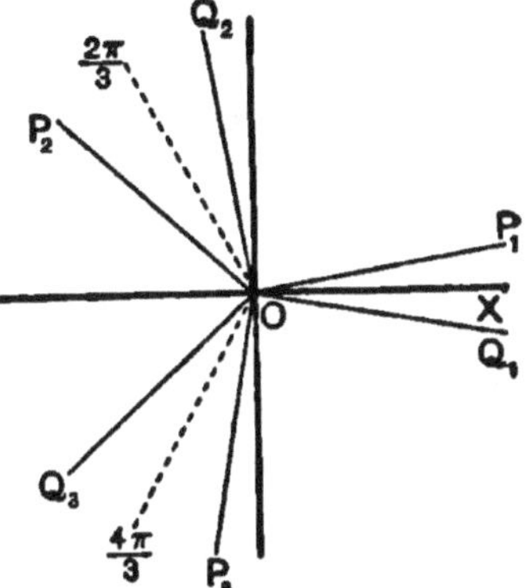

Let a be the smallest positive angle with the given cosine; then $A = 2n\pi \pm a$, and we have to find all the values of

$$\cos\frac{1}{3}(2n\pi \pm a).$$

Consider the angles denoted by the formula

$$\frac{1}{3}(2n\pi \pm a),$$

and ascribe to n in succession the values 0, 1, 2, 3, ...

When $n=0$, the angles are $\pm\frac{a}{3}$, bounded by OP_1 and OQ_1;

when $n=1$, the angles are $\frac{2\pi}{3} \pm \frac{a}{3}$, bounded by OP_2 and OQ_2

when $n=2$, the angles are $\frac{4\pi}{3} \pm \frac{a}{3}$, bounded by OP_3 and OQ_3.

By giving to n the values 3, 4, 5, ... we obtain a series of angles coterminal with those indicated in the figure.

Thus OP_1, OQ_1, OP_2, OQ_2, OP_3, OQ_3 bound all the angles included in the formula $\frac{1}{3}(2n\pi \pm a)$.

Now $\qquad \cos XOQ_1 = \cos XOP_1 = \cos\frac{a}{3}$;

$$\cos XOP_3 = \cos XOQ_2 = \cos\left(\frac{2\pi}{3} - \frac{a}{3}\right);$$

$$\cos XOQ_3 = \cos XOP_2 = \cos\left(\frac{2\pi}{3} + \frac{a}{3}\right).$$

Thus the values of $\cos\frac{A}{3}$ are $\cos\frac{a}{3}$, $\cos\frac{2\pi + a}{3}$, $\cos\frac{2\pi - a}{3}$.

EXAMPLES. XX. b.

1. If $A=320°$, prove that

$$\tan\frac{A}{2}=\frac{-1+\sqrt{1+\tan^2 A}}{\tan A}.$$

2. Shew that

$$\tan A=-\frac{1+\sqrt{1+\tan^2 2A}}{\tan 2A} \text{ when } A=110°.$$

3. Find $\tan A$ when $\cos 2A=\frac{12}{13}$ and A lies between 180° and 225°.

4. Find $\cot\frac{A}{2}$ when $\cos A=-\frac{4}{5}$ and A lies between 180° and 270°.

5. If $\cot 2\theta=\cot 2\alpha$, shew that $\cot\theta$ has the two values $\cot\alpha$ and $-\tan\alpha$.

6. Given that $\sin\theta=\sin\alpha$, shew that the values of $\sin\frac{\theta}{3}$ are

$$\sin\frac{\alpha}{3},\quad \sin\frac{\pi-\alpha}{3},\quad -\sin\frac{\pi+\alpha}{3}.$$

7. If $\tan\theta=\tan\alpha$, shew that the values of $\tan\frac{\theta}{3}$ are

$$\tan\frac{\alpha}{3},\quad \tan\frac{\pi+\alpha}{3},\quad -\tan\frac{\pi-\alpha}{3}.$$

8. Given that $\cos 3\theta=\cos 3\alpha$, shew that the values of $\sin\theta$ are

$$\pm\sin\alpha,\quad -\sin\left(\frac{\pi}{3}\pm\alpha\right),\quad \sin\left(\frac{2\pi}{3}\pm\alpha\right).$$

9. Given that $\sin 3\theta=\sin 3\alpha$, shew that the values of $\cos\theta$ are

$$\pm\cos\alpha,\quad \cos\left(\frac{\pi}{3}\pm\alpha\right),\quad \cos\left(\frac{2\pi}{3}\pm\alpha\right).$$

CHAPTER XXI.

LIMITS AND APPROXIMATIONS.

265. *If θ be the radian measure of an angle less than a right angle, to shew that* $\sin\theta$, θ, $\tan\theta$ *are in ascending order of magnitude.*

Let the angle θ be represented by AOP.

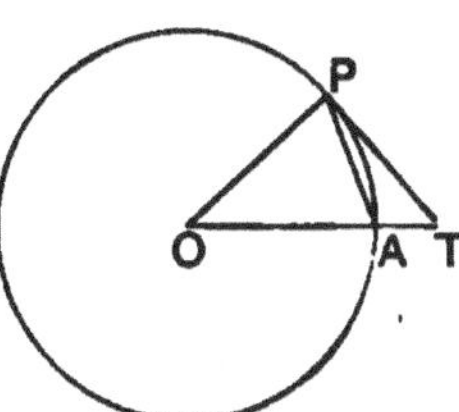

With centre O and radius OA describe a circle. Draw PT at right angles to OP to meet OA produced in T, and join PA.

Let r be the radius of the circle.

Area of $\Delta AOP = \frac{1}{2} AO \, . \, OP \sin AOP = \frac{1}{2} r^2 \sin\theta$;

area of sector $AOP = \frac{1}{2} r^2\theta$;

area of $\Delta OPT = \frac{1}{2} OP \, . \, PT = \frac{1}{2} r \, . \, r\tan\theta = \frac{1}{2} r^2 \tan\theta$.

But the areas of the triangle AOP, the sector AOP, and the triangle OTP are in ascending order of magnitude; that is,

$$\frac{1}{2} r^2 \sin\theta, \quad \frac{1}{2} r^2\theta, \quad \frac{1}{2} r^2 \tan\theta$$

are in ascending order of magnitude;

$\therefore$ $\sin\theta$, θ, $\tan\theta$ are in ascending order of magnitude.

266. *When θ is indefinitely diminished, to prove that $\frac{\sin\theta}{\theta}$ and $\frac{\tan\theta}{\theta}$ each have unity for their limit.*

In the last article, we have proved that $\sin\theta$, θ, $\tan\theta$ are in ascending order of magnitude. Divide each of these quantities by $\sin\theta$; then

$$1, \frac{\theta}{\sin\theta}, \frac{1}{\cos\theta} \text{ are in ascending order of magnitude;}$$

that is, $$\frac{\theta}{\sin\theta} \text{ lies between 1 and } \sec\theta.$$

But when θ is indefinitely diminished, the limit of $\sec\theta$ is 1; hence the limit of $\frac{\theta}{\sin\theta}$ is 1; that is, the limit of $\frac{\sin\theta}{\theta}$ is unity.

Again, by dividing each of the quantities $\sin\theta$, θ, $\tan\theta$ by $\tan\theta$, we find that $\cos\theta$, $\frac{\theta}{\tan\theta}$, 1 are in ascending order of magnitude. Hence the limit of $\frac{\tan\theta}{\theta}$ is unity.

These results are often written concisely in the forms

$$\underset{\theta=0}{Lt.}\left(\frac{\sin\theta}{\theta}\right)=1, \qquad \underset{\theta=0}{Lt.}\left(\frac{\tan\theta}{\theta}\right)=1.$$

Example. Find the limit of $n\sin\frac{\theta}{n}$ when $n=\infty$.

$$n\sin\frac{\theta}{n}=\theta\,.\,\frac{n}{\theta}\,.\,\sin\frac{\theta}{n}=\theta\left(\sin\frac{\theta}{n}\div\frac{\theta}{n}\right);$$

but since $\frac{\theta}{n}$ is indefinitely small, the limit of $\sin\frac{\theta}{n}\div\frac{\theta}{n}$ is unity;

$$\therefore \underset{n=\infty}{Lt.}\left(n\sin\frac{\theta}{n}\right)=\theta.$$

Similarly $$\underset{n=\infty}{Lt.}\left(n\tan\frac{\theta}{n}\right)=\theta.$$

267. It is important to remember that the conclusions of the foregoing articles only hold when the angle is expressed in radian measure. If any other system of measurement is used, the results will require modification.

Example. Find the value of $\underset{n=0}{Lt.}\left(\frac{\sin n^\circ}{n}\right)$.

Let θ be the number of radians in n°; then

$$\frac{n}{180}=\frac{\theta}{\pi}, \text{ and } n=\frac{180\theta}{\pi}; \text{ also } \sin n^\circ=\sin\theta;$$

$$\therefore \frac{\sin n^\circ}{n}=\frac{\pi\sin\theta}{180\theta}=\frac{\pi}{180}\cdot\frac{\sin\theta}{\theta}.$$

When n is indefinitely small, θ is indefinitely small;

$$\therefore \underset{n=0}{Lt.}\left(\frac{\sin n^\circ}{n}\right)=\frac{\pi}{180}\cdot\underset{\theta=0}{Lt.}\left(\frac{\sin\theta}{\theta}\right);$$

$$\therefore \underset{n=0}{Lt.}\left(\frac{\sin n^\circ}{n}\right)=\frac{\pi}{180}.$$

268. When θ is the radian measure of a very small angle, we have shewn that

$$\frac{\sin\theta}{\theta}=1, \quad \cos\theta=1, \quad \frac{\tan\theta}{\theta}=1;$$

that is, $\quad \sin\theta=\theta, \quad \cos\theta=1, \quad \tan\theta=\theta.$

Hence $r\tan\theta=r\theta$, and therefore in the figure of Art. 265, the tangent PT is equal to the arc PA, when $\angle AOP$ is very small.

In Art. 270, it will be shewn that these results hold so long as θ is so small that its square may be neglected. When this is the case, we have

$$\sin(a+\theta)=\sin a\cos\theta+\cos a\sin\theta$$
$$=\sin a+\theta\cos a;$$
$$\cos(a+\theta)=\cos a\cos\theta-\sin a\sin\theta$$
$$=\cos a-\theta\sin a.$$

Example 1. The inclination of a railway to the horizontal plane is 52′ 30″, find how many feet it rises in a mile.

Let OA be the horizontal plane, and OP a mile of the railway. Draw PN perpendicular to OA.

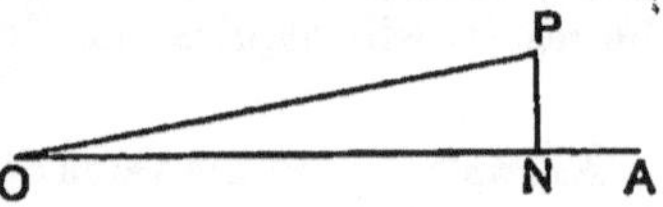

Let $PN = x$ feet, $\angle PON = \theta$;

then $$\frac{PN}{OP} = \sin\theta = \theta \text{ approximately.}$$

But θ = radian measure of $52'\,30'' = \dfrac{52\frac{1}{2}}{60} \times \dfrac{\pi}{180} = \dfrac{7}{8} \times \dfrac{\pi}{180}$;

$$\therefore \frac{x}{1760 \times 3} = \frac{7}{8} \times \frac{22}{7} \times \frac{1}{180};$$

$$\therefore x = \frac{1760 \times 3 \times 22}{8 \times 180} = \frac{242}{3} = 80\tfrac{2}{3}.$$

Thus the rise is $80\frac{2}{3}$ feet.

Example 2. A pole 6 ft. long stands on the top of a tower 54 ft. high: find the angle subtended by the pole at a point on the ground which is at a distance of 180 yds. from the foot of the tower.

Let A be the point on the ground, BC the tower, CD the pole.

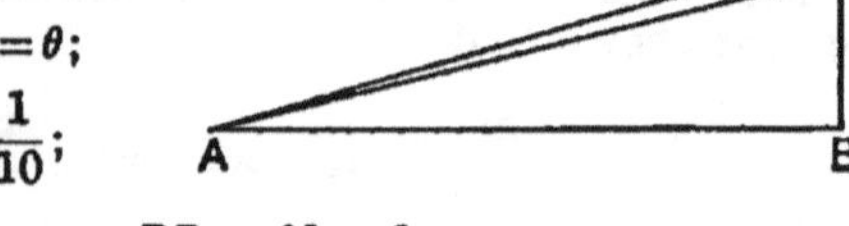

Let $\angle BAC = \alpha$, $\angle CAD = \theta$;

then $\tan\alpha = \dfrac{BC}{AB} = \dfrac{54}{540} = \dfrac{1}{10}$;

$$\tan(\alpha+\theta) = \frac{BD}{AB} = \frac{60}{540} = \frac{1}{9}.$$

But $$\tan(\alpha+\theta) = \frac{\tan\alpha + \tan\theta}{1 - \tan\alpha\tan\theta} = \frac{\tan\alpha + \theta}{1 - \theta\tan\alpha} \text{ approximately;}$$

$$\therefore \frac{1}{9} = \frac{\frac{1}{10} + \theta}{1 - \frac{\theta}{10}} = \frac{1 + 10\theta}{10 - \theta};$$

whence $\theta = \dfrac{1}{91}$; that is, the angle is $\dfrac{1}{91}$ of a radian, and therefore contains $\dfrac{1}{91} \times \dfrac{180}{\pi}$ degrees.

On reduction, we find that the angle is 37′ 46″ nearly.

269. *If θ be the number of radians in an acute angle, to prove that*

$$\cos\theta > 1 - \frac{\theta^2}{2}, \text{ and } \sin\theta > \theta - \frac{\theta^3}{4}.$$

Since $\cos\theta = 1 - 2\sin^2\frac{\theta}{2}$, and $\sin\frac{\theta}{2} < \frac{\theta}{2}$;

$$\therefore \cos\theta > 1 - 2\left(\frac{\theta}{2}\right)^2;$$

that is,

$$\cos\theta > 1 - \frac{\theta^2}{2}.$$

Again, $\sin\theta = 2\sin\frac{\theta}{2}\cos\frac{\theta}{2} = 2\tan\frac{\theta}{2}\cos^2\frac{\theta}{2}$;

but

$$\tan\frac{\theta}{2} > \frac{\theta}{2};$$

$$\therefore \sin\theta > 2\frac{\theta}{2}\cos^2\frac{\theta}{2};$$

$$\therefore \sin\theta > \theta\left(1 - \sin^2\frac{\theta}{2}\right).$$

But $\sin\frac{\theta}{2} < \frac{\theta}{2}$, and therefore

$$1 - \sin^2\frac{\theta}{2} > 1 - \left(\frac{\theta}{2}\right)^2;$$

$$\therefore \sin\theta > \theta\left\{1 - \left(\frac{\theta}{2}\right)^2\right\};$$

$$\therefore \sin\theta > \theta - \frac{\theta^3}{4}.$$

270. From the propositions established in this chapter, it follows that if θ is an acute angle,

$$\cos\theta \text{ lies between } 1 \text{ and } 1 - \frac{\theta^2}{2},$$

and

$$\sin\theta \text{ lies between } \theta \text{ and } \theta - \frac{\theta^3}{4}.$$

Thus $\cos\theta = 1 - k\theta^2$ and $\sin\theta = \theta - k'\theta^3$, where k and k' are proper fractions less than $\frac{1}{2}$ and $\frac{1}{4}$ respectively.

Hence if θ be so small that its square can be neglected,

$$\cos\theta=1, \quad \sin\theta=\theta.$$

Example. Find the approximate value of sin 10″.

The circular measure of 10″ is $\dfrac{10\pi}{180\times60\times60}$ or $\dfrac{\pi}{64800}$;

$$\therefore \sin 10'' < \frac{\pi}{64800} \text{ and } > \frac{\pi}{64800} - \frac{1}{4}\left(\frac{\pi}{64800}\right)^3.$$

But
$$\frac{\pi}{64800} = \frac{3{\cdot}1415926535...}{64800} = {\cdot}000048481368......;$$

$$\therefore \frac{\pi}{64800} < {\cdot}00005 \text{ and } \left(\frac{\pi}{64800}\right)^3 < {\cdot}000000000000125;$$

$$\therefore \sin 10'' < \frac{\pi}{64800} \text{ and } > \frac{\pi}{64800} - \frac{1}{4}({\cdot}000000000000125).$$

Hence to 12 places of decimals,

$$\sin 10'' = \frac{\pi}{64800} = {\cdot}000048481368....$$

271. *To shew that when* n *is an indefinitely large integer, the limit of*
$$\cos\frac{\theta}{2}\cos\frac{\theta}{4}\cos\frac{\theta}{8}...\cos\frac{\theta}{2^n} = \frac{\sin\theta}{\theta}.$$

We have
$$\sin\theta = 2\sin\frac{\theta}{2}\cos\frac{\theta}{2}$$
$$= 2^2\sin\frac{\theta}{4}\cos\frac{\theta}{4}\cos\frac{\theta}{2}$$
$$= 2^3\sin\frac{\theta}{8}\cos\frac{\theta}{8}\cos\frac{\theta}{4}\cos\frac{\theta}{2}$$
$$\dots\dots\dots\dots\dots\dots\dots\dots$$
$$= 2^n\sin\frac{\theta}{2^n}\cos\frac{\theta}{2^n}...\cos\frac{\theta}{8}\cos\frac{\theta}{4}\cos\frac{\theta}{2}.$$

$$\therefore \cos\frac{\theta}{2}\cos\frac{\theta}{4}\cos\frac{\theta}{8}...\cos\frac{\theta}{2^n} = \frac{\sin\theta}{2^n\sin\dfrac{\theta}{2^n}}.$$

But the limit of $2^n\sin\dfrac{\theta}{2^n}$ is θ, and thus the proposition is established. [See Art. 266.]

272. *To shew that* $\frac{\sin\theta}{\theta}$ *continually decreases from* 1 *to* $\frac{2}{\pi}$ *as* θ *continually increases from* 0 *to* $\frac{\pi}{2}$.

We shall first shew that the fraction

$$\frac{\sin\theta}{\theta} - \frac{\sin(\theta+h)}{\theta+h} \text{ is positive,}$$

h denoting the radian measure of a small positive angle.

$$\text{This fraction} = \frac{(\theta+h)\sin\theta - \theta(\sin\theta\cos h + \cos\theta\sin h)}{\theta(\theta+h)}$$

$$= \frac{\theta\sin\theta(1-\cos h) + (h\sin\theta - \theta\cos\theta\sin h)}{\theta(\theta+h)}.$$

Now $\tan\theta > \theta$, that is $\sin\theta > \theta\cos\theta$, and $h > \sin h$;

$$\therefore\ h\sin\theta > \theta\cos\theta\sin h.$$

Also $1-\cos h$ is positive; hence the numerator is positive, and therefore the fraction is positive;

$$\therefore\ \frac{\sin(\theta+h)}{\theta+h} < \frac{\sin\theta}{\theta};$$

$\therefore\ \frac{\sin\theta}{\theta}$ continually decreases as θ continually increases.

When $\theta=0$, $\frac{\sin\theta}{\theta}=1$; and when $\theta=\frac{\pi}{2}$, $\frac{\sin\theta}{\theta}=\frac{2}{\pi}$.

Thus the proposition is established.

EXAMPLES. XXI. a.

[*In this Exercise take* $\pi = \frac{22}{7}$.]

1. A tower 44 feet high subtends an angle of 35′ at a point A on the ground: find the distance of A from the tower.

2. From the top of a wall 7 ft. 4 in. high the angle of depression of an object on the ground is 24′ 30″: find its distance from the wall.

3. Find the height of an object whose angle of elevation at a distance of 840 yards is $1° 30'$.

4. Find the angle subtended by a pole 10 ft. 1 in. high at a distance of a mile.

5. Find the angle subtended by a circular target 4 feet in diameter at a distance of 1000 yards.

6. Taking the diameter of a penny as 1·25 inches, find at what distance it must be held from the eye so as just to hide the moon, supposing the diameter of the moon to be half a degree.

7. Find the distance at which a globe 11 inches in diameter subtends an angle of $5'$.

8. Two places on the same meridian are 11 miles apart: find the difference in their latitudes, taking the radius of the earth as 3960 miles.

9. A man 6 ft. high stands on a tower whose height is 120 ft.: shew that at a point 24 ft. from the tower the man subtends an angle of $31{\cdot}5'$ nearly.

10. A flagstaff standing on the top of a cliff 490 feet high subtends an angle of ·04 radians at a point 980 feet from the base of the cliff: find the height of the flagstaff.

11. When $n=0$, find the limit of

$$(1)\quad \frac{\sin n'}{n}; \qquad (2)\quad \frac{\sin n''}{n}.$$

12. When $n=\infty$, find the limit of $\frac{1}{2}nr^2\sin\frac{2\pi}{n}$.

When $\theta=0$, find the limit of

13. $\dfrac{1-\cos\theta}{\theta\sin\theta}$.

14. $\dfrac{m\sin m\theta-n\sin n\theta}{\tan m\theta+\tan n\theta}$.

15. If $\theta=\cdot01$ of a radian, calculate $\cos\left(\frac{\pi}{3}+\theta\right)$.

16. Find the value of $\sin 30° 10' 30''$.

17. Given $\cos\left(\frac{\pi}{3}+\theta\right)=\cdot49$, find the sexagesimal value of θ.

Distance and Dip of the Visible Horizon.

273. Let A be a point above the earth's surface, BCD a section of the earth by a plane passing through its centre E and A.

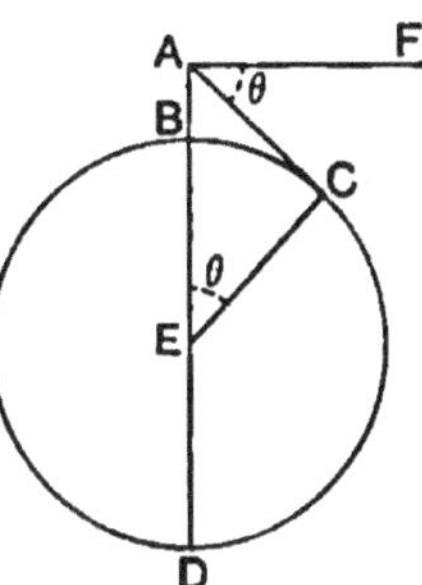

Let AE cut the circumference in B and D.

From A draw AC to touch the circle BCD in C, and join EC.

Draw AF at right angles to AD; then $\angle FAC$ is called the **dip of the horizon** as seen from A.

Thus the *dip of the horizon* is the angle of depression of any point on the horizon visible from A.

274. *To find the distance of the horizon.*

In the figure of the last article, let

$$AB = h, \quad EB = ED = r, \quad AC = x;$$

then by Euc. III. 36, $\quad AC^2 = AB . AD$;

that is, $\quad x^2 = h(2r + h) = 2hr + h^2.$

For ordinary altitudes h^2 is very small in comparison with $2hr$; hence approximately,

$$x^2 = 2hr \quad \text{and} \quad x = \sqrt{2hr}.$$

In this formula, suppose the measurements are made in *miles*, and let a be the number of *feet* in AB; then

$$a = 1760 \times 3 \times h.$$

By taking $r = 3960$, we have

$$x^2 = \frac{2 \times 3960 \times a}{1760 \times 3} = \frac{3a}{2}.$$

Thus we have the following rule:

Twice the square of the distance of the horizon measured in miles *is equal to three times the height of the place of observation measured in* feet.

Hence a man whose eye is 6 feet from the ground can see to a distance of 3 miles on a horizontal plane.

Example. The top of a ship's mast is $66\frac{2}{3}$ ft. above the sea-level, and from it the lamp of a lighthouse can just be seen. After the ship has sailed directly towards the lighthouse for half-an-hour the lamp can be seen from the deck, which is 24 ft. above the sea. Find the rate at which the ship is sailing.

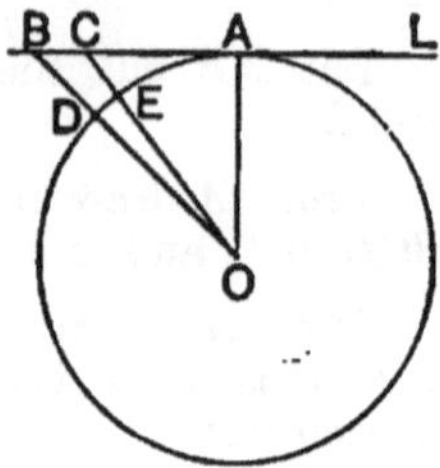

Let L denote the lamp, D and E the two positions of the ship, B the top of the mast, C the point on the deck from which the lamp is seen; then LCB is a tangent to the earth's surface at A.

[In problems like this some of the lines must necessarily be greatly out of proportion.]

Let AB and AC be expressed in miles; then since $DB = 66\frac{2}{3}$ feet and $EC = 24$ feet, we have by the rule

$$AB^2 = \frac{3}{2} \times 66\tfrac{2}{3} = 100;$$

$$\therefore\ AB = 10 \text{ miles.}$$

$$AC^2 = \frac{3}{2} \times 24 = 36;$$

$$\therefore\ AC = 6 \text{ miles.}$$

But the angles subtended by AB and AC at O the centre of the earth are very small;

$$\therefore\ \text{arc } AD = AB, \text{ and arc } AC = AE. \qquad \text{[Art. 268.]}$$

$$\therefore\ \text{arc } DE = AD - AE = AB - AC = 4 \text{ miles.}$$

Thus the ship sails 4 miles in half-an-hour, or 8 miles per hour.

275. Let θ be the number of radians in the dip of the horizon; then with the figure of Art. 273, we have

$$\cos\theta = \frac{EC}{EA} = \frac{r}{h+r} = \left(1 + \frac{h}{r}\right)^{-1};$$

$$\therefore\ 1 - 2\sin^2\frac{\theta}{2} = 1 - \frac{h}{r} + \frac{h^2}{r^2} - \dots;$$

$$\therefore\ 2\sin^2\frac{\theta}{2} = \frac{h}{r} - \frac{h^2}{r^2} + \dots.$$

Since θ and $\frac{h}{r}$ are small, we may replace $\sin\frac{\theta}{2}$ by $\frac{\theta}{2}$ and neglect the terms on the right after the first.

Thus $$\frac{\theta^2}{2}=\frac{h}{r}, \text{ or } \theta=\sqrt{\frac{2h}{r}}.$$

Let N be the number of degrees in θ radians; then

$$N=\frac{180\theta}{\pi}=\frac{180}{\pi}\sqrt{\frac{2h}{r}}.$$

Now $\sqrt{r}=63$ nearly; hence we have approximately

$$N=\frac{180\times 7\times\sqrt{2h}}{22\times 63},$$

or $$N=\frac{10}{11}\sqrt{2h},$$

a formula connecting the dip of the horizon in degrees and the height of the place of observation in miles.

EXAMPLES. XXI. b.

[Here $\pi=\frac{22}{7}$, *and radius of earth*$=3960$ *miles.*]

1. Find the greatest distance at which the lamp of a lighthouse can be seen, the light being 96 feet above the sea-level.

2. If the lamp of a lighthouse begins to be seen at a distance of 15 miles, find its height above the sea-level.

3. The tops of the masts of two ships are 32 ft. 8 in. and 42 ft. 8 in. above the sea-level: find the greatest distance at which one mast can be seen from the other.

4. Find the height of a ship's mast which is just visible at a distance of 20 miles from a point on the mast of another ship which is 54 ft. above the sea-level.

5. From the mast of a ship 73 ft. 6 in. high the lamp of a lighthouse is just visible at a distance of 28 miles: find the height of the lamp.

6. Find the sexagesimal measure of the dip of the horizon from a hill 2640 feet high.

7. Along a straight coast there are lighthouses at intervals of 24 miles: find at what height the lamp must be placed so that the light of one at least may be visible at a distance of $3\frac{1}{2}$ miles from any point of the coast.

8. From the top of a mountain the dip of the horizon is $1\cdot\dot{8}\dot{1}^\circ$: find its height in feet.

9. The distance of the horizon as seen from the top of a hill is 30·25 miles: find the height of the hill and the dip of the horizon.

10. If x miles be the distance of the visible horizon and N degrees the dip, shew that

$$N = \frac{x}{66}\sqrt{\frac{10}{11}}.$$

When $\theta = 0$, find the limit of

11. $\dfrac{\sin 4\theta \cot \theta}{\text{vers } 2\theta \cot^2 2\theta}$. **12.** $\dfrac{1 - \cos\theta + \sin\theta}{1 - \cos\theta - \sin\theta}$.

13. When $\theta = \alpha$, find the limit of

(1) $\dfrac{\sin\theta - \sin\alpha}{\theta - \alpha}$; (2) $\dfrac{\cos\theta - \cos\alpha}{\theta - \alpha}$.

14. Two sides of a triangle are 31 and 32, and they include a right angle: find the other angles.

15. A person walks directly towards a distant object P, and observes that at the three points A, B, C, the elevations of P are α, 2α, 3α respectively: shew that $AB = 3BC$ nearly.

16. Shew that $\dfrac{\tan\theta}{\theta}$ continually increases from 1 to ∞ as θ continually increases from 0 to $\dfrac{\pi}{2}$.

CHAPTER XXII.

GEOMETRICAL PROOFS.

276. *To find the expansion of* tan $(A+B)$ *geometrically.*

Let $\angle LOM = A$, and $\angle MON = B$; then $\angle LON = A+B$.

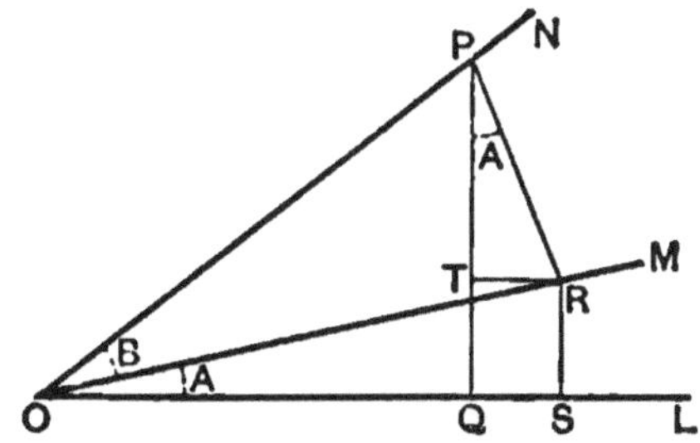

In ON take any point P, and draw PQ and PR perpendicular to OL and OM respectively. Also draw RS and RT perpendicular to OL and PQ respectively.

$$\tan(A+B) = \frac{PQ}{OQ} = \frac{RS+PT}{OS-TR}$$

$$= \frac{\dfrac{RS}{OS} + \dfrac{PT}{OS}}{1 - \dfrac{TR}{OS}} = \frac{\dfrac{RS}{OS} + \dfrac{PT}{OS}}{1 - \dfrac{TR}{TP} \cdot \dfrac{TP}{OS}}.$$

Now $\qquad \dfrac{RS}{OS} = \tan A$, and $\dfrac{TR}{TP} = \tan A$;

also the triangles ROS and TPR are similar, and therefore

$$\frac{TP}{OS} = \frac{PR}{OR} = \tan B.$$

$$\therefore \ \tan(A+B) = \frac{\tan A + \tan B}{1 - \tan A \tan B}.$$

In like manner, with the help of the figure on page 95, we may obtain the expansion of $\tan(A-B)$ geometrically.

277. *To prove geometrically the formulæ for transformation of sums into products.*

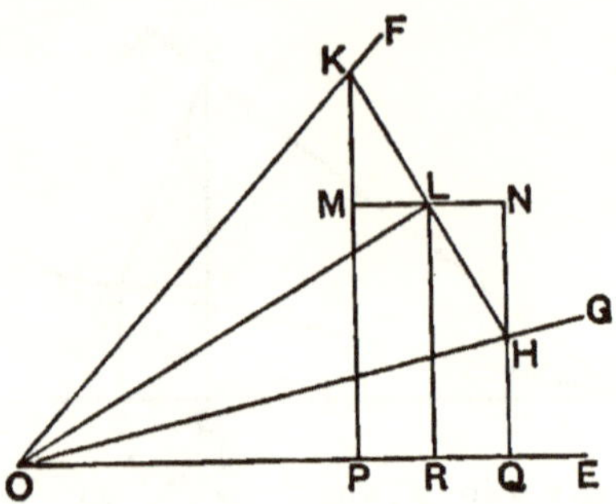

Let $\angle EOF$ be denoted by A, and $\angle EOG$ by B.

With centre O and any radius describe an arc of a circle meeting OG in H and OF in K.

Bisect $\angle KOH$ by OL; then OL bisects HK at right angles.

Draw KP, HQ, LR perpendicular to OE, and through L draw MLN parallel to OE meeting KP in M and QH in N.

It is easy to prove that the triangles MKL and NHL are equal in all respects, so that $KM=NH$, $ML=LN$, $PR=RQ$.

Also $\angle GOF=A-B$, and therefore

$$\angle HOL=\angle KOL=\frac{A-B}{2};$$

$$\therefore\ \angle EOL=B+\frac{A-B}{2}=\frac{A+B}{2}.$$

$$\sin A+\sin B=\frac{KP}{OK}+\frac{HQ}{OH}=\frac{KP+HQ}{OK}$$

$$=\frac{(KM+LR)+(LR-NH)}{OK}=2\,\frac{LR}{OK};$$

$$\therefore \sin A + \sin B = 2\frac{LR}{OL}\cdot\frac{OL}{OK} = 2\sin ROL \cos KOL$$

$$= 2\sin\frac{A+B}{2}\cos\frac{A-B}{2}.$$

$$\cos A + \cos B = \frac{OP}{OK} + \frac{OQ}{OH} = \frac{OP+OQ}{OK}$$

$$= \frac{(OR-PR)+(OR+RQ)}{OK} = 2\frac{OR}{OK}$$

$$= 2\frac{OR}{OL}\cdot\frac{OL}{OK} = 2\cos ROL \cos KOL$$

$$= 2\cos\frac{A+B}{2}\cos\frac{A-B}{2}.$$

$$\sin A - \sin B = \frac{KP}{OK} - \frac{HQ}{OH} = \frac{KP-HQ}{OK}$$

$$= \frac{(KM+LR)-(LR-NH)}{OK} = 2\frac{KM}{OK}$$

$$= 2\frac{KM}{KL}\cdot\frac{KL}{OK} = 2\cos LKM \sin KOL$$

$$= 2\cos\frac{A+B}{2}\sin\frac{A-B}{2},$$

since $\angle LKM$ = comp^t of $\angle KLM = \angle MLO = \angle LOE = \frac{A+B}{2}$.

$$\cos B - \cos A = \frac{OQ}{OH} - \frac{OP}{OK} = \frac{OQ-OP}{OK}$$

$$= \frac{(OR+RQ)-(OR-PR)}{OK} = 2\frac{PR}{OK} = 2\frac{ML}{OK}$$

$$= 2\frac{ML}{KL}\cdot\frac{KL}{OK} = 2\sin LKM \sin KOL$$

$$= 2\sin\frac{A+B}{2}\sin\frac{A-B}{2}.$$

278. *Geometrical proof of the 2A formulæ.*

Let BPD be a semicircle, BD the diameter, C the centre.

On the circumference, take any point P, and join PB, PC, PD.

Draw PN perpendicular to BD.

Let $\angle PBD = A$, then

$$\angle PCD = 2A.$$

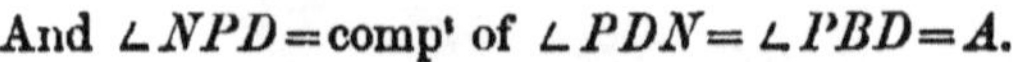

And $\angle NPD =$ comp^t of $\angle PDN = \angle PBD = A$.

$$\sin 2A = \frac{PN}{CP} = \frac{2PN}{2CP} = \frac{2PN}{BD} = 2\,\frac{PN}{BP} \cdot \frac{BP}{BD}$$

$$= 2 \sin PBN \cos PBD$$

$$= 2 \sin A \cos A.$$

$$\cos 2A = \frac{CN}{CP} = \frac{2CN}{BD} = \frac{CN + CN}{BD}$$

$$= \frac{(BN - BC) + (CD - ND)}{BD} = \frac{BN - ND}{BD}$$

$$= \frac{BN}{BP} \cdot \frac{BP}{BD} - \frac{ND}{PD} \cdot \frac{PD}{BD}$$

$$= \cos A \,.\, \cos A - \sin A \,.\, \sin A$$

$$= \cos^2 A - \sin^2 A.$$

$$\cos 2A = \frac{CN}{CP} = \frac{CD - DN}{CP} = 1 - \frac{DN}{CP} = 1 - \frac{2DN}{BD}$$

$$= 1 - 2\,\frac{DN}{DP} \cdot \frac{DP}{BD} = 1 - 2 \sin A \,.\, \sin A$$

$$= 1 - 2 \sin^2 A.$$

$$\cos 2A = \frac{CN}{CP} = \frac{BN - BC}{CP} = \frac{BN}{CP} - 1 = \frac{2BN}{BD} - 1$$

$$= 2\,\frac{BN}{BP} \cdot \frac{BP}{BD} - 1 = 2 \cos A \,.\, \cos A - 1$$

$$= 2 \cos^2 A - 1.$$

$$\tan 2A = \frac{PN}{CN} = \frac{2PN}{2CN} = \frac{2PN}{BN - ND}$$

$$= \frac{2\frac{PN}{BN}}{1 - \frac{ND}{BN}} = \frac{2\frac{PN}{BN}}{1 - \frac{ND}{PN} \cdot \frac{PN}{BN}}$$

$$= \frac{2 \tan A}{1 - \tan A \,.\, \tan A}$$

$$= \frac{2 \tan A}{1 - \tan^2 A}.$$

279. *To find the value of* sin 18° *geometrically.*

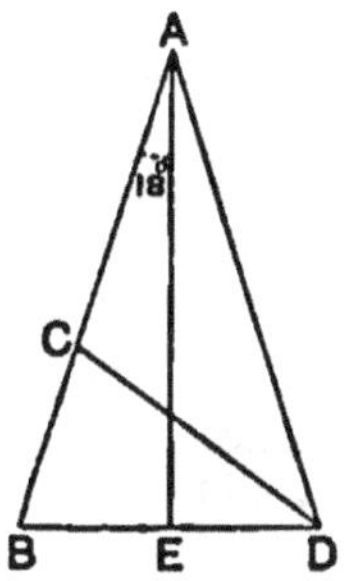

Let ABD be an isosceles triangle in which each angle at the base BD is double the vertical angle A; then

$$A + 2A + 2A = 180°,$$

and therefore $A = 36°$.

Bisect $\angle BAD$ by AE; then AE bisects BD at right angles;

$$\therefore\ \angle BAE = 18°.$$

Thus $\qquad \sin 18° = \frac{BE}{AB} = \frac{x}{a},$

where $\qquad AB = a$, and $BE = x$.

From the construction given in Euc. IV. 10,

$$AC = BD = 2BE = 2x,$$

and

$$AB \,.\, BC = AC^2;$$

$$\therefore\ a(a - 2x) = (2x)^2;$$

$$\therefore\ 4x^2 + 2ax - a^2 = 0;$$

$$\therefore\ x = \frac{-2a \pm \sqrt{20a^2}}{8} = \frac{-1 \pm \sqrt{5}}{4} a.$$

The upper sign must be taken, since x is positive. Thus

$$\sin 18° = \frac{\sqrt{5} - 1}{4}.$$

Proofs by Projection.

280. DEFINITION. If from any two points A and B, lines AC and BD are drawn perpendicular to OX, then the intercept CD is called the **projection** of AB upon OX.

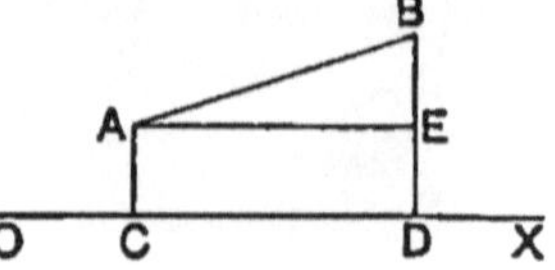

Through A draw AE parallel to OX; then

$$CD = AE = AB \cos BAE;$$

that is, $$CD = AB \cos \alpha,$$

where α is the angle of inclination of the lines AB and OX.

281. *To shew that the projection of a straight line is equal to the projection of an equal and parallel straight line drawn from a fixed point.*

Let AB be any straight line, O a fixed point, which we shall call the origin, OP a straight line equal and parallel to AB.

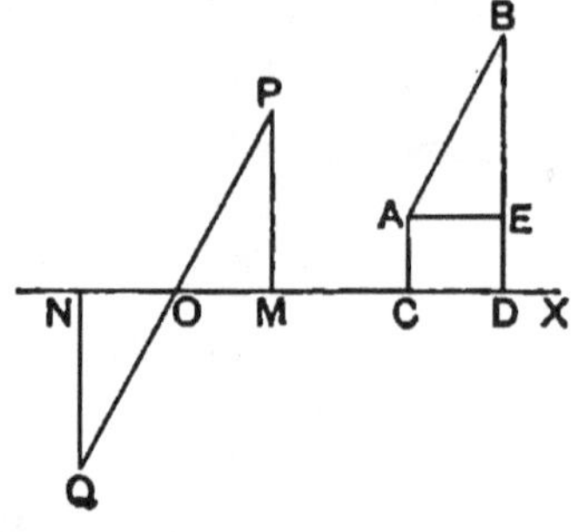

Let CD and OM be the projections of AB and OP upon any straight line OX drawn through the origin.

Draw AE parallel to OX.

The two triangles AEB and OMP are identically equal;

$$\therefore\ OM = AE = CD;$$

that is, projection of OP = projection of AB.

282. In the figure of the last article, *two* straight lines OP and OQ can be drawn from O equal and parallel to AB; it is therefore necessary to have some means of fixing the *direction* in which the line from O is to be drawn. Accordingly it is agreed to consider that

the direction of a line is fixed by the order of the letters.

Thus AB denotes a line drawn from A to B, and BA denotes a line drawn from B to A.

Hence OP denotes a line drawn from the origin parallel to AB, and OQ denotes a line drawn from the origin parallel to BA.

Similarly the direction of a projected line is fixed by the order of the letters.

Thus CD is drawn to the right from C to D and is positive, while DC is drawn to the left from D to C and is negative.

Hence in sign as well as in magnitude

$$OM = CD, \text{ and } ON = DC;$$

that is, projection of OP = projection of AB,

and projection of OQ = projection of BA.

Thus *the projection of a straight line can be represented both in sign and magnitude by the projection of an equal and parallel straight line drawn from the origin.*

283. Whatever be the direction of AB, the line OP will fall within one of the four quadrants.

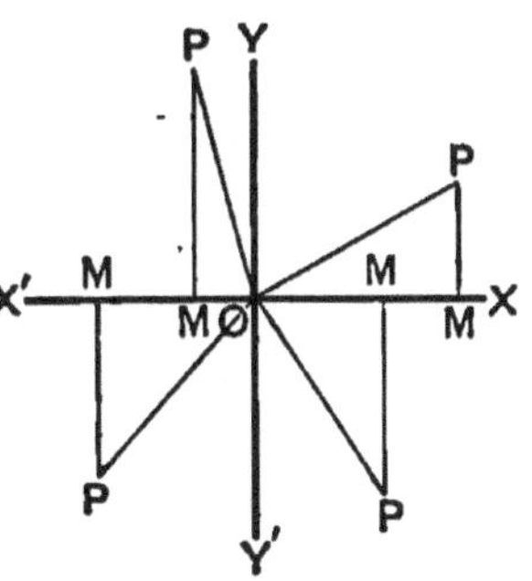

Also from the definitions given in Art. 75, we have

$$\frac{OM}{OP} = \cos XOP,$$

that is,

$$OM = OP \cos XOP,$$

whatever be the magnitude of the angle XOP. We shall always suppose, unless the contrary is stated, that the angles are measured in the positive direction.

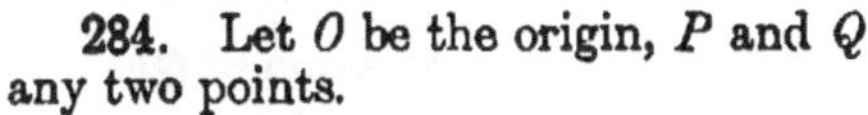

284. Let O be the origin, P and Q any two points.

Join OP, OQ, PQ, and draw PM and QN perpendicular to OX.

We have

$$OM = ON + NM,$$

since the line NM is to be regarded as negative; that is,

the projection of OP = projection of OQ + projection of QP.

Hence, *the projection of one side of a triangle is equal to the sum of the projections of the other two sides taken in order.* Thus

projection of OQ = projection of OP + projection of PQ;

projection of QP = projection of QO + projection of OP.

General Proof of the Addition Formulæ.

285. In Fig. 1, let a line starting from OX revolve until it has traced the angle A, taking up the position OM, and then let it further revolve until it has traced the angle B, taking up the final position ON. Thus XON is the angle $A+B$.

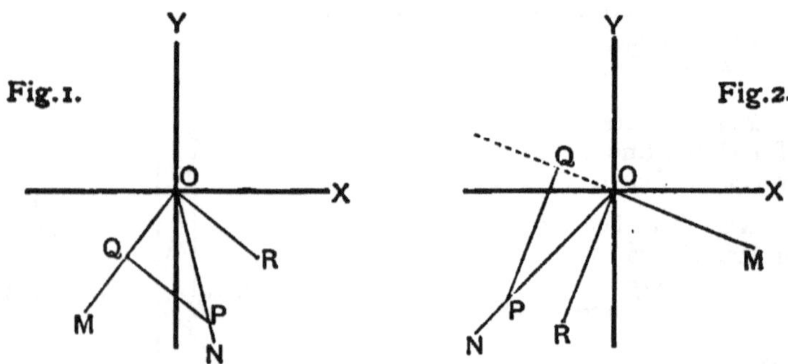

In ON take any point P, and draw PQ perpendicular to OM; also draw OR equal and parallel to QP.

Projecting upon OX, we have

projection of OP = projection of OQ + projection of QP

= projection of OQ + projection of OR.

$$\therefore\ OP\cos XOP = OQ\cos XOQ + OR\cos XOR \quad\text{............(1)}$$

$$= OP\cos B\cos XOQ + OP\sin B\cos XOR;$$

$$\therefore\ \cos XOP = \cos B\cos XOQ + \sin B\cos XOR;$$

that is, $\cos(A+B) = \cos B\cos A + \sin B\cos(90°+A)$

$$= \cos A\cos B - \sin A\sin B.$$

Projecting upon OY, we have only to write Y for X in (1);

thus $OP \cos YOP = OQ \cos YOQ + OR \cos YOR$

$$= OP \cos B \cos YOQ + OP \sin B \cos YOR;$$

$$\therefore \cos YOP = \cos B \cos YOQ + \sin B \cos YOR;$$

that is,

$$\cos(A + B - 90°) = \cos B \cos(A - 90°) + \sin B \cos A;$$

$$\therefore \sin(A + B) = \sin A \cos B + \cos A \sin B.$$

In Fig. 2, let a line starting from OX revolve until it has traced the angle A, taking up the position OM, and then let it revolve *back again* until it has traced the angle B, taking up the final position ON. Thus XON is the angle $A - B$.

In ON take any point P, and draw PQ perpendicular to MO produced; also draw OR equal and parallel to QP.

Projecting upon OX, we have as in the previous case

$$OP \cos XOP = OQ \cos XOQ + OR \cos XOR$$

$$= OP \cos(180° - B) \cos XOQ$$

$$+ OP \sin(180° - B) \cos XOR;$$

$$\therefore \cos XOP = -\cos B \cos XOQ + \sin B \cos XOR;$$

that is,

$$\cos(A - B) = -\cos B \cos(A - 180°) + \sin B \cos(A - 90°)$$

$$= -\cos B(-\cos A) + \sin B \sin A$$

$$= \cos A \cos B + \sin A \sin B.$$

Projecting upon OY, we have

$$OP \cos YOP = OQ \cos YOQ + OR \cos YOR;$$

$$= OP \cos(180° - B) \cos YOQ$$

$$+ OP \sin(180° - B) \cos YOR;$$

$$\therefore \cos YOP = -\cos B \cos YOQ + \sin B \cos YOR;$$

that is,

$$\cos(A - B - 90°) = -\cos B \cos(A - 270°) + \sin B \cos(A - 180°);$$

$$\therefore \sin(A - B) = -\cos B(-\sin A) + \sin B(-\cos A)$$

$$= \sin A \cos B - \cos A \sin B.$$

286. The above method of proof is applicable to every case, and therefore the Addition Formulæ are universally established.

The universal truth of the Addition Formulæ may also be deduced from the special geometrical investigations of Arts. 110 and 111 by analysis, as in the next article.

287. When each of the angles A, B, $A+B$ is less than 90°, we have shewn that

$$\cos(A+B)=\cos A\cos B-\sin A\sin B\text{............(1)}.$$

But $\cos(A+B)=\sin(\overline{A+B}+90^\circ)=\sin(\overline{A+90^\circ}+B)$;

also $$\cos A=\sin(A+90^\circ),$$

and $$-\sin A=\cos(A+90^\circ).\quad\text{[Art. 98.]}$$

Hence by substitution in (1), we have

$$\sin(\overline{A+90^\circ}+B)=\sin(A+90^\circ)\cos B+\cos(A+90^\circ)\sin B.$$

In like manner, it may be proved that

$$\cos(\overline{A+90^\circ}+B)=\cos(A+90^\circ)\cos B-\sin(A+90^\circ)\sin B.$$

Thus the formulæ for the sine and cosine of $A+B$ hold when A is increased by 90°. Similarly we may shew that they hold when B is increased by 90°.

By repeated applications of the same process it may be proved that the formulæ are true when either or both of the angles A and B is increased by any multiple of 90°.

Again, $\cos(A+B)=\cos A\cos B-\sin A\sin B$............(1).

But $\cos(A+B)=-\sin(\overline{A+B}-90^\circ)=-\sin(\overline{A-90^\circ}+B)$;

also $$\cos A=-\sin(A-90^\circ),$$

and $$\sin A=\cos(A-90^\circ).\quad\text{[Arts. 99 and 102.]}$$

Hence by substitution in (1), we have

$$\sin(\overline{A-90^\circ}+B)=\sin(A-90^\circ)\cos B+\cos(A-90^\circ)\sin B.$$

Similarly we may shew that

$$\cos(\overline{A-90^\circ}+B)=\cos(A-90^\circ)\cos B-\sin(A-90^\circ)\sin B.$$

Thus the formulæ for the sine and cosine of $A+B$ hold when A is diminished by 90°. In like manner we may prove that they are true when B is diminished by 90°.

By repeated applications of the same process it may be shewn that the formulæ hold when either or both of the angles A and B is diminished by any multiple of 90°. Further, it will be seen that the formulæ are true if either of the angles A or B is increased by a multiple of 90° and the other is diminished by a multiple of 90°.

Thus $\sin(P+Q)=\sin P\cos Q+\cos P\sin Q,$

and $\cos(P+Q)=\cos P\cos Q-\sin P\sin Q,$

where $P=A\pm m\,.\,90^\circ$, and $Q=B\pm n\,.\,90^\circ$,

m and n being any positive integers, and A and B any acute angles.

Thus the Addition Formulæ are true for the algebraical sum of any two angles.

MISCELLANEOUS EXAMPLES. H.

1. If the sides of a right-angled triangle are

$$\cos 2\alpha+\cos 2\beta+2\cos(\alpha+\beta) \text{ and } \sin 2\alpha+\sin 2\beta+\sin 2(\alpha+\beta),$$

shew that the hypotenuse is $4\cos^2\frac{\alpha-\beta}{2}$.

2. If the in-centre and circum-centre be at equal distances from BC, prove that

$$\cos B+\cos C=1.$$

3. The shadow of a tower is observed to be half the known height of the tower, and some time afterwards to be equal to the height: how much will the sun have gone down in the interval? Given log 2,

$$L\tan 63^\circ\,26'=10{\cdot}3009994, \text{ diff. for } 1'=3159.$$

4. If $(1+\sin\alpha)(1+\sin\beta)(1+\sin\gamma)$

$$=(1-\sin\alpha)(1-\sin\beta)(1-\sin\gamma),$$

shew that each expression is equal to $\pm\cos\alpha\cos\beta\cos\gamma$.

5. Two parallel chords of a circle lying on the same side of the centre subtend 72° and 144° at the centre: prove that the distance between them is one-half of the radius.

Also shew that the sum of the squares of the chords is equal to five times the square of the radius.

6. Two straight railways are inclined at an angle of 60°. From their point of intersection two trains A and B start at the same time, one along each line. A travels at the rate of 48 miles per hour, at what rate must B travel so that after one hour they shall be 43 miles apart?

7. If $$\alpha=\cos^{-1}\frac{x}{a}+\cos^{-1}\frac{y}{b},$$

shew that $$\sin^2\alpha=\frac{x^2}{a^2}-\frac{2xy}{ab}\cos\alpha+\frac{y^2}{b^2}.$$

8. If p, q, r denote the sides of the ex-central triangle, prove that

$$\frac{a^2}{p^2}+\frac{b^2}{q^2}+\frac{c^2}{r^2}+\frac{2abc}{pqr}=1.$$

9. A tower is situated within the angle formed by two straight roads OA and OB, and subtends angles α and β at the points A and B where the roads are nearest to it. If $OA=a$, and $OB=b$, shew that the height of the tower is

$$\sqrt{a^2-b^2}\sin\alpha\sin\beta\,/\,\sqrt{\sin(\alpha+\beta)\sin(\alpha-\beta)}.$$

10. In a triangle, shew that

$$r^2+r_1^2+r_2^2+r_3^2=16R^2-a^2-b^2-c^2.$$

11. If AD be a median of the triangle ABC, shew that

(1) $\cot BAD=2\cot A+\cot B$;

(2) $2\cot ADC=\cot B-\cot C$.

12. If p, q, r are the distances of the orthocentre from the sides, prove that

$$4\left(\frac{a}{p}+\frac{b}{q}+\frac{c}{r}\right)=\left(\frac{a}{p}+\frac{b}{q}-\frac{c}{r}\right)\left(\frac{b}{q}+\frac{c}{r}-\frac{a}{p}\right)\left(\frac{c}{r}+\frac{a}{p}-\frac{b}{q}\right).$$

Graphical Representation of the Circular Functions.

288. Definition. Let $f(x)$ be a function of x which has a single value for all values of x, and let the values of x be represented by lines measured from O along OX or OX', and the values of $f(x)$ by lines drawn perpendicular to XX'. Then with the figure of the next article, if OM represent any value of x, and MP the corresponding value of $f(x)$, the curve traced out by the point P is called the **Graph** of $f(x)$.

Graphs of sin θ and cos θ.

289. Suppose that the unit of length is chosen to represent a radian; then any angle of θ radians will be represented by a line OM which contains θ units of length.

Graph of sin θ.

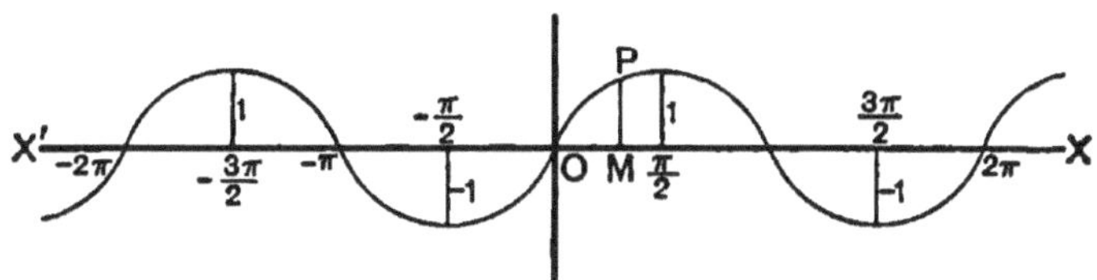

Let MP, drawn perpendicular to OX, represent the value of $\sin\theta$ corresponding to the value OM of θ; then the curve traced out by the point P represents the graph of $\sin\theta$.

As OM or θ increases from 0 to $\frac{\pi}{2}$, MP or $\sin\theta$ increases from 0 to 1, which is its greatest value.

As OM increases from $\frac{\pi}{2}$ to π, MP decreases from 0 to 1.

As OM increases from π to $\frac{3\pi}{2}$, MP increases numerically from 0 to -1.

As OM increases from $\frac{3\pi}{2}$ to 2π, MP decreases numerically from -1 to 0.

As OM increases from 2π to 4π, from 4π to 6π, from 6π to 8π, ..., MP passes through the same series of values as when OM increases from 0 to 2π.

Since $\sin(-\theta) = -\sin\theta$, the values of MP lying to the left of O are equal in magnitude but are of opposite sign to values of MP lying at an equal distance to the right of O.

Thus the graph of $\sin\theta$ is a *continuous* waving line extending to an infinite distance on each side of O.

The graph of $\cos\theta$ is the same as that of $\sin\theta$, the origin being at the point marked $\frac{\pi}{2}$ in the figure.

Graphs of tan θ and cot θ.

290. As before, suppose that the unit of length is chosen to represent a radian; then any angle of θ radians will be represented by a line OM which contains θ units of length.

Let MP, drawn perpendicular to OX, represent the value of $\tan\theta$ corresponding to the value OM of θ; then the curve traced out by the point P represents the graph of $\tan\theta$.

By tracing the changes in the value of $\tan\theta$ as θ varies from 0 to 2π, from 2π to 4π,......, it will be seen that the graph of $\tan\theta$ consists of an infinite number of *discontinuous* equal branches as represented in the figure below. The part of each branch beneath XX' is convex towards XX', and the part of each branch above XX' is also convex towards XX'; hence at the point where any branch cuts XX' there is what is called a *point of inflexion*, where the direction of curvature changes. The proof of these statements is however beyond the range of the present work.

The various branches touch the dotted lines passing through the points marked

$$\pm\frac{\pi}{2},\quad \pm\frac{3\pi}{2},\quad \pm\frac{5}{2},\ldots\ldots,$$

at an infinite distance from XX'.

Graph of tan θ.

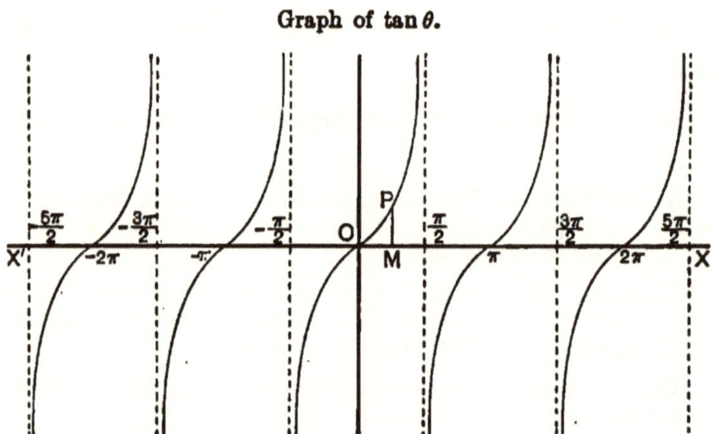

The student should draw the graph of $\cot\theta$, which is very similar to that of $\tan\theta$.

Graphs of sec θ and cosec θ.

291. The graph of sec θ is represented in the figure below. It consists of an infinite number of equal festoons lying alternately above and below XX', the vertex of each being at the unit of distance from XX'. The various festoons touch the dotted lines passing through the points marked

$$\pm\frac{\pi}{2},\quad \pm\frac{3\pi}{2},\quad \pm\frac{5\pi}{2},\ldots\ldots,$$

at an infinite distance from XX'.

Graph of sec θ.

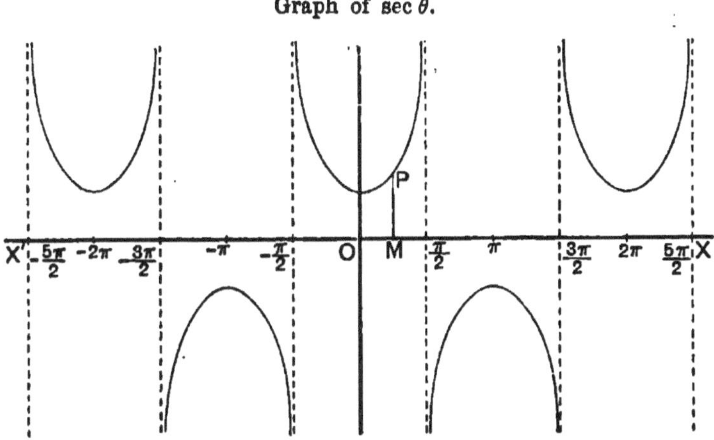

The graph of cosec θ is the same as that of sec θ, the origin being at the point marked $-\frac{\pi}{2}$ in the figure.

CHAPTER XXIII.

SUMMATION OF FINITE SERIES.

292. An expression in which the successive terms are formed by some regular law is called a **series**. If the series ends at some assigned term it is called a **finite series**; if the number of terms is unlimited it is called an **infinite series**.

A series may be denoted by an expression of the form

$$u_1+u_2+u_3+\ldots+u_{n-1}+u_n+u_{n+1}+\ldots,$$

where u_{n+1}, the $(n+1)^{\text{th}}$ term, is obtained from u_n, the nth term, by replacing n by $n+1$.

Thus if $u_n=\cos(\alpha+n\beta)$, then $u_{n+1}=\cos\{\alpha+(n+1)\beta\}$;
and if $u_n=\cot 2^{n-1}\alpha$, then $u_{n+1}=\cot 2^n\alpha$.

293. If the rth term of a series can be expressed as the difference of two quantities one of which is the same function of r that the other is of $r+1$, the sum of the series may be readily found.

For let the series be denoted by

$$u_1+u_2+u_3+\ldots\ldots+u_n,$$

and its sum by S, and suppose that any term

$$u_r=v_{r+1}-v_r;$$

then $S=(v_2-v_1)+(v_3-v_2)+(v_4-v_3)+\ldots+(v_n-v_{n-1})+(v_{n+1}-v_n)$
$\quad=v_{n+1}-v_1.$

Example. Find the sum of the series

$$\operatorname{cosec}\alpha+\operatorname{cosec}2\alpha+\operatorname{cosec}4\alpha+\ldots\ldots+\operatorname{cosec}2^{n-1}\alpha.$$

$$\operatorname{cosec}\alpha=\frac{1}{\sin\alpha}=\frac{\sin\frac{\alpha}{2}}{\sin\frac{\alpha}{2}\sin\alpha}=\frac{\sin\left(\alpha-\frac{\alpha}{2}\right)}{\sin\frac{\alpha}{2}\sin\alpha}.$$

Hence $$\operatorname{cosec} \alpha = \cot\frac{\alpha}{2} - \cot \alpha.$$

If we replace α by 2α, we obtain

$$\operatorname{cosec} 2\alpha = \cot \alpha - \cot 2\alpha.$$

Similarly, $$\operatorname{cosec} 4\alpha = \cot 2\alpha - \cot 4\alpha,$$

....

$$\operatorname{cosec} 2^{n-1}\alpha = \cot 2^{n-2}\alpha - \cot 2^{n-1}\alpha.$$

By addition, $$S = \cot\frac{\alpha}{2} - \cot 2^{n-1}\alpha.$$

294. *To find the sum of the sines of a series of* n *angles which are in arithmetical progression.*

Let the sine-series be denoted by

$$\sin\alpha + \sin(\alpha+\beta) + \sin(\alpha+2\beta) + \ldots\ldots + \sin\{\alpha+(n-1)\beta\}.$$

We have the identities

$$2\sin\alpha\sin\frac{\beta}{2} = \cos\left(\alpha-\frac{\beta}{2}\right) - \cos\left(\alpha+\frac{\beta}{2}\right),$$

$$2\sin(\alpha+\beta)\sin\frac{\beta}{2} = \cos\left(\alpha+\frac{\beta}{2}\right) - \cos\left(\alpha+\frac{3\beta}{2}\right),$$

$$2\sin(\alpha+2\beta)\sin\frac{\beta}{2} = \cos\left(\alpha+\frac{3\beta}{2}\right) - \cos\left(\alpha+\frac{5\beta}{2}\right),$$

....

$$2\sin\{\alpha+(n-1)\beta\}\sin\frac{\beta}{2} = \cos\left(\alpha+\frac{2n-3}{2}\beta\right) - \cos\left(\alpha+\frac{2n-1}{2}\beta\right).$$

By addition,

$$2S\sin\frac{\beta}{2} = \cos\left(\alpha-\frac{\beta}{2}\right) - \cos\left(\alpha+\frac{2n-1}{2}\beta\right)$$

$$= 2\sin\left(\alpha+\frac{n-1}{2}\beta\right)\sin\frac{n\beta}{2};$$

$$\therefore\ S = \frac{\sin\frac{n\beta}{2}}{\sin\frac{\beta}{2}}\sin\left(\alpha+\frac{n-1}{2}\beta\right).$$

295. In like manner we may shew that the sum of the cosine-series

$$\cos\alpha+\cos(\alpha+\beta)+\cos(\alpha+2\beta)+\ldots+\cos\{\alpha+(n-1)\beta\}$$

$$=\frac{\sin\frac{n\beta}{2}}{\sin\frac{\beta}{2}}\cos\left(\alpha+\frac{n-1}{2}\beta\right).$$

296. The formulæ of the two last articles may be expressed verbally as follows.

The sum of the sines of a series of n *angles in* A.P.

$$=\frac{\sin\frac{\text{n } diff.}{2}}{\sin\frac{diff.}{2}}\sin\frac{first\ angle+last\ angle}{2}.$$

The sum of the cosines of a series of n *angles in* A.P.

$$=\frac{\sin\frac{\text{n } diff.}{2}}{\sin\frac{diff.}{2}}\cos\frac{first\ angle+last\ angle}{2}.$$

Example. Find the sum of the series

$$\cos\alpha+\cos 3\alpha+\cos 5\alpha+\ldots\ldots+\cos(2n-1)\alpha.$$

Here the common difference of the angles is 2α;

$$\therefore\ S=\frac{\sin n\alpha}{\sin\alpha}\cos\frac{\alpha+(2n-1)\alpha}{2}$$

$$=\frac{\sin n\alpha\cos n\alpha}{\sin\alpha}=\frac{\sin 2n\alpha}{2\sin\alpha}.$$

297. If $\sin\frac{n\beta}{2}=0$, each of the expressions found in Arts. 294 and 295 for the sum vanishes. In this case

$$\frac{n\beta}{2}=k\pi,\quad\text{or}\quad\beta=\frac{2k\pi}{n},\ \text{where } k \text{ is any integer.}$$

Hence *the sum of the sines and the sum of the cosines of* n *angles in arithmetical progression are each equal to zero, when the common difference of the angles is an even multiple of* $\frac{\pi}{n}$.

298. Some series may be brought under the rule of Art. 296 by a simple transformation.

Example 1. Find the sum of n terms of the series

$$\cos\alpha-\cos(\alpha+\beta)+\cos(\alpha+2\beta)-\cos(\alpha+3\beta)+\ldots\ldots.$$

This series is equal to

$$\cos\alpha+\cos(\alpha+\beta+\pi)+\cos(\alpha+2\beta+2\pi)+\cos(\alpha+3\beta+3\pi)+\ldots\ldots,$$

a series in which the common difference of the angles is $\beta+\pi$, and the last angle is $\alpha+(n-1)(\beta+\pi)$.

$$\therefore S=\frac{\sin\frac{n(\beta+\pi)}{2}}{\sin\frac{\beta+\pi}{2}}\cos\left\{\alpha+\frac{(n-1)(\beta+\pi)}{2}\right\}.$$

Example 2. Find the sum of n terms of the series

$$\sin\alpha+\cos(\alpha+\beta)-\sin(\alpha+2\beta)-\cos(\alpha+3\beta)+\sin(\alpha+4\beta)+\ldots\ldots.$$

This series is equal to

$$\sin\alpha+\sin\left(\alpha+\beta+\frac{\pi}{2}\right)+\sin(\alpha+2\beta+\pi)+\sin\left(\alpha+3\beta+\frac{3\pi}{2}\right)+\ldots\ldots,$$

a series in which the common difference of the angles is $\beta+\frac{\pi}{2}$.

$$\therefore S=\frac{\sin\frac{n(2\beta+\pi)}{4}}{\sin\frac{2\beta+\pi}{4}}\sin\left\{\alpha+\frac{(n-1)(2\beta+\pi)}{4}\right\}.$$

EXAMPLES. XXIII. a.

Sum each of the following series to n terms :

1. $\sin\alpha+\sin 3\alpha+\sin 5\alpha+\ldots\ldots.$

2. $\cos\alpha+\cos(\alpha-\beta)+\cos(\alpha-2\beta)+\ldots\ldots.$

3. $\sin\alpha+\sin\left(\alpha-\frac{\pi}{n}\right)+\sin\left(\alpha-\frac{2\pi}{n}\right)+\ldots\ldots.$

4. $\cos\frac{\pi}{k}+\cos\frac{2\pi}{k}+\cos\frac{3\pi}{k}+\ldots\ldots.$

Find the sum of each of the following series:

5. $\cos\frac{\pi}{19}+\cos\frac{3\pi}{19}+\cos\frac{5\pi}{19}+\ldots\ldots+\cos\frac{17\pi}{19}$.

6. $\cos\frac{2\pi}{21}+\cos\frac{4\pi}{21}+\cos\frac{6\pi}{21}+\ldots\ldots+\cos\frac{20\pi}{21}$.

7. $\sin\frac{\pi}{n}+\sin\frac{2\pi}{n}+\sin\frac{3\pi}{n}+\ldots\ldots$ to $n-1$ terms.

8. $\cos\frac{\pi}{n}+\cos\frac{3\pi}{n}+\cos\frac{5\pi}{n}+\ldots\ldots$ to $2n-1$ terms.

9. $\sin n\alpha+\sin(n-1)\alpha+\sin(n-2)\alpha+\ldots\ldots$ to $2n$ terms.

Sum each of the following series to n terms:

10. $\sin\theta-\sin 2\theta+\sin 3\theta-\sin 4\theta+\ldots\ldots$.

11. $\cos\alpha-\cos(\alpha-\beta)+\cos(\alpha-2\beta)-\cos(\alpha-3\beta)+\ldots\ldots$.

12. $\cos\alpha-\sin(\alpha-\beta)-\cos(\alpha-2\beta)+\sin(\alpha-3\beta)+\ldots\ldots$.

13. $\sin 2\theta\sin\theta+\sin 3\theta\sin 2\theta+\sin 4\theta\sin 3\theta+\ldots\ldots$.

14. $\sin\alpha\cos 3\alpha+\sin 3\alpha\cos 5\alpha+\sin 5\alpha\cos 7\alpha+\ldots\ldots$.

15. $\sec\alpha\sec 2\alpha+\sec 2\alpha\sec 3\alpha+\sec 3\alpha\sec 4\alpha+\ldots\ldots$.

16. $\operatorname{cosec}\theta\operatorname{cosec}3\theta+\operatorname{cosec}3\theta\operatorname{cosec}5\theta$
$+\operatorname{cosec}5\theta\operatorname{cosec}7\theta+\ldots\ldots$.

17. $\tan\frac{\alpha}{2}\sec\alpha+\tan\frac{\alpha}{2^2}\sec\frac{\alpha}{2}+\tan\frac{\alpha}{2^3}\sec\frac{\alpha}{2^2}+\ldots\ldots$.

18. $\cos 2\alpha\operatorname{cosec}3\alpha+\cos 6\alpha\operatorname{cosec}9\alpha+\cos 18\alpha\operatorname{cosec}27\alpha+\ldots\ldots$.

19. $\sin\alpha\sec 3\alpha+\sin 3\alpha\sec 9\alpha+\sin 9\alpha\sec 27\alpha+\ldots\ldots$.

20. The circumference of a semicircle of radius a is divided into n equal arcs. Shew that the sum of the distances of the several points of section from either extremity of the diameter is

$$a\left(\cot\frac{\pi}{4n}-1\right).$$

21. From the angular points of a regular polygon, perpendiculars are drawn to XX' and YY' the horizontal and vertical diameter of the circumscribing circle: shew that the algebraical sums of each of the two sets of perpendiculars are equal to zero.

299. By means of the identities

$$2\sin^2 a = 1 - \cos 2a, \qquad 2\cos^2 a = 1 + \cos 2a,$$

$$4\sin^3 a = 3\sin a - \sin 3a, \qquad 4\cos^3 a = 3\cos a + \cos 3a,$$

we can find the sum of the squares and cubes of the sines and cosines of a series of angles in arithmetical progression.

Example 1. Find the sum of n terms of the series

$$\sin^2 a + \sin^2(a+\beta) + \sin^2(a+2\beta) + \ldots\ldots.$$

$$2S = \{1 - \cos 2a\} + \{1 - \cos(2a+2\beta)\} + \{1 - \cos(2a+4\beta)\} + \ldots\ldots$$

$$= n - \{\cos 2a + \cos(2a+2\beta) + \cos(2a+4\beta) + \ldots\ldots\};$$

$$= n - \frac{\sin n\beta}{\sin\beta}\cos\frac{2a + \{2a+(n-1)\,2\beta\}}{2};$$

$$\therefore\ S = \frac{n}{2} - \frac{\sin n\beta}{2\sin\beta}\cos\{2a+(n-1)\beta\}.$$

Example 2. Find the sum of the series

$$\cos^3 a + \cos^3 3a + \cos^3 5a + \ldots\ldots + \cos^3(2n-1)\,a.$$

$$4S = (3\cos a + \cos 3a) + (3\cos 3a + \cos 9a) + (3\cos 5a + \cos 15a) + \ldots\ldots$$

$$= 3(\cos a + \cos 3a + \cos 5a + \ldots\ldots) + (\cos 3a + \cos 9a + \cos 15a + \ldots\ldots)$$

$$= \frac{3\sin na}{\sin a}\cos\left\{\frac{a+(2n-1)\,a}{2}\right\} + \frac{\sin 3na}{\sin 3a}\cos\left\{\frac{3a+(2n-1)\,3a}{2}\right\};$$

$$\therefore\ S = \frac{3\sin na\cos na}{4\sin a} + \frac{\sin 3na\cos 3na}{4\sin 3a}.$$

300. The following further examples illustrate the principle of Art. 293.

Example 1. Find the sum of the series

$$\tan^{-1}\frac{x}{1+1\,.\,2\,.\,x^2} + \tan^{-1}\frac{x}{1+2\,.\,3\,.\,x^2} + \ldots\ldots + \tan^{-1}\frac{x}{1+n(n+1)\,x^2}.$$

As in Art. 249, we have

$$\tan^{-1}\frac{x}{1+r(r+1)\,x^2} = \tan^{-1}(r+1)\,x - \tan^{-1}rx;$$

$$\therefore\ S = \tan^{-1}(n+1)\,x - \tan^{-1}x.$$

Example 2. Find the sum of n terms of the series

$$\tan a+\frac{1}{2}\tan\frac{a}{2}+\frac{1}{2^2}\tan\frac{a}{2^2}+\frac{1}{2^3}\tan\frac{a}{2^3}+\dots\dots.$$

We have $$\tan a=\cot a-2\cot 2a.$$

Replacing a by $\frac{a}{2}$ and dividing by 2, we obtain

$$\frac{1}{2}\tan\frac{a}{2}=\frac{1}{2}\cot\frac{a}{2}-\cot a.$$

Similarly, $$\frac{1}{2^2}\tan\frac{a}{2^2}=\frac{1}{2^2}\cot\frac{a}{2^2}-\frac{1}{2}\cot\frac{a}{2};$$

....

$$\frac{1}{2^{n-1}}\tan\frac{a}{2^{n-1}}=\frac{1}{2^{n-1}}\cot\frac{a}{2^{n-1}}-\frac{1}{2^{n-2}}\cot\frac{a}{2^{n-2}}.$$

By addition, $$S=\frac{1}{2^{n-1}}\cot\frac{a}{2^{n-1}}-2\cot 2a.$$

EXAMPLES. XXIII. b.

Sum each of the following series to n terms:

1. $\cos^2\theta+\cos^2 3\theta+\cos^2 5\theta+\dots\dots.$

2. $\sin^2 a+\sin^2\left(a+\frac{\pi}{n}\right)+\sin^2\left(a+\frac{2\pi}{n}\right)+\dots\dots.$

3. $\cos^2 a+\cos^2\left(a-\frac{\pi}{n}\right)+\cos^2\left(a-\frac{2\pi}{n}\right)+\dots\dots.$

4. $\sin^3\theta+\sin^3 2\theta+\sin^3 3\theta+\dots\dots.$

5. $\sin^3 a+\sin^3\left(a+\frac{2\pi}{n}\right)+\sin^3\left(a+\frac{4\pi}{n}\right)+\dots\dots.$

6. $\cos^3 a+\cos^3\left(a-\frac{2\pi}{n}\right)+\cos^3\left(a-\frac{4\pi}{n}\right)+\dots\dots.$

7. $\tan\theta+2\tan 2\theta+2^2\tan 2^2\theta+\dots\dots$

8. $\frac{1}{\cos a+\cos 3a}+\frac{1}{\cos a+\cos 5a}+\frac{1}{\cos a+\cos 7a}+\dots\dots.$

9. $\sin^2\theta\sin 2\theta+\frac{1}{2}\sin^2 2\theta\sin 4\theta+\frac{1}{4}\sin^2 4\theta\sin 8\theta+\ldots\ldots$

10. $2\cos\theta\sin^2\frac{\theta}{2}+2^2\cos\frac{\theta}{2}\sin^2\frac{\theta}{2^2}+2^3\cos\frac{\theta}{2^2}\sin^2\frac{\theta}{2^3}+\ldots\ldots$

11. $\tan^{-1}\frac{x}{1.2+x^2}+\tan^{-1}\frac{x}{2.3+x^2}+\tan^{-1}\frac{x}{3.4+x^2}+\ldots\ldots$

12. $\tan^{-1}\frac{1}{1+1+1^2}+\tan^{-1}\frac{1}{1+2+2^2}+\tan^{-1}\frac{1}{1+3+3^2}+\ldots\ldots$

13. $\tan^{-1}\frac{2}{2+1^2+1^4}+\tan^{-1}\frac{4}{2+2^2+2^4}+\tan^{-1}\frac{6}{2+3^2+3^4}+\ldots$

14. $\tan^{-1}\frac{2}{1-1^2+1^4}+\tan^{-1}\frac{4}{1-2^2+2^4}+\tan^{-1}\frac{6}{1-3^2+3^4}+\ldots$

15. From any point on the circumference of a circle of radius r, chords are drawn to the angular points of the regular inscribed polygon of n sides: shew that the sum of the squares of the chords is $2nr^2$.

16. From a point P within a regular polygon of $2n$ sides, perpendiculars PA_1, PA_2, PA_3, ...PA_{2n} are drawn to the sides: shew that

$$PA_1+PA_3+\ldots+PA_{2n-1}=PA_2+PA_4+\ldots+PA_{2n}=nr,$$

where r is the radius of the inscribed circle.

17. If $A_1A_2A_3\ldots A_{2n+1}$ is a regular polygon and P a point on the circumscribed circle lying on the arc A_1A_{2n+1}, shew that

$$PA_1+PA_2+\ldots+PA_{2n+1}=PA_2+PA_4+\ldots+PA_{2n}.$$

18. From any point on the circumference of a circle, perpendiculars are drawn to the sides of the regular circumscribing polygon of n sides: shew that

(1) the sum of the squares of the perpendiculars is $\frac{3nr^2}{2}$;

(2) the sum of the cubes of the perpendiculars is $\frac{5nr^2}{2}$.

CHAPTER XXIV.

MISCELLANEOUS TRANSFORMATIONS AND IDENTITIES.

Symmetrical Expressions.

301. An expression is said to be *symmetrical* with respect to certain of the letters it contains, if the value of the expression remains unaltered when any pair of these letters are interchanged. Thus

$$\cos\alpha+\cos\beta+\cos\gamma, \qquad \sin\alpha\sin\beta\sin\gamma,$$
$$\tan(\alpha-\theta)+\tan(\beta-\theta)+\tan(\gamma-\theta),$$

are expressions which are symmetrical with respect to the letters α, β, γ.

302. A symmetrical expression involving the *sum* of a number of quantities may be concisely denoted by writing down one of the *terms* and prefixing the symbol Σ. Thus $\Sigma\cos\alpha$ stands for the sum of all the terms of which $\cos\alpha$ is the type, $\Sigma\sin\alpha\sin\beta$ stands for the sum of all the terms of which $\sin\alpha\sin\beta$ is the type; and so on.

For instance, if the expression is symmetrical with respect to the three letters α, β, γ,

$$\Sigma\cos\beta\cos\gamma=\cos\beta\cos\gamma+\cos\gamma\cos\alpha+\cos\alpha\cos\beta;$$
$$\Sigma\sin(\alpha-\theta)=\sin(\alpha-\theta)+\sin(\beta-\theta)+\sin(\gamma-\theta).$$

303. A symmetrical expression involving the *product* of a number of quantities may be denoted by writing down one of the *factors* and prefixing the symbol Π. Thus $\Pi\sin\alpha$ stands for the product of all the factors of which $\sin\alpha$ is the type.

For instance, if the expression is symmetrical with respect to the three letters α, β, γ,

$$\Pi\tan(\alpha+\theta)=\tan(\alpha+\theta)\tan(\beta+\theta)\tan(\gamma+\theta);$$

$$\Pi(\cos\beta+\cos\gamma)=(\cos\beta+\cos\gamma)(\cos\gamma+\cos\alpha)(\cos\alpha+\cos\beta).$$

304. With the notation just explained, certain theorems in Chap. XII. involving the three angles A, B, C, which are connected by the relation $A+B+C=180°$, may be written more concisely. For instance

$$\Sigma \sin 2A = 4\Pi \sin A;$$

$$\Sigma \sin A = 4\Pi \cos\frac{A}{2};$$

$$\Sigma \tan A = \Pi \tan A;$$

$$\Sigma \tan\frac{B}{2}\tan\frac{C}{2} = 1.$$

Example 1. Find the ratios of $a : b : c$ from the equations

$$a\cos\theta + b\sin\theta = c \quad\text{and}\quad a\cos\phi + b\sin\phi = c.$$

From the given equations, we have

$$a\cos\theta + b\sin\theta - c = 0,$$

and

$$a\cos\phi + b\sin\phi - c = 0;$$

whence by *cross multiplication*

$$\frac{a}{\sin\phi - \sin\theta} = \frac{b}{\cos\theta - \cos\phi} = \frac{c}{\sin\phi\cos\theta - \cos\phi\sin\theta};$$

$$\therefore \frac{a}{2\cos\frac{\phi+\theta}{2}\sin\frac{\phi-\theta}{2}} = \frac{b}{2\sin\frac{\phi+\theta}{2}\sin\frac{\phi-\theta}{2}} = \frac{c}{\sin(\phi-\theta)}.$$

Dividing each denominator by $2\sin\frac{\phi-\theta}{2}$, we have

$$\frac{a}{\cos\frac{\theta+\phi}{2}} = \frac{b}{\sin\frac{\theta+\phi}{2}} = \frac{c}{\cos\frac{\theta-\phi}{2}}.$$

NOTE. This result is important in Analytical Geometry.

It should be remarked that $\cos(\theta-\phi)$ is a symmetrical function of θ and ϕ, for $\cos(\theta-\phi) = \cos(\phi-\theta)$; hence the values obtained for $a : b : c$ involve θ and ϕ symmetrically.

Example 2. If α and β are two different values of θ which satisfy the equation $a\cos\theta + b\sin\theta = c$, find the values of

$$4\cos^2\frac{\alpha}{2}\cos^2\frac{\beta}{2}, \qquad \sin\alpha + \sin\beta, \qquad \sin\alpha\ \sin\beta.$$

From the given equation, by transposing and squaring,

$$(a \cos \theta - c)^2 = b^2 \sin^2 \theta = b^2 (1 - \cos^2 \theta);$$

$$\therefore (a^2 + b^2) \cos^2 \theta - 2ac \cos \theta + c^2 - b^2 = 0.$$

The roots of this quadratic in $\cos \theta$ are $\cos \alpha$ and $\cos \beta$;

$$\therefore \cos \alpha + \cos \beta = \frac{2ac}{a^2 + b^2} \text{ ....(1)},$$

and

$$\cos \alpha \cos \beta = \frac{c^2 - b^2}{a^2 + b^2} \text{ ....(2)}.$$

And

$$4 \cos^2 \frac{\alpha}{2} \cos^2 \frac{\beta}{2} = (1 + \cos \alpha)(1 + \cos \beta)$$

$$= 1 + \frac{2ac}{a^2 + b^2} + \frac{c^2 - b^2}{a^2 + b^2}$$

$$= \frac{(a + c)^2}{a^2 + b^2}.$$

From the data, we see that $\frac{\pi}{2} - \alpha$ and $\frac{\pi}{2} - \beta$ are values of θ which satisfy the equation $\quad a \sin \theta + b \cos \theta = c.$

By writing a for b and b for a, equation (1) becomes

$$\cos \left(\frac{\pi}{2} - \alpha \right) + \cos \left(\frac{\pi}{2} - \beta \right) = \frac{2bc}{b^2 + a^2},$$

or

$$\sin \alpha + \sin \beta = \frac{2bc}{a^2 + b^2}.$$

Similarly, from equation (2) we have

$$\sin \alpha \sin \beta = \frac{c^2 - a^2}{a^2 + b^2}.$$

These last two results may also be derived from the equation

$$(b \sin \theta - c)^2 = a^2 \cos^2 \theta = a^2 (1 - \sin^2 \theta).$$

Example 3. If α and β are two different values of θ which satisfy the equation $a \cos \theta + b \sin \theta = c$, prove that $\tan \frac{\alpha + \beta}{2} = \frac{b}{a}$. Also if the values of α and β are equal, shew that $a^2 + b^2 = c^2$.

By substituting $\cos \theta = \dfrac{1 - \tan^2 \frac{\theta}{2}}{1 + \tan^2 \frac{\theta}{2}}$ and $\sin \theta = \dfrac{2 \tan \frac{\theta}{2}}{1 + \tan^2 \frac{\theta}{2}}$

in the given equation $a \cos\theta + b \sin\theta = c$, we have

$$a\left(1 - \tan^2\frac{\theta}{2}\right) + 2b\tan\frac{\theta}{2} = c\left(1 + \tan^2\frac{\theta}{2}\right);$$

that is,
$$(c+a)\tan^2\frac{\theta}{2} - 2b\tan\frac{\theta}{2} + (c-a) = 0 \quad \text{..............(1)}.$$

The roots of this equation are $\tan\frac{\alpha}{2}$ and $\tan\frac{\beta}{2}$;

$$\therefore \tan\frac{\alpha}{2} + \tan\frac{\beta}{2} = \frac{2b}{c+a}, \quad \text{and} \quad \tan\frac{\alpha}{2}\tan\frac{\beta}{2} = \frac{c-a}{c+a};$$

$$\therefore \tan\frac{\alpha+\beta}{2} = \frac{2b}{c+a} \Big/ \left(1 - \frac{c-a}{c+a}\right) = \frac{b}{a}.$$

If the roots of equation (1) are equal, we have

$$b^2 = (c+a)(c-a);$$

whence
$$a^2 + b^2 = c^2.$$

NOTE. The substitution here employed is frequently used in Analytical Geometry.

Example 4. If $\cos\theta + \cos\phi = a$ and $\sin\theta + \sin\phi = b$, find the values of $\cos(\theta+\phi)$ and $\sin 2\theta + \sin 2\phi$.

From the given equations, we have

$$\frac{\sin\theta + \sin\phi}{\cos\theta + \cos\phi} = \frac{b}{a};$$

$$\therefore \tan\frac{\theta+\phi}{2} = \frac{b}{a}.$$

For shortness write t instead of $\tan\frac{\theta+\phi}{2}$; then

$$\cos(\theta+\phi) = \frac{1-t^2}{1+t^2} = \frac{a^2-b^2}{a^2+b^2},$$

and
$$\sin(\theta+\phi) = \frac{2t}{1+t^2} = \frac{2ab}{a^2+b^2}.$$

Multiplying the two given equations together, we have

$$\sin 2\theta + \sin 2\phi + 2\sin(\theta+\phi) = 2ab;$$

$$\therefore \sin 2\theta + \sin 2\phi = 2ab\left(1 - \frac{2}{a^2+b^2}\right).$$

Example 5. Resolve into factors the expression

$$\cos^2\alpha+\cos^2\beta+\cos^2\gamma+2\cos\alpha\cos\beta\cos\gamma-1,$$

and shew that it vanishes if any one of the four angles $\alpha\pm\beta\pm\gamma$ is an odd multiple of two right angles.

$$\begin{aligned}\text{The expression} &= \cos^2\alpha+(\cos^2\beta+\cos^2\gamma-1)+2\cos\alpha\cos\beta\cos\gamma\\ &= \cos^2\alpha+(\cos^2\beta-\sin^2\gamma)+2\cos\alpha\cos\beta\cos\gamma\\ &= \cos^2\alpha+\cos(\beta+\gamma)\cos(\beta-\gamma)+\cos\alpha\left\{\cos(\beta+\gamma)+\cos(\beta-\gamma)\right\}\\ &= \left\{\cos\alpha+\cos(\beta+\gamma)\right\}\left\{\cos\alpha+\cos(\beta-\gamma)\right\}\\ &= 4\cos\frac{\alpha+\beta+\gamma}{2}\cos\frac{\alpha-\beta-\gamma}{2}\cos\frac{\alpha+\beta-\gamma}{2}\cos\frac{\alpha-\beta+\gamma}{2}.\end{aligned}$$

The expression vanishes if one of the quantities $\cos\dfrac{\alpha\pm\beta\pm\gamma}{2}=0$;

that is, if one of the four angles $\dfrac{\alpha\pm\beta\pm\gamma}{2}=(2n+1)\dfrac{\pi}{2}$;

that is, if $\alpha\pm\beta\pm\gamma=(2n+1)\pi$, where n is any integer.

Example 6. If $\tan\theta=\dfrac{\sin\alpha\sin\beta}{\cos\alpha+\cos\beta}$,

prove that one value of $\tan\dfrac{\theta}{2}$ is $\tan\dfrac{\alpha}{2}\tan\dfrac{\beta}{2}$.

From the given equation, we have

$$\begin{aligned}\sec^2\theta=1+\frac{\sin^2\alpha\sin^2\beta}{(\cos\alpha+\cos\beta)^2} &= \frac{(\cos\alpha+\cos\beta)^2+(1-\cos^2\alpha)(1-\cos^2\beta)}{(\cos\alpha+\cos\beta)^2}\\ &= \frac{1+2\cos\alpha\cos\beta+\cos^2\alpha\cos^2\beta}{(\cos\alpha+\cos\beta)^2}.\end{aligned}$$

Taking the positive root, $\sec\theta=\dfrac{1+\cos\alpha\cos\beta}{\cos\alpha+\cos\beta}$;

$$\therefore\ \cos\theta=\frac{\cos\alpha+\cos\beta}{1+\cos\alpha\cos\beta}.$$

$$\therefore\ \frac{1-\cos\theta}{1+\cos\theta}=\frac{1-\cos\alpha-\cos\beta+\cos\alpha\cos\beta}{1+\cos\alpha+\cos\beta+\cos\alpha\cos\beta}=\frac{(1-\cos\alpha)(1-\cos\beta)}{(1+\cos\alpha)(1+\cos\beta)};$$

$$\therefore\ \tan^2\frac{\theta}{2}=\tan^2\frac{\alpha}{2}\tan^2\frac{\beta}{2};$$

and therefore one value of $\tan\dfrac{\theta}{2}$ is $\tan\dfrac{\alpha}{2}\tan\dfrac{\beta}{2}$.

Example 7. In any triangle, shew that

$$\Sigma a^3 \cos A = abc\,(1 + 4\Pi \cos A).$$

Let
$$k = \frac{a}{\sin A} = \frac{b}{\sin B} = \frac{c}{\sin C};$$

so that
$$a = k \sin A, \quad b = k \sin B, \quad c = k \sin C.$$

By substituting these values in the given identity, and dividing by k^3, we have to prove that

$$\Sigma \sin^3 A \cos A = \sin A \sin B \sin C\,(1 + 4\Pi \cos A).$$

Now
$$8\Sigma \sin^3 A \cos A = 4\Sigma \sin^2 A \sin 2A$$
$$= 2\Sigma\,(1 - \cos 2A) \sin 2A$$
$$= 2\Sigma \sin 2A - \Sigma \sin 4A;$$

and it has been shewn in Example 1, Art. 135, that

$$\Sigma \sin 2A = 4\Pi \sin A;$$

and it is easy to prove that

$$\Sigma \sin 4A = -4\Pi \sin 2A = -32\Pi \sin A \,.\, \Pi \cos A;$$

$$\therefore\ 8\,\Sigma \sin^3 A \cos A = 8\Pi \sin A + 32\Pi \sin A \,.\, \Pi \cos A;$$

$$\therefore\ \Sigma \sin^3 A \cos A = \Pi \sin A\,(1 + 4\Pi \cos A).$$

EXAMPLES. XXIV. a.

1. If $\theta = \alpha$, and $\theta = \beta$ satisfy the equation

$$\frac{1}{a}\cos\theta + \frac{1}{b}\sin\theta = \frac{1}{c},$$

prove that
$$a \cos\frac{\alpha+\beta}{2} = b \sin\frac{\alpha+\beta}{2} = c \cos\frac{\alpha-\beta}{2}.$$

Solve the simultaneous equations :

2. $\dfrac{x}{a}\cos\alpha + \dfrac{y}{b}\sin\alpha = 1, \qquad \dfrac{x}{a}\cos\beta + \dfrac{y}{b}\sin\beta = 1.$

3. $\dfrac{x}{a}\cos\alpha + \dfrac{y}{b}\sin\alpha = 1, \qquad \dfrac{x}{a}\sin\alpha - \dfrac{y}{b}\cos\alpha = 1.$

If α and β are two different solutions of $a\cos\theta+b\sin\theta=c$, prove that

4. $\cos(\alpha+\beta)=\dfrac{a^2-b^2}{a^2+b^2}$.   5. $\cos^2\dfrac{\alpha-\beta}{2}=\dfrac{c^2}{a^2+b^2}$.

6. $\sin 2\alpha+\sin 2\beta=\dfrac{4ab(2c^2-a^2-b^2)}{(a^2+b^2)^2}$.

7. $\sin^2\alpha+\sin^2\beta=\dfrac{2a^2(a^2+b^2)-2c^2(a^2-b^2)}{(a^2+b^2)^2}$.

8. If $a\cos\alpha+b\sin\alpha=a\cos\beta+b\sin\beta=c$, prove that

$$\sin(\alpha+\beta)=\frac{2ab}{a^2+b^2},\quad\text{and}\quad\cot\alpha+\cot\beta=\frac{2ab}{c^2-a^2}.$$

If $\cos\theta+\cos\phi=a$ and $\sin\theta+\sin\phi=b$, prove that

9. $\cos\theta\cos\phi=\dfrac{(a^2+b^2)^2-4b^2}{4(a^2+b^2)}$.

10. $\cos 2\theta+\cos 2\phi=\dfrac{(a^2-b^2)(a^2+b^2-2)}{a^2+b^2}$.

11. $\tan\theta+\tan\phi=\dfrac{8ab}{(a^2+b^2)^2-4b^2}$.

12. $\tan\dfrac{\theta}{2}+\tan\dfrac{\phi}{2}=\dfrac{4b}{a^2+b^2+2a}$.

13. Express

$$1-\cos^2\alpha-\cos^2\beta-\cos^2\gamma+2\cos\alpha\cos\beta\cos\gamma$$

as the product of four sines, and shew that it vanishes if any one of the four angles $\alpha\pm\beta\pm\gamma$ is zero or an even multiple of π.

14. Express

$$\sin^2\alpha+\sin^2\beta-\sin^2\gamma+2\sin\alpha\sin\beta\cos\gamma$$

as the product of two sines and two cosines.

15. Express

$$\sin^2\alpha+\sin^2\beta+\sin^2\gamma-2\sin\alpha\sin\beta\sin\gamma-1$$

as the product of four cosines.

16. If $$\cos\theta = \frac{\cos\alpha - \cos\beta}{1 - \cos\alpha\cos\beta},$$

prove that one value of $\tan\frac{\theta}{2}$ is $\tan\frac{\alpha}{2}\cot\frac{\beta}{2}$.

17. If $$\tan^2\theta\cos^2\frac{\alpha+\beta}{2} = \sin\alpha\sin\beta,$$

prove that one value of $\tan^2\frac{\theta}{2}$ is $\tan\frac{\alpha}{2}\tan\frac{\beta}{2}$.

18. If $$\tan\theta(\cos\alpha + \sin\beta) = \sin\alpha\cos\beta,$$
prove that one value of $\tan\frac{\theta}{2}$ is $\tan\frac{\alpha}{2}\tan\left(\frac{\pi}{4} - \frac{\beta}{2}\right)$.

In any triangle, shew that

19. $\Sigma a^3 \sin B \sin C = 2abc(1 + \cos A \cos B \cos C)$.

20. $\Sigma a \cos^3 A = \frac{abc}{4R^2}(1 - 4\cos A\cos B\cos C)$.

21. $\Sigma a^3 \cos(B - C) = 3abc$.

22. If α and β are roots of the equation $a\cos\theta + b\sin\theta = c$, form the equations whose roots are

(1) $\sin\alpha$ and $\sin\beta$; (2) $\cos 2\alpha$ and $\cos 2\beta$.

Alternating Expressions.

305. An expression is said to be *alternating* with respect to certain of the letters it contains, if the sign of the expression but not its numerical value is altered when any pair of these letters are interchanged.

Thus $\cos\alpha - \cos\beta$, $\sin(\alpha - \beta)$, $\tan(\alpha - \beta)$,
$$\cos^2\alpha\sin(\beta-\gamma) + \cos^2\beta\sin(\gamma-\alpha) + \cos^2\gamma\sin(\alpha-\beta)$$
are alternating expressions.

306. Alternating expressions may be abridged by means of the symbols Σ and Π. Thus

$$\Sigma\sin^2\alpha\sin(\beta-\gamma) = \sin^2\alpha\sin(\beta-\gamma) + \sin^2\beta\sin(\gamma-\alpha) + \sin^2\gamma\sin(\alpha-\beta);$$

$$\Pi\tan(\beta-\gamma) = \tan(\beta-\gamma)\tan(\gamma-\alpha)\tan(\alpha-\beta).$$

We shall confine our attention chiefly to alternating expressions involving the three letters α, β, γ, and we shall adopt the *cyclical arrangement* $\beta-\gamma$, $\gamma-\alpha$, $\alpha-\beta$ in which β follows α, γ follows β, and α follows γ.

Example 1. Prove that $\Sigma \cos(\alpha+\theta)\sin(\beta-\gamma)=0$.

$$\begin{aligned}\Sigma\cos(\alpha+\theta)\sin(\beta-\gamma) &= \Sigma(\cos\alpha\cos\theta-\sin\alpha\sin\theta)\sin(\beta-\gamma)\\ &= \cos\theta\,\Sigma\cos\alpha\sin(\beta-\gamma)-\sin\theta\,\Sigma\sin\alpha\sin(\beta-\gamma)\\ &= 0,\end{aligned}$$

since $\Sigma\cos\alpha\sin(\beta-\gamma)=0$ and $\Sigma\sin\alpha\sin(\beta-\gamma)=0$.

Example 2. Shew that $\Sigma\sin 2(\beta-\gamma)=-4\Pi\sin(\beta-\gamma)$.

$$\begin{aligned}&\sin 2(\beta-\gamma)+\sin 2(\gamma-\alpha)+\sin 2(\alpha-\beta)\\ &\quad=2\sin(\beta-\alpha)\cos(\alpha+\beta-2\gamma)+2\sin(\alpha-\beta)\cos(\alpha-\beta)\\ &\quad=2\sin(\alpha-\beta)\{\cos(\alpha-\beta)-\cos(\alpha+\beta-2\gamma)\}\\ &\quad=4\sin(\alpha-\beta)\sin(\alpha-\gamma)\sin(\beta-\gamma)\\ &\quad=-4\Pi\sin(\beta-\gamma).\end{aligned}$$

Example 3. Prove that

(1) $\Sigma\tan(\beta-\gamma)=\Pi\tan(\beta-\gamma)$;

(2) $\Sigma\tan\beta\tan\gamma\tan(\beta-\gamma)=-\Pi\tan(\beta-\gamma)$.

(1) From Art. 118, if $A+B+C=0$, we see that

$$\tan A+\tan B+\tan C=\tan A\tan B\tan C.$$

Hence by writing $A=\beta-\gamma$, $B=\gamma-\alpha$, $C=\alpha-\beta$, we have

$$\Sigma\tan(\beta-\gamma)=\Pi\tan(\beta-\gamma).$$

(2) From the formulæ for $\tan(\beta-\gamma)$, $\tan(\gamma-\alpha)$, $\tan(\alpha-\beta)$, we have

$$\Sigma(1+\tan\beta\tan\gamma)\tan(\beta-\gamma)=\Sigma(\tan\beta-\tan\gamma)=0;$$

whence by transposition

$$\begin{aligned}\Sigma\tan\beta\tan\gamma\tan(\beta-\gamma)&=-\Sigma\tan(\beta-\gamma)\\ &=-\Pi\tan(\beta-\gamma).\end{aligned}$$

Example 4. Shew that

$$\Sigma \cos 3\alpha \sin (\beta-\gamma)=4 \cos (\alpha+\beta+\gamma)\, \Pi \sin (\beta-\gamma).$$

Since $\quad 2 \cos 3\alpha \sin (\beta-\gamma)=\sin (3\alpha+\beta-\gamma)-\sin (3\alpha-\beta+\gamma)$,

we have

$$2\Sigma \cos 3\alpha \sin (\beta-\gamma)=\sin (3\alpha+\beta-\gamma)-\sin (3\alpha-\beta+\gamma)+\sin (3\beta+\gamma-\alpha)$$
$$-\sin (3\beta-\gamma+\alpha)+\sin (3\gamma+\alpha-\beta)-\sin (3\gamma-\alpha+\beta).$$

Combining the second and third terms, the fourth and fifth terms, the sixth and first terms, and dividing by 2, we have

$$\Sigma \cos 3\alpha \sin (\beta-\gamma)$$
$$=\cos (\alpha+\beta+\gamma)\{\sin 2(\beta-\alpha)+\sin 2(\gamma-\beta)+\sin 2(\alpha-\gamma)\}$$
$$=4 \cos (\alpha+\beta+\gamma)\, \Pi \sin (\beta-\gamma).$$ [See Example 2.]

307. The following example is given as a specimen of a concise solution.

Example. If $(y+z) \tan \alpha+(z+x) \tan \beta+(x+y) \tan \gamma=0$,

and $\quad x \tan \beta \tan \gamma+y \tan \gamma \tan \alpha+z \tan \alpha \tan \beta=x+y+z$,

prove that $\quad x \sin 2\alpha+y \sin 2\beta+z \sin 2\gamma=0$.

From the given equations, we have

$$x(1-\tan \beta \tan \gamma)+y(1-\tan \gamma \tan \alpha)+z(1-\tan \alpha \tan \beta)=0,$$

and $$x(\tan \beta+\tan \gamma)+y(\tan \gamma+\tan \alpha)+z(\tan \alpha+\tan \beta)=0.$$

If we find the values of $x : y : z$ by cross multiplication, the denominator of x

$$=(1-\tan \gamma \tan \alpha)(\tan \alpha+\tan \beta)-(1-\tan \alpha \tan \beta)(\tan \gamma+\tan \alpha)$$
$$=(\tan \beta-\tan \gamma)+\tan^2 \alpha(\tan \beta-\tan \gamma)$$
$$=(1+\tan^2 \alpha)(\tan \beta-\tan \gamma)$$
$$=\sec^2 \alpha(\tan \beta-\tan \gamma)$$
$$=\frac{\sec \alpha \sin (\beta-\gamma)}{\cos \alpha \cos \beta \cos \gamma}.$$

Hence $$\frac{x}{\sec \alpha \sin (\beta-\gamma)}=\frac{y}{\sec \beta \sin (\gamma-\alpha)}=\frac{z}{\sec \gamma \sin (\alpha-\beta)}=k \text{ say.}$$

$$\therefore\ x \sin 2\alpha+y \sin 2\beta+z \sin 2\gamma=k \Sigma \sin 2\alpha \sec \alpha \sin (\beta-\gamma)$$
$$=2k \Sigma \sin \alpha \sin (\beta-\gamma)$$
$$=0.$$

Allied formulæ in Algebra and Trigonometry.

308. From well-known algebraical identities we can deduce some interesting trigonometrical identities.

Example 1. In the identity

$$(x-a)(b-c)+(x-b)(c-a)+(x-c)(a-b)=0,$$

put $\qquad x=\cos 2\theta,\quad a=\cos 2\alpha,\quad b=\cos 2\beta,\quad c=\cos 2\gamma;$

then $\qquad x-a=\cos 2\theta-\cos 2\alpha=2\sin(\alpha+\theta)\sin(\alpha-\theta),$

and $\qquad b-c=\cos 2\beta-\cos 2\gamma=-2\sin(\beta+\gamma)\sin(\beta-\gamma);$

$$\therefore\ \Sigma\sin(\alpha+\theta)\sin(\alpha-\theta)\sin(\beta+\gamma)\sin(\beta-\gamma)=0.$$

Example 2. In the identity

$$\Sigma a^2(b-c)=-\Pi(b-c),$$

put $\qquad a=\sin^2\alpha,\quad b=\sin^2\beta,\quad c=\sin^2\gamma;$

then $\qquad b-c=\sin^2\beta-\sin^2\gamma=\sin(\beta+\gamma)\sin(\beta-\gamma);$

$$\therefore\ \Sigma\sin^4\alpha\sin(\beta+\gamma)\sin(\beta-\gamma)=-\Pi\sin(\beta+\gamma)\,.\,\Pi\sin(\beta-\gamma).$$

Example 3. In the identity

$$\Sigma a^3(b-c)=-(a+b+c)\,\Pi(b-c),$$

put $\qquad a=\cos\alpha,\quad b=\cos\beta,\quad c=\cos\gamma;$

$$\therefore\ \Sigma\cos^3\alpha(\cos\beta-\cos\gamma)=-(\cos\alpha+\cos\beta+\cos\gamma)\,\Pi(\cos\beta-\cos\gamma).$$

But $\qquad \Sigma\cos\alpha(\cos\beta-\cos\gamma)=0;$

$$\therefore\ \Sigma(4\cos^3\alpha-3\cos\alpha)(\cos\beta-\cos\gamma)$$
$$=-4(\cos\alpha+\cos\beta+\cos\gamma)\,\Pi(\cos\beta-\cos\gamma);$$

that is,

$$\Sigma\cos 3\alpha(\cos\beta-\cos\gamma)=-4(\cos\alpha+\cos\beta+\cos\gamma)\,\Pi(\cos\beta-\cos\gamma).$$

Example 4. If $a+b+c=0$, then $a^3+b^3+c^3=3abc$.

Here a, b, c may be any three quantities whose sum is zero; this condition is satisfied if we put $a=\cos(\alpha+\theta)\sin(\beta-\gamma)$, and b and c equal to corresponding quantities.

Thus $\ \Sigma\cos^3(\alpha+\theta)\sin^3(\beta-\gamma)=3\Pi\cos(\alpha+\theta)\sin(\beta-\gamma).$

309. An algebraical identity may sometimes be established by the aid of Trigonometry.

Example. If $x+y+z=xyz$, prove that

$$x(1-y^2)(1-z^2)+y(1-z^2)(1-x^2)+z(1-x^2)(1-y^2)=4xyz.$$

By putting $x=\tan\alpha$, $y=\tan\beta$, $z=\tan\gamma$, we have

$$\tan\alpha+\tan\beta+\tan\gamma=\tan\alpha\tan\beta\tan\gamma;$$

whence
$$\tan\alpha=-\frac{\tan\beta+\tan\gamma}{1-\tan\beta\tan\gamma}=-\tan(\beta+\gamma);$$

$$\therefore\ \alpha=n\pi-(\beta+\gamma),\text{ where } n \text{ is an integer};$$

$$\therefore\ \alpha+\beta+\gamma=n\pi;$$

$$\therefore\ 2\alpha+2\beta+2\gamma=2n\pi.$$

From this relation it is easy to shew that

$$\tan 2\alpha+\tan 2\beta+\tan 2\gamma=\tan 2\alpha\tan 2\beta\tan 2\gamma;$$

$$\therefore\ \frac{2x}{1-x^2}+\frac{2y}{1-y^2}+\frac{2z}{1-z^2}=\frac{8xyz}{(1-x^2)(1-y^2)(1-z^2)};$$

$$\therefore\ x(1-y^2)(1-z^2)+y(1-z^2)(1-x^2)+z(1-x^2)(1-y^2)=4xyz.$$

EXAMPLES. XXIV. b.

Prove the following identities:

1. $\Sigma\sin(\alpha-\theta)\sin(\beta-\gamma)=0.$

2. $\Sigma\cos\beta\cos\gamma\sin(\beta-\gamma)=\Sigma\sin\beta\sin\gamma\sin(\beta-\gamma).$

3. $\Sigma\sin(\beta-\gamma)\cos(\beta+\gamma+\theta)=0.$

4. $\Sigma\cos 2(\beta-\gamma)=4\Pi\cos(\beta-\gamma)-1.$

5. $\Sigma\sin\beta\sin\gamma\sin(\beta-\gamma)=-\Pi\sin(\beta-\gamma).$

6. $\Sigma\cot(\alpha-\beta)\cot(\alpha-\gamma)+1=0.$

7. $\Sigma\sin 3\alpha\sin(\beta-\gamma)=4\sin(\alpha+\beta+\gamma)\,\Pi\sin(\beta-\gamma).$

8. $\Sigma\cos^3\alpha\sin(\beta-\gamma)=\cos(\alpha+\beta+\gamma)\,\Pi\sin(\beta-\gamma).$

9. $\Sigma\cos(\theta+\alpha)\cos(\beta+\gamma)\sin(\theta-\alpha)\sin(\beta-\gamma)=0.$

10. $\Sigma\sin^2\beta\sin^2\gamma\sin(\beta+\gamma)\sin(\beta-\gamma)$
$$=-\Pi\sin(\beta+\gamma)\,.\,\Pi\sin(\beta-\gamma).$$

Prove the following identities:

11. $\Sigma \cos 2\beta \cos 2\gamma \sin(\beta+\gamma)\sin(\beta-\gamma)$
$$= -4\Pi \sin(\beta+\gamma) . \Pi \sin(\beta-\gamma).$$

12. $\Sigma \cos 4\alpha \sin(\beta+\gamma)\sin(\beta-\gamma)$
$$= -8\Pi \sin(\beta+\gamma) . \Pi \sin(\beta-\gamma).$$

13. $\Sigma \sin 3\alpha(\sin\beta - \sin\gamma)$
$$= 4(\sin\alpha+\sin\beta+\sin\gamma)\,\Pi(\sin\beta-\sin\gamma).$$

14. $\Sigma \sin^3(\beta+\gamma)\sin^3(\beta-\gamma) = 3\Pi \sin(\beta+\gamma) . \Pi \sin(\beta-\gamma).$

15. $\Sigma \cos^3(\beta+\gamma+\theta)\sin^3(\beta-\gamma)$
$$= 3\Pi \cos(\beta+\gamma+\theta) . \Pi \sin(\beta-\gamma).$$

16. If $x+y+z=xyz$, prove that
$$\Sigma \frac{3x-x^3}{1-3x^2} = \Pi \frac{3x-x^3}{1-3x^2}.$$

17. If $yz+zx+xy=1$, prove that
$$\Sigma x(1-y^2)(1-z^2) = 4xyz.$$

310. From a trigonometrical identity many others may be derived by various substitutions.

For instance, if A, B, C are *any* angles, positive or negative, connected by the relation $A+B+C=\pi$, we know that
$$\sin A + \sin B + \sin C = 4\cos\frac{A}{2}\cos\frac{B}{2}\cos\frac{C}{2}.$$

Let $\quad A=\pi-2\alpha, \quad B=\pi-2\beta, \quad C=\pi-2\gamma;$

then $\quad \sin A = \sin 2\alpha$, and $\cos\frac{A}{2} = \sin\alpha.$

Also $\quad 2(\alpha+\beta+\gamma) = 3\pi - (A+B+C) = 2\pi;$
$$\therefore \alpha+\beta+\gamma=\pi,$$
and $\quad \sin 2\alpha + \sin 2\beta + \sin 2\gamma = 4\sin\alpha\sin\beta\sin\gamma.$

Again, let $\quad A=\frac{\pi}{2}-\frac{\alpha}{2}, \quad B=\frac{\pi}{2}-\frac{\beta}{2}, \quad C=\frac{\pi}{2}-\frac{\gamma}{2};$

then $\quad \sin A = \cos\frac{\alpha}{2}$, and $\cos\frac{A}{2} = \cos\frac{\pi-\alpha}{4}.$

Also $\alpha+\beta+\gamma=3\pi-2(A+B+C)=3\pi-2\pi$;

$$\therefore\ \alpha+\beta+\gamma=\pi,$$

and $$\cos\frac{\alpha}{2}+\cos\frac{\beta}{2}+\cos\frac{\gamma}{2}=4\cos\frac{\pi-\alpha}{4}\cos\frac{\pi-\beta}{4}\cos\frac{\pi-\gamma}{4}.$$

Example. If $A+B+C=\pi$, shew that

$$\cos\frac{A}{2}+\cos\frac{B}{2}-\cos\frac{C}{2}=4\cos\frac{\pi+A}{4}\cos\frac{\pi+B}{4}\cos\frac{\pi-C}{4}.$$

Put $$\frac{\pi+A}{4}=\frac{\alpha}{2},\quad \frac{\pi+B}{4}=\frac{\beta}{2},\quad \frac{C-\pi}{4}=\frac{\gamma}{2};$$

then $\cos\frac{A}{2}=\cos\left(\alpha-\frac{\pi}{2}\right)=\sin\alpha$, and $\cos\frac{C}{2}=\cos\left(\gamma+\frac{\pi}{2}\right)=-\sin\gamma$,

so that the above identity becomes

$$\sin\alpha+\sin\beta+\sin\gamma=4\cos\frac{\alpha}{2}\cos\frac{\beta}{2}\cos\frac{\gamma}{2},$$

which is clearly true since

$$\alpha+\beta+\gamma=\frac{\pi}{2}+\frac{A+B+C}{2}=\frac{\pi}{2}+\frac{\pi}{2}=\pi.$$

311. When $A+B+C=n\pi$,

$$\tan(A+B)=\tan(n\pi-C)=-\tan C;$$

whence we obtain $\Sigma\tan A=\Pi\tan A$.

When $n=0$, the given condition is satisfied in the case of any three angles whose sum is 0; as for instance if

$$A=\beta+\gamma-2\alpha,\quad B=\gamma+\alpha-2\beta,\quad C=\alpha+\beta-2\gamma.$$

Hence $\Sigma\tan(\beta+\gamma-2\alpha)=\Pi\tan(\beta+\gamma-2\alpha)$.

Example. If $\alpha+\beta+\gamma=0$, shew that

$$\Sigma\cot(\gamma+\alpha-\beta)\cot(\alpha+\beta-\gamma)=1.$$

Put $\beta+\gamma-\alpha=A,\quad \gamma+\alpha-\beta=B,\quad \alpha+\beta-\gamma=C$;

then, by addition,

$$A+B+C=\alpha+\beta+\gamma=0;$$

$$\therefore\ \cot(A+B)=-\cot C;$$

whence $\Sigma\cot A\cot B=1$,

that is, $\Sigma\cot(\gamma+\alpha-\beta)\cot(\alpha+\beta-\gamma)=1$.

312. The following example is a further illustration of the manner in which an identity may be established by appropriate substitutions in some simpler identity.

Example. Prove that

$$2\Pi \cos(\beta+\gamma) + \Pi \cos 2\alpha = \Sigma \cos 2\alpha \cos^2(\beta+\gamma).$$

In Example 5, Art. 133, we have proved that

$$4 \cos\alpha \cos\beta \cos\gamma = \Sigma \cos(\beta+\gamma-\alpha) + \cos(\alpha+\beta+\gamma).$$

In this identity first replace α, β, γ by $\beta+\gamma, \gamma+\alpha, \alpha+\beta$ respectively, and secondly replace α, β, γ by $2\alpha, 2\beta, 2\gamma$ respectively.

Thus $\quad 8\Pi \cos(\beta+\gamma) = 2\Sigma \cos 2\alpha + 2\cos 2(\alpha+\beta+\gamma),$

and $\quad 4\Pi \cos 2\alpha = \Sigma \cos 2(\beta+\gamma-\alpha) + \cos 2(\alpha+\beta+\gamma);$

whence by addition

$$\begin{aligned} 8\Pi \cos(\beta+\gamma) + 4\Pi \cos 2\alpha \\ &= 2\Sigma \cos 2\alpha + \Sigma \cos 2(\beta+\gamma-\alpha) + 3\cos 2(\alpha+\beta+\gamma) \\ &= 2\Sigma \cos 2\alpha + \Sigma \{\cos 2(\beta+\gamma-\alpha) + \cos 2(\alpha+\beta+\gamma)\} \\ &= 2\Sigma \cos 2\alpha + 2\Sigma \cos 2(\beta+\gamma) \cos 2\alpha \\ &= 2\Sigma \cos 2\alpha \{1 + \cos 2(\beta+\gamma)\} \\ &= 4\Sigma \cos 2\alpha \cos^2(\beta+\gamma); \end{aligned}$$

$$\therefore\ 2\Pi \cos(\beta+\gamma) + \Pi \cos 2\alpha = \Sigma \cos 2\alpha \cos^2(\beta+\gamma).$$

313. Suppose that $A'B'C'$ is the pedal triangle of ABC, and let the sides and angles of the pedal triangle be denoted by a', b', c', and A', B', C', and its circum-radius by R'. Then from Arts. 224 and 225, we have

$$a' = a\cos A, \quad b' = b\cos B, \quad c' = c\cos C, \quad R' = \frac{R}{2},$$
$$A' = 180° - 2A, \quad B' = 180° - 2B, \quad C' = 180° - 2C.$$

By means of these relations, we may from any identity proved for the triangle ABC derive another, as in the following case.

In the triangle ABC, we know that

$$\Sigma a \cos A = 4R \sin A \sin B \sin C;$$

hence in the pedal triangle $A'B'C'$,

$$\Sigma a' \cos A' = 4R' \sin A' \sin B' \sin C';$$

$$\therefore\ \Sigma a \cos A \cos(180° - 2A) = 2R\Pi \sin(180° - 2A);$$

that is, $\quad -\Sigma a \cos A \cos 2A = 2R \sin 2A \sin 2B \sin 2C.$

Example. In any triangle ABC, shew that

$$\frac{a^2\cos^2 A - b^2\cos^2 B - c^2\cos^2 C}{2bc\cos B\cos C} = \cos 2A.$$

In the pedal triangle $A'B'C'$, we have

$$\frac{b'^2+c'^2-a'^2}{2b'c'} = \cos A';$$

hence, by substituting the equivalents of a', b', c', A', we have

$$\frac{b^2\cos^2 B + c^2\cos^2 C - a^2\cos^2 A}{2bc\cos B\cos C} = \cos(180° - 2A) = -\cos 2A;$$

whence the required identity follows at once.

314. If $A_1B_1C_1$ be the ex-central triangle of ABC, we may, as in the preceding article, from any identity proved for the triangle ABC derive another by means of the relations

$$a_1 = a\operatorname{cosec}\frac{A}{2}, \quad b_1 = b\operatorname{cosec}\frac{B}{2}, \quad c_1 = c\operatorname{cosec}\frac{C}{2}, \quad R_1 = 2R,$$

$$A_1 = 90° - \frac{A}{2}, \quad B_1 = 90° - \frac{B}{2}, \quad C_1 = 90° - \frac{C}{2}.$$

315. The following Exercise consists of miscellaneous questions on the subject of this Chapter.

EXAMPLES. XXIV. c.

1. Shew that

$$\Sigma\cot(2\alpha+\beta-3\gamma)\cot(2\beta+\gamma-3\alpha) = 1.$$

2. Shew that

(1) $2\Pi\sin(\beta+\gamma) + \Pi\sin 2\alpha = \Sigma\sin 2\alpha\sin^2(\beta+\gamma)$;

(2) $\Pi\sin(\beta+\gamma-\alpha) + 2\Pi\sin\alpha = \Sigma\sin^2\alpha\sin(\beta+\gamma-\alpha)$.

3. In any triangle, prove that

(1) $a^2\cos^2 A - b^2\cos^2 B = Rc\cos C\sin 2(B-A)$;

(2) $a^2\operatorname{cosec}^2\frac{A}{2} - b^2\operatorname{cosec}^2\frac{B}{2} = 4Rc\operatorname{cosec}\frac{C}{2}\sin\frac{B-A}{2}$;

(3) $\Sigma(b\cos B + c\cos C)\cot A = -2R\Sigma\cos 2A$.

4. If $$\sin 2\theta = 2\sin\alpha\sin\gamma,$$
and $$\cos 2\theta = \cos 2\alpha\cos 2\beta = \cos 2\gamma\cos 2\delta,$$
prove that one value of $\tan\theta$ is $\tan\beta\tan\delta$.

5. If $$\tan\frac{\theta}{2}\tan\frac{\phi}{2} = \tan\frac{\gamma}{2},$$
and $$\sec\alpha\cos\theta = \sec\beta\cos\phi = \cos\gamma,$$
prove that $$\sin^2\gamma = (\sec\alpha - 1)(\sec\beta - 1).$$

6. If $$\frac{\cos\theta - \cos\alpha}{\cos\theta - \cos\beta} = \frac{\sin^2\alpha\cos\beta}{\sin^2\beta\cos\alpha},$$
prove that one value of $\tan\frac{\theta}{2}$ is $\tan\frac{\alpha}{2}\tan\frac{\beta}{2}$.

7. If $\sin\theta = \cot\alpha\tan\gamma$ and $\tan\theta = \cos\alpha\tan\beta$, prove that one value of $\cos\theta$ is $\cos\beta\sec\gamma$.

8. If α and β are two different values of θ which satisfy
$$bc\cos\theta\cos\phi + ac\sin\theta\sin\phi = ab,$$
prove that
$$(b^2+c^2-a^2)\cos\alpha\cos\beta + (c^2+a^2-b^2)\sin\alpha\sin\beta = a^2+b^2-c^2.$$

9. If β and γ are two different values of θ which satisfy
$$\sin\alpha\cos\theta + \cos\alpha\sin\theta = \cos\alpha\sin\alpha,$$
prove that $$\frac{\cos\beta\cos\gamma}{\cos^2\alpha} + \frac{\sin\beta\sin\gamma}{\sin^2\alpha} = 1.$$

10. If β and γ are two different values of θ which satisfy
$$k^2\cos\alpha\cos\theta + k(\sin\alpha + \sin\theta) + 1 = 0,$$
prove that $$k^2\cos\beta\cos\gamma + k(\sin\beta + \sin\gamma) + 1 = 0.$$

11. If β and γ are two different values of θ which satisfy
$$\frac{\cos\theta\cos\phi}{\cos^2\alpha} + \frac{\sin\theta\sin\phi}{\sin^2\alpha} + 1 = 0,$$
prove that $$\frac{\cos\beta\cos\gamma}{\cos^2\alpha} + \frac{\sin\beta\sin\gamma}{\sin^2\alpha} + 1 = 0.$$

CHAPTER XXV.

MISCELLANEOUS THEOREMS AND EXAMPLES.

Inequalities. Maxima and Minima.

316. THE methods of proving trigonometrical inequalities are in many cases identical with those by which algebraical inequalities are established.

Example 1. Shew that $a^2\tan^2\theta + b^2\cot^2\theta > 2ab$.

We have $a^2\tan^2\theta + b^2\cot^2\theta = (a\tan\theta - b\cot\theta)^2 + 2ab$;

$$\therefore\ a^2\tan^2\theta + b^2\cot^2\theta > 2ab,$$

unless $a\tan\theta - b\cot\theta = 0$, or $a\tan^2\theta = b$.

In this case the inequality becomes an equality.

This proposition may be otherwise expressed by saying that the *minimum value* of $a^2\tan^2\theta + b^2\cot^2\theta$ is $2ab$.

Example 2. Shew that

$$1 + \sin^2\alpha + \sin^2\beta > \sin\alpha + \sin\beta + \sin\alpha\sin\beta.$$

Since $(1 - \sin\alpha)^2$ is positive,

$$1 + \sin^2\alpha > 2\sin\alpha;$$

similarly $1 + \sin^2\beta > 2\sin\beta$,

and $\sin^2\alpha + \sin^2\beta > 2\sin\alpha\sin\beta$.

Adding and dividing by 2, we have

$$1 + \sin^2\alpha + \sin^2\beta > \sin\alpha + \sin\beta + \sin\alpha\sin\beta.$$

Example 3. When is $12\sin\theta - 9\sin^2\theta$ a maximum?

The expression $= 4 - (2 - 3\sin\theta)^2$, and is therefore a maximum when $2 - 3\sin\theta = 0$, so that its maximum value is 4.

317. *To find the numerically greatest values of*

$$a \cos\theta + b \sin\theta.$$

Let $a = r\cos\alpha$ and $b = r\sin\alpha$,

so that $r^2 = a^2 + b^2$ and $\tan\alpha = \frac{b}{a}$;

then $a\cos\theta + b\sin\theta = r(\cos\theta\cos\alpha + \sin\theta\sin\alpha)$

$$= r\cos(\theta - \alpha).$$

Thus the expression is numerically greatest when

$$\cos(\theta - \alpha) = \pm 1;$$

that is, the greatest positive value $= r = \sqrt{a^2+b^2}$,
and the numerically greatest negative value $= -r = -\sqrt{a^2+b^2}$.

Hence, if $c^2 > a^2 + b^2$,

the maximum value of $a\cos\theta + b\sin\theta + c$ is $c + \sqrt{a^2+b^2}$,
and the minimum value is $c - \sqrt{a^2+b^2}$.

318. The expression $a\cos(\alpha+\theta) + b\cos(\beta+\theta)$

$$= (a\cos\alpha + b\cos\beta)\cos\theta - (a\sin\alpha + b\sin\beta)\sin\theta;$$

and therefore its numerically greatest values are equal to the positive and negative square roots of

$$(a\cos\alpha + b\cos\beta)^2 + (a\sin\alpha + b\sin\beta)^2;$$

that is, are equal to

$$\pm\sqrt{a^2 + b^2 + 2ab\cos(\alpha - \beta)}.$$

In like manner, we may find the maximum and minimum values of the sum of any number of expressions of the form $a\cos(\alpha+\theta)$ or $a\sin(\alpha+\theta)$.

319. *If* α *and* β *are two angles, each lying between* 0 *and* $\frac{\pi}{2}$, *whose sum is given, to find the maximum value of* $\cos\alpha\cos\beta$ *and of* $\cos\alpha + \cos\beta$.

Suppose that $\alpha + \beta = \sigma$;

then $2\cos\alpha\cos\beta = \cos(\alpha+\beta) + \cos(\alpha-\beta)$

$$= \cos\sigma + \cos(\alpha - \beta),$$

and is therefore a maximum when $\alpha-\beta=0$, or $\alpha=\beta=\frac{\sigma}{2}$.

Thus the maximum value of $\cos\alpha\cos\beta$ is $\cos^2\frac{\sigma}{2}$.

Again, $$\cos\alpha+\cos\beta=2\cos\frac{\alpha+\beta}{2}\cos\frac{\alpha-\beta}{2}$$
$$=2\cos\frac{\sigma}{2}\cos\frac{\alpha-\beta}{2},$$

and is therefore a maximum when $\alpha=\beta=\frac{\sigma}{2}$.

Thus the maximum value of $\cos\alpha+\cos\beta$ is $2\cos\frac{\sigma}{2}$.

Similar theorems hold in case of the sine.

Example 1. If A, B, C are the angles of a triangle, find the maximum value of

$$\sin A+\sin B+\sin C \text{ and of } \sin A\sin B\sin C.$$

Let us suppose that C remains constant, while A and B vary.

$$\sin A+\sin B+\sin C=2\sin\frac{A+B}{2}\cos\frac{A-B}{2}+\sin C$$
$$=2\cos\frac{C}{2}\cos\frac{A-B}{2}+\sin C.$$

This expression is a maximum when $A=B$.

Hence, so long as any two of the angles A, B, C are unequal, the expression $\sin A+\sin B+\sin C$ is not a maximum; that is, the expression is a maximum when $A=B=C=60°$.

Thus the maximum value $=3\sin 60°=\frac{3\sqrt{3}}{2}$.

Again,
$$2\sin A\sin B\sin C=\{\cos(A-B)-\cos(A+B)\}\sin C$$
$$=\{\cos(A-B)+\cos C\}\sin C.$$

This expression is a maximum when $A=B$.

Hence, by reasoning as before, $\sin A\sin B\sin C$ has its maximum value when $A=B=C=60°$.

Thus the maximum value $=\sin^3 60°=\frac{3\sqrt{3}}{8}$.

Example 2. If α and β are two angles, each lying between 0 and $\frac{\pi}{2}$, whose sum is constant, find the minimum value of $\sec\alpha+\sec\beta$.

We have $$\sec\alpha+\sec\beta=\frac{1}{\cos\alpha}+\frac{1}{\cos\beta}=\frac{\cos\alpha+\cos\beta}{\cos\alpha\cos\beta}$$

$$=\frac{4\cos\frac{\alpha+\beta}{2}\cos\frac{\alpha-\beta}{2}}{\cos(\alpha+\beta)+\cos(\alpha-\beta)}=\frac{2\cos\frac{\alpha+\beta}{2}\cos\frac{\alpha-\beta}{2}}{\cos^2\frac{\alpha-\beta}{2}-\sin^2\frac{\alpha+\beta}{2}}$$

$$=\cos\frac{\alpha+\beta}{2}\left(\frac{1}{\cos\frac{\alpha-\beta}{2}+\sin\frac{\alpha+\beta}{2}}+\frac{1}{\cos\frac{\alpha-\beta}{2}-\sin\frac{\alpha+\beta}{2}}\right).$$

Since $\alpha+\beta$ is constant, this expression is least when the denominators are greatest; that is, when $\alpha=\beta=\frac{\alpha+\beta}{2}$.

Thus the minimum value is $2\sec\frac{\alpha+\beta}{2}$.

320. *If $\alpha, \beta, \gamma, \delta, \ldots\ldots$ are n angles, each lying between* 0 *and $\frac{\pi}{2}$, whose sum is constant, to find the maximum value of*

$$\cos\alpha\cos\beta\cos\gamma\cos\delta\ldots\ldots$$

Let $$\alpha+\beta+\gamma+\delta+\ldots\ldots=\sigma.$$

Suppose that any two of the angles, say α and β, are unequal; then if in the given product we replace the two unequal factors $\cos\alpha$ and $\cos\beta$ by the two equal factors $\cos\frac{\alpha+\beta}{2}$ and $\cos\frac{\alpha+\beta}{2}$, the value of the product is increased while the sum of the angles remains unaltered. Hence so long as any two of the angles $\alpha, \beta, \gamma, \delta, \ldots$ are unequal the product is not a maximum; that is, the product is a maximum when all the angles are equal. In this case each angle $=\frac{\sigma}{n}$.

Thus the maximum value is $\cos^n\frac{\sigma}{n}$.

In like manner we may shew that

the maximum value of $\cos\alpha+\cos\beta+\cos\gamma+\ldots\ldots=n\cos\frac{\sigma}{n}$.

321. The methods of solution used in the following examples are worthy of notice.

Example 1. Shew that $\tan 3\alpha \cot \alpha$ cannot lie between 3 and $\frac{1}{3}$.

We have $\tan 3\alpha \cot \alpha = \frac{\tan 3\alpha}{\tan \alpha} = \frac{3 - \tan^2 \alpha}{1 - 3\tan^2 \alpha} = n$ say;

$$\therefore \tan^2 \alpha = \frac{n-3}{3n-1} = \frac{3-n}{1-3n}.$$

These two fractional values of $\tan^2 \alpha$ must be positive, and therefore n must be greater than 3 or less than $\frac{1}{3}$.

Example 2. If a and b are positive quantities, of which a is the greater, find the minimum value of $a \sec \theta - b \tan \theta$.

Denote the expression by x, and put $\tan \theta = t$;

then
$$x = a\sqrt{1+t^2} - bt;$$

$$\therefore b^2t^2 + 2bxt + x^2 = a^2(1+t^2);$$

$$\therefore t^2(b^2 - a^2) + 2bxt + x^2 - a^2 = 0.$$

In order that the values of t found from this equation may be real,

$$b^2x^2 > (b^2 - a^2)(x^2 - a^2);$$

$$\therefore 0 > a^2(a^2 - b^2 - x^2);$$

$$\therefore x^2 > a^2 - b^2.$$

Thus the minimum value is $\sqrt{a^2 - b^2}$.

Example 3. If a, b, c, k are constant quantities and α, β, γ variable quantities subject to the relation $a \tan \alpha + b \tan \beta + c \tan \gamma = k$, find the minimum value of $\tan^2 \alpha + \tan^2 \beta + \tan^2 \gamma$.

By multiplying out and re-arranging the terms, we have

$$(a^2+b^2+c^2)(\tan^2 \alpha + \tan^2 \beta + \tan^2 \gamma) - (a \tan \alpha + b \tan \beta + c \tan \gamma)^2$$
$$= (b \tan \gamma - c \tan \beta)^2 + (c \tan \alpha - a \tan \gamma)^2 + (a \tan \beta - b \tan \alpha)^2.$$

But the minimum value of the right side of this equation is zero; hence the minimum value of

$$(a^2+b^2+c^2)(\tan^2 \alpha + \tan^2 \beta + \tan^2 \gamma) - k^2 = 0;$$

that is, the minimum value of

$$\tan^2 \alpha + \tan^2 \beta + \tan^2 \gamma = \frac{k^2}{a^2+b^2+c^2}.$$

EXAMPLES. XXV. a.

When θ is variable find the minimum value of the following expressions:

1. $p\cot\theta+q\tan\theta$.
2. $4\sin^2\theta+\operatorname{cosec}^2\theta$.
3. $8\sec^2\theta+18\cos^2\theta$.
4. $3-2\cos\theta+\cos^2\theta$.

Prove the following inequalities:

5. $\tan^2\alpha+\tan^2\beta+\tan^2\gamma>\tan\beta\tan\gamma+\tan\gamma\tan\alpha+\tan\alpha\tan\beta$.
6. $\sin^2\alpha+\sin^2\beta>2(\sin\alpha+\sin\beta-1)$.

When θ is variable, find the maximum value of

7. $\sin\theta+\cos\theta$.
8. $\cos\theta+\sqrt{3}\sin\theta$.
9. $a\cos(\alpha+\theta)+b\sin\theta$.
10. $p\cos\theta+q\sin(\alpha+\theta)$.

If $\sigma=\alpha+\beta$, where α and β are two angles each lying between 0 and $\frac{\pi}{2}$, and σ is constant, find the maximum or minimum value of

11. $\sin\alpha+\sin\beta$.
12. $\sin\alpha\sin\beta$.
13. $\tan\alpha+\tan\beta$.
14. $\operatorname{cosec}\alpha+\operatorname{cosec}\beta$.

If A, B, C are the angles of a triangle, find the maximum or minimum value of

15. $\cos A\cos B\cos C$.
16. $\cot A+\cot B+\cot C$.
17. $\sin^2\frac{A}{2}+\sin^2\frac{B}{2}+\sin^2\frac{C}{2}$.
18. $\sec A+\sec B+\sec C$.
19. $\tan^2\frac{A}{2}+\tan^2\frac{B}{2}+\tan^2\frac{C}{2}$. $\left[\textit{Use } \Sigma\tan\frac{B}{2}\tan\frac{C}{2}=1.\right]$
20. $\cot^2 A+\cot^2 B+\cot^2 C$. [*Use* $\Sigma\cot B\cot C=1$.]
21. If $b^2<4ac$, find the maximum and minimum values of
$$a\sin^2\theta+b\sin\theta\cos\theta+c\cos^2\theta.$$

22. If α, β, γ lie between 0 and $\frac{\pi}{2}$, shew that

$$\sin\alpha+\sin\beta+\sin\gamma>\sin(\alpha+\beta+\gamma).$$

23. If a and b are two positive quantities of which a is the greater, shew that $a\operatorname{cosec}\theta>b\cot\theta+\sqrt{a^2-b^2}$.

24. Shew that $\frac{\sec^2\theta-\tan\theta}{\sec^2\theta+\tan\theta}$ lies between 3 and $\frac{1}{3}$.

25. Find the maximum value of $\frac{\tan^2\theta-\cot^2\theta+1}{\tan^2\theta+\cot^2\theta-1}$.

26. If a, b, c, k are constant positive quantities, and α, β, γ variable quantities subject to the relation

$$a\cos\alpha+b\cos\beta+c\cos\gamma=k,$$

find the minimum value of

$$\cos^2\alpha+\cos^2\beta+\cos^2\gamma \text{ and of } a\cos^2\alpha+b\cos^2\beta+c\cos^2\gamma.$$

Elimination.

322. No general rules can be given for the elimination of some assigned quantity or quantities from two or more trigonometrical equations. The form of the equations will often suggest special methods, and in addition to the usual algebraical artifices we shall always have at our disposal the identical relations subsisting between the trigonometrical functions. Thus suppose it is required to eliminate θ from the equations

$$x\cos\theta=a, \quad y\cot\theta=b.$$

Here $$\sec\theta=\frac{x}{a}, \text{ and } \tan\theta=\frac{y}{b};$$

but for all values of θ, we have

$$\sec^2\theta-\tan^2\theta=1.$$

$\therefore$ by substitution,

$$\frac{x^2}{a^2}-\frac{y^2}{b^2}=1.$$

From this example we see that since θ satisfies *two* equations (either of which is sufficient to determine θ) there is a relation, independent of θ, which subsists between the coefficients and constants of the equations. To determine this relation we *eliminate* θ, and the result is called the *eliminant* of the given equations.

323. The following examples will illustrate some useful methods of elimination.

Example 1. Eliminate θ between the equations

$$l\cos\theta+m\sin\theta+n=0 \text{ and } p\cos\theta+q\sin\theta+r=0.$$

From the given equations, we have by cross multiplication

$$\frac{\cos\theta}{mr-nq}=\frac{\sin\theta}{np-lr}=\frac{1}{lq-mp};$$

$$\therefore \cos\theta=\frac{mr-nq}{lq-mp}, \text{ and } \sin\theta=\frac{np-lr}{lq-mp};$$

whence by squaring, adding, and clearing of fractions, we obtain

$$(mr-nq)^2+(np-lr)^2=(lq-mp)^2.$$

The particular instance in which $q=l$ and $p=-m$ is of frequent occurrence in Analytical Geometry. In this case the eliminant may be written down at once; for we have

$$l\cos\theta+m\sin\theta=-n,$$

and

$$l\sin\theta-m\cos\theta=-r;$$

whence by squaring and adding, we obtain

$$l^2+m^2=n^2+r^2.$$

Example 2. Eliminate θ between the equations

$$\frac{ax}{\cos\theta}-\frac{by}{\sin\theta}=c^2 \text{ and } l\tan\theta=m.$$

From the second equation, we have

$$\frac{\sin\theta}{m}=\frac{\cos\theta}{l}=\frac{\sqrt{\sin^2\theta+\cos^2\theta}}{\sqrt{m^2+l^2}}=\frac{1}{\sqrt{m^2+l^2}};$$

$$\therefore \sin\theta=\frac{m}{\sqrt{m^2+l^2}}, \text{ and } \cos\theta=\frac{l}{\sqrt{m^2+l^2}}.$$

By substituting in the first equation, we obtain

$$\frac{ax}{l}-\frac{by}{m}=\frac{c^2}{\sqrt{m^2+l^2}}.$$

Example 3. Eliminate θ between the equations

$$x=\cot\theta+\tan\theta \text{ and } y=\sec\theta-\cos\theta.$$

From the given equations, we have

$$x=\frac{1}{\tan\theta}+\tan\theta=\frac{1+\tan^2\theta}{\tan\theta}$$

$$=\frac{\sec^2\theta}{\tan\theta},$$

and

$$y=\sec\theta-\frac{1}{\sec\theta}=\frac{\sec^2\theta-1}{\sec\theta}$$

$$=\frac{\tan^2\theta}{\sec\theta}.$$

From these values of x and y we obtain

$$x^2y=\sec^3\theta \text{ and } xy^2=\tan^3\theta.$$

But

$$\sec^2\theta-\tan^2\theta=1;$$

$$\therefore (x^2y)^{\frac{2}{3}}-(xy^2)^{\frac{2}{3}}=1;$$

that is,

$$x^{\frac{4}{3}}y^{\frac{2}{3}}-x^{\frac{2}{3}}y^{\frac{4}{3}}=1.$$

Example 4. Eliminate θ from the equations

$$\frac{x}{a}=\cos\theta+\cos 2\theta \text{ and } \frac{y}{b}=\sin\theta+\sin 2\theta.$$

From the given equations, we have

$$\frac{x}{a}=2\cos\frac{3\theta}{2}\cos\frac{\theta}{2},$$

and

$$\frac{y}{b}=2\sin\frac{3\theta}{2}\cos\frac{\theta}{2};$$

whence by squaring and adding, we obtain

$$\frac{x^2}{a^2}+\frac{y^2}{b^2}=4\cos^2\frac{\theta}{2}.$$

But

$$\frac{x}{a}=2\cos\frac{\theta}{2}\left(4\cos^3\frac{\theta}{2}-3\cos\frac{\theta}{2}\right)$$

$$=2\cos^2\frac{\theta}{2}\left(4\cos^2\frac{\theta}{2}-3\right);$$

$$\therefore \frac{2x}{a}=\left(\frac{x^2}{a^2}+\frac{y^2}{b^2}\right)\left(\frac{x^2}{a^2}+\frac{y^2}{b^2}-3\right).$$

324. The following examples are instances of the elimination of two quantities.

Example 1. Eliminate θ and ϕ from the equations

$$a\sin^2\theta+b\cos^2\theta=m, \quad b\sin^2\phi+a\cos^2\phi=n, \quad a\tan\theta=b\tan\phi.$$

From the first equation, we have

$$a\sin^2\theta+b\cos^2\theta=m(\sin^2\theta+\cos^2\theta);$$

$$\therefore\ (a-m)\sin^2\theta=(m-b)\cos^2\theta;$$

$$\therefore\ \tan^2\theta=\frac{m-b}{a-m}.$$

From the second equation, we have

$$b\sin^2\phi+a\cos^2\phi=n(\sin^2\phi+\cos^2\phi);$$

$$\therefore\ \tan^2\phi=\frac{n-a}{b-n}.$$

From the third equation,

$$a^2\tan^2\theta=b^2\tan^2\phi;$$

$$\therefore\ \frac{a^2(m-b)}{a-m}=\frac{b^2(n-a)}{b-n};$$

$$\therefore\ a^2(bm-b^2-mn+bn)=b^2(an-a^2-mn+am);$$

$$\therefore\ mab(a-b)+nab(a-b)=mn(a^2-b^2);$$

$$\therefore\ mab+nab=mn(a+b);$$

$$\therefore\ \frac{1}{n}+\frac{1}{m}=\frac{1}{a}+\frac{1}{b}.$$

Example 2. Eliminate θ and ϕ from the equations

$$x\cos\theta+y\sin\theta=x\cos\phi+y\sin\phi=2a, \quad 2\sin\frac{\theta}{2}\sin\frac{\phi}{2}=1.$$

From the data, we see that θ and ϕ are the roots of the equation

$$x\cos\alpha+y\sin\alpha=2a;$$

$$\therefore\ (x\cos\alpha-2a)^2=y^2\sin^2\alpha=y^2(1-\cos^2\alpha);$$

$$\therefore\ (x^2+y^2)\cos^2\alpha-4ax\cos\alpha+4a^2-y^2=0,$$

which is a quadratic in $\cos\alpha$ with roots $\cos\theta$ and $\cos\phi$.

But $$1=4\sin^2\frac{\theta}{2}\sin^2\frac{\phi}{2}=(1-\cos\theta)(1-\cos\phi);$$

whence $$\cos\theta+\cos\phi=\cos\theta\cos\phi;$$

$$\therefore\ \frac{4ax}{x^2+y^2}=\frac{4a^2-y^2}{x^2+y^2};$$

$$\therefore\ y^2=4a(a-x).$$

325. The method exhibited in the following example is one frequently used in Analytical Geometry.

Example. If a, b, c are unequal, find the relations that hold between the coefficients, when

$$a\cos\theta+b\sin\theta=c,$$

and $$a\cos^2\theta+2a\cos\theta\sin\theta+b\sin^2\theta=c.$$

The required relation will be obtained by eliminating θ from the given equations. This is most conveniently done by making each equation homogeneous in $\sin\theta$ and $\cos\theta$.

From the first equation, we have

$$a\cos\theta+b\sin\theta=c\sqrt{\cos^2\theta+\sin^2\theta};$$

whence, by squaring and transposing,

$$(a^2-c^2)\cos^2\theta+2ab\cos\theta\sin\theta+(b^2-c^2)\sin^2\theta=0\ \ldots\ldots(1).$$

From the second equation, we have

$$a\cos^2\theta+2a\cos\theta\sin\theta+b\sin^2\theta=c(\cos^2\theta+\sin^2\theta);$$

$$\therefore\ (a-c)\cos^2\theta+2a\cos\theta\sin\theta+(b-c)\sin^2\theta=0\ \ldots\ldots(2).$$

From (1) and (2) we have by cross-multiplication,

$$\frac{\cos^2\theta}{2ab(b-c)-2a(b^2-c^2)}=\frac{\cos\theta\sin\theta}{(b^2-c^2)(a-c)-(a^2-c^2)(b-c)}$$

$$=\frac{\sin^2\theta}{2a(a^2-c^2)-2ab(a-c)};$$

or $$\frac{\cos^2\theta}{-2ac(b-c)}=\frac{\cos\theta\sin\theta}{(b-c)(a-c)(b-a)}=\frac{\sin^2\theta}{2a(a-c)(a+c-b)};$$

$$\therefore\ -4a^2c(b-c)(a-c)(a+c-b)=(b-c)^2(a-c)^2(b-a)^2.$$

By supposition, the quantities a, b, c are unequal; hence dividing by $(b-c)(a-c)$, we obtain

$$4a^2c(a+c-b)+(b-c)(a-c)(a-b)^2=0.$$

EXAMPLES. XXV. b.

Eliminate θ between the equations:

1. $\frac{x}{a}\cos\theta+\frac{y}{b}\sin\theta=1,\quad \frac{x}{a}\sin\theta-\frac{y}{b}\cos\theta=1.$

2. $a\sec\theta-x\tan\theta=y,\quad b\sec\theta+y\tan\theta=x.$

3. $\cos\theta+\sin\theta=a,\quad \cos 2\theta=b,$

4. $x=\sin\theta+\cos\theta,\quad y=\tan\theta+\cot\theta.$

5. $a=\cot\theta+\cos\theta,\quad b=\cot\theta-\cos\theta.$

Find the eliminant in each of the following cases:

6. $x=\cot\theta+\tan\theta,\quad y=\operatorname{cosec}\theta-\sin\theta.$

7. $\operatorname{cosec}\theta-\sin\theta=a^3,\quad \sec\theta-\cos\theta=b^3.$

8. $4x=3a\cos\theta+a\cos 3\theta,\quad 4y=3a\sin\theta-a\sin 3\theta.$

9. $x=\tan^2\theta(a\tan\theta-x),\quad y=\sec^2\theta(y-a\sec\theta).$

10. $x=a\cos\theta(2\cos 2\theta-1),\quad y=b\sin\theta(4\cos^2\theta-1).$

11. If $\cos(\theta-\alpha)=a$, and $\sin(\theta-\beta)=b$,
shew that $a^2-2ab\sin(\alpha-\beta)+b^2=\cos^2(\alpha-\beta).$

Find the relation that must hold between x and y if

12. $x+y=3-\cos 4\theta,\quad x-y=4\sin 2\theta.$

13. $x=\sin\theta+\cos\theta\sin 2\theta,\quad y=\cos\theta+\sin\theta\sin 2\theta.$

14. If $\sin\theta+\cos\theta=a$, and $\sin 2\theta+\cos 2\theta=b$,
shew that $(a^2-b-1)^2=a^2(2-a^2).$

15. If $\cos\theta-\sin\theta=b$, and $\cos 3\theta+\sin 3\theta=a$,
shew that $a=3b-2b^3.$

16. Eliminate θ from the equations:
$$a\cos\theta-b\sin\theta=c,\quad 2ab\cos 2\theta+(a^2-b^2)\sin 2\theta=2c^2.$$

17. If $x=a\cos\theta+b\cos 2\theta$, and $y=a\sin\theta+b\sin 2\theta$,
shew that $a^2\{(x+b)^2+y^2\}=(x^2+y^2-b^2)^2.$

18. If $\dfrac{\tan(\theta+\alpha)}{\tan(\theta-\alpha)}=\dfrac{a+b}{a-b}$, and $a\cos 2\alpha+b\cos 2\theta=c$,

shew that $$a^2+c^2-2ac\cos 2\alpha=b^2.$$

19. If $x=a(\sin 3\theta-\sin\theta)$, and $y=a(\cos\theta-\cos 3\theta)$,

shew that $$(x^2+y^2)(2a^2-x^2-y^2)^2=4a^4x^2.$$

Eliminate θ from the equations:

20. $x\cos\theta-y\sin\theta=a\cos 2\theta,\quad x\sin\theta+y\cos\theta=2a\sin 2\theta.$

21. $x\sin\theta-y\cos\theta=\sqrt{x^2+y^2},\quad \dfrac{\cos^2\theta}{a^2}+\dfrac{\sin^2\theta}{b^2}=\dfrac{1}{x^2+y^2}.$

22. $\dfrac{x\cos\theta}{a}+\dfrac{y\sin\theta}{b}=1,\quad x\sin\theta-y\cos\theta=\sqrt{a^2\sin^2\theta+b^2\cos^2\theta}.$

23. If $\cos(\alpha-3\theta)=m\cos^3\theta$, and $\sin(\alpha-3\theta)=m\sin^3\theta$,

shew that $$m^2+m\cos\alpha=2.$$

Eliminate θ and ϕ from the equations:

24. $\tan\theta+\tan\phi=x,\quad \cot\theta+\cot\phi=y,\quad \theta+\phi=\alpha.$

25. $\sin\theta+\sin\phi=a,\quad \cos\theta+\cos\phi=b,\quad \theta-\phi=\alpha.$

26. $a\sin^2\theta+b\cos^2\theta=a\cos^2\phi+b\sin^2\phi=1,\quad a\tan\theta=b\tan\phi.$

27. If $\dfrac{x}{a}\cos\theta+\dfrac{y}{b}\sin\theta=\dfrac{x}{a}\cos\phi+\dfrac{y}{b}\sin\phi=1$, and $\theta-\phi=\alpha$,

shew that $$\frac{x^2}{a^2}+\frac{y^2}{b^2}=\sec^2\frac{\alpha}{2}.$$

28. If $\tan\theta+\tan\phi=a,\quad \cot\theta+\cot\phi=b,\quad \theta-\phi=\alpha$,

shew that $$ab(ab-4)=(a+b)^2\tan^2\alpha.$$

Eliminate θ and ϕ between the equations:

29. $a\cos^2\theta+b\sin^2\theta=m\cos^2\phi,\quad a\sin^2\theta+b\cos^2\theta=n\sin^2\phi,$
$$m\tan^2\theta-n\tan^2\phi=0.$$

30. $x\cos\theta+y\sin\theta=2a\sqrt{3},\quad x\cos(\theta+\phi)+y\sin(\theta+\phi)=4a,$
$$x\cos(\theta-\phi)+y\sin(\theta-\phi)=2a.$$

31. $c\sin\theta=a\sin(\theta+\phi),\quad a\sin\phi=b\sin\theta,\quad \cos\theta-\cos\phi=2m.$

Application of Trigonometry to the Theory of Equations.

326. In the Theory of Equations it is shewn that the solution of *any* cubic equation may be made to depend on the solution of a cubic equation of the form $x^3+ax+b=0$. In certain cases the solution is very conveniently obtained by Trigonometry.

327. Consider the equation

$$x^3-qx-r=0 \quad \text{.........................(1)},$$

in which each of the letters q and r represents a positive quantity.

From the identity $\cos 3\theta=4\cos^3\theta-3\cos\theta$,

we have
$$\cos^3\theta-\frac{3}{4}\cos\theta-\frac{\cos 3\theta}{4}=0 \quad \text{.................(2)}.$$

Let $x=y\cos\theta$, where y is a positive quantity; then from (1),

$$\cos^3\theta-\frac{q}{y^2}\cos\theta-\frac{r}{y^3}=0 \quad \text{.....................(3)}.$$

If the equations (2) and (3) are identical, we have $\frac{q}{y^2}=\frac{3}{4}$, so that $y=+\sqrt{\frac{4q}{3}}$, since y is positive; and

$$\frac{\cos 3\theta}{4}=\frac{r}{y^3}=\sqrt{\frac{27r^2}{64q^3}};$$

whence
$$\cos 3\theta=\sqrt{\frac{27r^2}{4q^3}}.$$

Hence the values of θ are real if $27r^2<4q^3$;

that is, if
$$\left(\frac{r}{2}\right)^2<\left(\frac{q}{3}\right)^3.$$

Let α be the smallest angle whose cosine is equal to $\sqrt{\frac{27r^2}{4q^3}}$; then $\cos 3\theta=\cos\alpha$; whence $3\theta=2n\pi\pm\alpha$.

Thus the values of $\cos\theta$ are

$$\cos\frac{\alpha}{3},\quad \cos\frac{2\pi+\alpha}{3},\quad \cos\frac{2\pi-\alpha}{3}.$$ [See Art. 264.]

But $$x = y\cos\theta = \sqrt{\frac{4q}{3}}\cos\theta,$$

and therefore the roots of $x^3 - qx - r = 0$ are

$$\sqrt{\frac{4q}{3}}\cos a,\quad \sqrt{\frac{4q}{3}}\cos\frac{2\pi + a}{3},\quad \sqrt{\frac{4q}{3}}\cos\frac{2\pi - a}{3}.$$

328. Following the method explained in the preceding article, we may use the identity

$$\sin^3\theta - \frac{3}{4}\sin\theta + \frac{\sin 3\theta}{4} = 0$$

to obtain the solution of the equation

$$x^3 - qx + r = 0,$$

each of the quantities represented by q and r being positive.

Example. Solve the equation $x^3 - 12x + 8 = 0$.

We have $$\sin^3\theta - \frac{3}{4}\sin\theta + \frac{\sin 3\theta}{4} = 0.$$

In the given equation put $x = y\sin\theta$, where y is positive; then

$$\sin^3\theta - \frac{12}{y^2}\sin\theta + \frac{8}{y^3} = 0.$$

$$\therefore\ \frac{3}{4} = \frac{12}{y^2};\ \text{whence } y = 4;$$

and $$\frac{\sin 3\theta}{4} = \frac{8}{y^3} = \frac{1}{8};\ \text{whence } \sin 3\theta = \frac{1}{2}.$$

Suppose that θ is estimated in sexagesimal measure; then

$$3\theta = n\,.\,180^\circ + (-1)^n 30^\circ.$$

By ascribing to n the values 0, 1, 2, 3, 4 we obtain

$$\theta = 10^\circ,\quad \theta = 50^\circ,\quad \theta = 130^\circ,\quad \theta = 170^\circ,\quad \theta = 250^\circ;$$

and by further ascribing to n the values 5, 6, 7,... it will easily be seen that the values of $\sin\theta$ are equal to some one of the three quantities

$$\sin 10^\circ,\quad \sin 50^\circ,\quad -\sin 70^\circ.$$

But $x = y\sin\theta = 4\sin\theta$, and therefore the roots are

$$4\sin 10^\circ,\quad 4\sin 50^\circ,\quad -4\sin 70^\circ.$$

Application of the Theory of Equations to Trigonometry.

329. In the Theory of Equations it is shewn that the equation whose roots are $a_1, a_2, a_3, \ldots\ldots, a_n$ is

$$(x-a_1)(x-a_2)(x-a_3)\ldots\ldots\ldots\ldots(x-a_n)=0,$$

or $$x^n - S_1x^{n-1} + S_2x^{n-2} - S_3x^{n-3} + \ldots\ldots + (-1)^nS_n = 0,$$

where S_1 = sum of the roots;

S_2 = sum of the products of the roots taken two at a time;

S_3 = sum of the products of the roots taken three at a time;

....

S_n = product of the roots.

[See Hall and Knight's *Higher Algebra*, Art. 538 and Art. 539.]

Example 1. If α, β, γ are the values of θ which satisfy the equation

$$a\tan^3\theta + (2a-x)\tan\theta + y = 0 \quad \ldots\ldots\ldots\ldots\ldots\ldots(1),$$

shew that (i) if $\tan\alpha + \tan\beta = h$, then $ah^3 + (2a-x)h = y$;

(ii) if $\tan\alpha\tan\beta = k$, then $y^2 + (2a-x)ak^2 = a^2k^3$.

(i) From the theory of equations, we have from (1),

$$\tan\alpha + \tan\beta + \tan\gamma = 0;$$

$$\therefore\ h + \tan\gamma = 0, \text{ or } \tan\gamma = -h.$$

But $$a\tan^3\gamma + (2a-x)\tan\gamma + y = 0;$$

$$\therefore\ ah^3 + (2a-x)h - y = 0.$$

(ii) From the theory of equations, we have from (1),

$$\tan\alpha\tan\beta\tan\gamma = -\frac{y}{a};$$

$$\therefore\ k\tan\gamma = -\frac{y}{a}, \text{ or } \tan\gamma = -\frac{y}{ak}.$$

Substituting in $a\tan^3\gamma + (2a-x)\tan\gamma + y = 0$, we have

$$-\frac{ay^3}{a^3k^3} - (2a-x)\frac{y}{ak} + y = 0;$$

$$\therefore\ y^2 + (2a-x)ak^2 - a^2k^3 = 0.$$

Example 2. Shew that

$$\cos^2 a + \cos^2\left(\frac{2\pi}{3} + a\right) + \cos^2\left(\frac{2\pi}{3} - a\right) = \frac{3}{2}.$$

Suppose that $\cos 3\theta = k$;

then $4\cos^3\theta - 3\cos\theta = \cos 3\theta = k$;

$$\therefore \cos^3\theta - \frac{3}{4}\cos\theta - \frac{k}{4} = 0.$$

The roots of this cubic in $\cos\theta$ are

$$\cos a,\ \cos\left(\frac{2\pi}{3} + a\right),\ \text{and}\ \cos\left(\frac{2\pi}{3} - a\right),$$

where a is any angle which satisfies the equation $\cos 3a = k$. For shortness, denote the roots by a, b, c; then

$$a^2 + b^2 + c^2 = (a + b + c)^2 - 2(bc + ca + ab)$$

$$= 0 - 2\left(-\frac{3}{4}\right);$$

$$\therefore \cos^2 a + \cos^2\left(\frac{2\pi}{3} + a\right) + \cos^2\left(\frac{2\pi}{3} - a\right) = \frac{3}{2}.$$

330. If $5\theta = 2n\pi$, where n is any integer, we have

$$3\theta = 2n\pi - 2\theta;$$

$$\therefore \sin 3\theta = -\sin 2\theta.$$

The values of $\sin\theta$ found from this equation are

$$0,\ \sin\frac{2\pi}{5},\ \sin\frac{4\pi}{5},\ \sin\frac{6\pi}{5},\ \sin\frac{8\pi}{5},$$

being obtained by giving to n the values 0, 1, 2, 3, 4. It will easily be seen that no new values of $\sin\theta$ are obtained by ascribing to n the values 5, 6, 7, ...

But $\sin\frac{6\pi}{5} = -\sin\frac{4\pi}{5} = -\sin\frac{\pi}{5}$,

and $\sin\frac{8\pi}{5} = -\sin\frac{2\pi}{5}$;

hence rejecting the zero solution, the values of $\sin\theta$ found from the equation $\sin 3\theta = -\sin 2\theta$ are

$$\pm\sin\frac{\pi}{5},\ \text{and}\ \pm\sin\frac{2\pi}{5}.$$

If we put $\sin\theta = x$, the equation $\sin 3\theta = -\sin 2\theta$ becomes

$$3x - 4x^3 = -2x\sqrt{1-x^2}.$$

Dividing by x, and thus removing the solution $x=0$, we have

$$(3-4x^2)^2 = 4(1-x^2),$$

or

$$16x^4 - 20x^2 + 5 = 0.$$

This is a quadratic in x^2, and as we have just seen the values of x^2 are

$$\sin^2\frac{\pi}{5} \text{ and } \sin^2\frac{2\pi}{5}.$$

From the theory of quadratic equations, we have

$$\sin^2\frac{\pi}{5} + \sin^2\frac{2\pi}{5} = \frac{20}{16} = \frac{5}{4};$$

$$\sin^2\frac{\pi}{5}\sin^2\frac{2\pi}{5} = \frac{5}{16}.$$

Example. Shew that

$$\sin\frac{2\pi}{7} + \sin\frac{4\pi}{7} + \sin\frac{8\pi}{7} = \frac{1}{2}\sqrt{7}.$$

If $7\theta = 2n\pi$, where n is any integer, we have

$$\sin 4\theta = -\sin 3\theta.$$

The values of $\sin\theta$ found from this equation are

$$0,\quad \pm\sin\frac{2\pi}{7},\quad \pm\sin\frac{4\pi}{7},\quad \pm\sin\frac{8\pi}{7},$$

since

$$\sin\frac{6\pi}{7} = -\sin\frac{8\pi}{7}.$$

If $\sin\theta = x$, the equation $\sin 4\theta = -\sin 3\theta$ becomes

$$4x(1-2x^2)\sqrt{1-x^2} = 4x^3 - 3x;$$

whence rejecting the solution $x=0$, we obtain

$$16(1-4x^2+4x^4)(1-x^2) = 16x^4 - 24x^2 + 9,$$

or

$$64x^6 - 112x^4 + 56x^2 - 7 = 0 \quad\text{......................(1).}$$

The values of x^2 found from this equation are

$$\sin^2\frac{2\pi}{7},\quad \sin^2\frac{4\pi}{7},\quad \sin^2\frac{8\pi}{7};$$

hence $$\sin^2\frac{2\pi}{7}+\sin^2\frac{4\pi}{7}+\sin^2\frac{8\pi}{7}=\frac{112}{64}=\frac{7}{4}.$$

But $$\sin\frac{2\pi}{7}\sin\frac{4\pi}{7}+\sin\frac{2\pi}{7}\sin\frac{8\pi}{7}+\sin\frac{4\pi}{7}\sin\frac{8\pi}{7}$$

$$=\frac{1}{2}\left\{\left(\cos\frac{2\pi}{7}-\cos\frac{6\pi}{7}\right)+\left(\cos\frac{6\pi}{7}-\cos\frac{10\pi}{7}\right)+\left(\cos\frac{4\pi}{7}-\cos\frac{12\pi}{7}\right)\right\}$$

$$=0.$$

$$\therefore\ \left(\sin\frac{2\pi}{7}+\sin\frac{4\pi}{7}+\sin\frac{8\pi}{7}\right)^2=\sin^2\frac{2\pi}{7}+\sin^2\frac{4\pi}{7}+\sin^2\frac{8\pi}{7}=\frac{7}{4};$$

$$\therefore\ \sin\frac{2\pi}{7}+\sin\frac{4\pi}{7}+\sin\frac{8\pi}{7}=\frac{1}{2}\sqrt{7}.$$

331. If $7\theta=2n\pi$, where n is any integer, we have

$$4\theta=2n\pi-3\theta;$$

$$\therefore\ \cos 4\theta=\cos 3\theta.$$

By giving to n the values 0, 1, 2, 3, the values of $\cos\theta$ obtained from this equation are

$$1,\ \cos\frac{2\pi}{7},\ \cos\frac{4\pi}{7},\ \cos\frac{6\pi}{7}.$$

It will easily be seen that no new values of $\cos\theta$ are found by ascribing to n the values 4, 5, 6, 7, ...; for

$$\cos\frac{8\pi}{7}=\cos\frac{6\pi}{7},\ \cos\frac{10\pi}{7}=\cos\frac{4\pi}{7},\ \ldots\ldots\ldots\ldots$$

Now $$\cos 4\theta=8\cos^4\theta-8\cos^2\theta+1,$$

and therefore if $x=\cos\theta$, the equation $\cos 4\theta=\cos 3\theta$ becomes

$$8x^4-8x^2+1=4x^3-3x,$$

or $$8x^4-4x^3-8x^2+3x+1=0.$$

Removing the factor $x-1$, which corresponds to the root $\cos\theta=1$, we obtain

$$8x^3+4x^2-4x-1=0,$$

the roots of which equation are

$$\cos\frac{2\pi}{7},\ \cos\frac{4\pi}{7},\ \cos\frac{6\pi}{7}.$$

Example 1. Find the values of

$$\tan^2\frac{\pi}{7}+\tan^2\frac{2\pi}{7}+\tan^2\frac{3\pi}{7} \text{ and } \tan\frac{\pi}{7}\tan\frac{2\pi}{7}\tan\frac{3\pi}{7}.$$

If $7\theta=n\pi$, where n is any integer, we have

$$\tan 4\theta=-\tan 3\theta.$$

By writing $\tan\theta=t$, this equation becomes

$$\frac{4t-4t^3}{1-6t^2+t^4}=-\frac{3t-t^3}{1-3t^2},$$

or

$$t^6-21t^4+35t^2-7=0.$$

The roots of this cubic in t^2 are

$$\tan^2\frac{\pi}{7},\quad \tan^2\frac{2\pi}{7},\quad \tan^2\frac{3\pi}{7}.$$

$$\therefore \tan^2\frac{\pi}{7}+\tan^2\frac{2\pi}{7}+\tan^2\frac{3\pi}{7}=21,$$

and

$$\tan\frac{\pi}{7}\tan\frac{2\pi}{7}\tan\frac{3\pi}{7}=\sqrt{7},$$

the positive value of the square root being taken, since each of the factors on the left is positive.

Example 2. Shew that

$$\cos^4\frac{\pi}{7}+\cos^4\frac{2\pi}{7}+\cos^4\frac{3\pi}{7}=\frac{13}{16};$$

and

$$\sec^4\frac{\pi}{7}+\sec^4\frac{2\pi}{7}+\sec^4\frac{3\pi}{7}=416.$$

Let y denote any one of the quantities

$$\cos^2\frac{\pi}{7},\quad \cos^2\frac{2\pi}{7},\quad \cos^2\frac{3\pi}{7};$$

then $2y=1+x$, where x denotes one of the quantities

$$\cos\frac{2\pi}{7},\quad \cos\frac{4\pi}{7},\quad \cos\frac{6\pi}{7}.$$

From Art. 331, the equation whose roots are

$$\cos\frac{2\pi}{7},\quad \cos\frac{4\pi}{7},\quad \cos\frac{6\pi}{7}$$

is

$$8x^3+4x^2-4x-1=0;$$

whence by substituting $x=2y-1$, it follows that

$$\cos^2\frac{\pi}{7},\quad \cos^2\frac{2\pi}{7},\quad \cos^2\frac{3\pi}{7}$$

are the roots of the equation

$$8(2y-1)^3+4(2y-1)^2-4(2y-1)-1=0,$$

or

$$64y^3-80y^2+24y-1=0.$$

$$\therefore\ \cos^2\frac{\pi}{7}+\cos^2\frac{2\pi}{7}+\cos^2\frac{3\pi}{7}=\frac{80}{64}=\frac{5}{4};$$

and

$$\Sigma\cos^2\frac{\pi}{7}\cos^2\frac{2\pi}{7}=\frac{24}{64}=\frac{3}{8}.$$

By squaring the first of these equations and subtracting twice the second equation, we have

$$\cos^4\frac{\pi}{7}+\cos^4\frac{2\pi}{7}+\cos^4\frac{3\pi}{7}=\frac{13}{16}.$$

By putting $z=\frac{1}{y}$, we see that

$$\sec^2\frac{\pi}{7},\quad \sec^2\frac{2\pi}{7},\quad \sec^2\frac{3\pi}{7}$$

are the roots of the equation

$$z^3-24z^2+80z-64=0;$$

$$\therefore\ \sec^4\frac{\pi}{7}+\sec^4\frac{2\pi}{7}+\sec^4\frac{3\pi}{7}=(24)^2-(2\times 80)=416.$$

332. To find $\cos 5\theta$ and $\sin 5\theta$, we may proceed as follows:

$$\begin{aligned}\cos 5\theta+\cos\theta&=2\cos 3\theta\cos 2\theta\\&=(4\cos^3\theta-3\cos\theta)(4\cos^2\theta-2);\end{aligned}$$

$$\therefore\ \cos 5\theta=16\cos^5\theta-20\cos^3\theta+5\cos\theta.$$

$$\begin{aligned}\sin 5\theta+\sin\theta&=2\sin 3\theta\cos 2\theta\\&=(3\sin\theta-4\sin^3\theta)(2-4\sin^2\theta);\end{aligned}$$

$$\therefore\ \sin 5\theta=16\sin^5\theta-20\sin^3\theta+5\sin\theta.$$

It is easy to prove that

$$\cos 6\theta=32\cos^6\theta-48\cos^4\theta+18\cos^2\theta-1,$$

and

$$\sin 6\theta=\cos\theta\,(32\sin^5\theta-32\sin^3\theta+6\sin\theta).$$

EXAMPLES. XXV. c.

Solve the following equations:

1. $x^3-3x-1=0$.
2. $x^3-3x+1=0$.
3. $x^3-3x-\sqrt{3}=0$.
4. $8x^3-6x+\sqrt{2}=0$.
5. $8a^3x^3-6ax+2\sin 3A=0$.
6. $x^3-3a^2x-2a^3\cos 3A=0$.

7. If $\sin\alpha$ and $\sin\beta$ are the roots of the equation

$$a\sin^2\theta+b\sin\theta+c=0,$$

shew that (1) if $\sin\alpha+2\sin\beta=1$, then $a^2+2b^2+3ab+ac=0$,

(2) if $c\sin\alpha=a\sin\beta$, then $a+c=\pm b$.

8. If $\tan\alpha$ and $\tan\beta$ are the roots of the equation

$a\tan^2\theta-b\tan\theta+c=0$, and if $a\tan\alpha+b\tan\beta=2b$,

shew that $$b^2(2a-b)+c(a-b)^2=0.$$

9. If $\tan\alpha$, $\tan\beta$, $\tan\gamma$ are the roots of the equation

$$a\tan^3\theta+(2a-x)\tan\theta+y=0,$$

and if $a(\tan^2\alpha+\tan^2\beta)=2x-5a$, shew that $x\pm y=3a$.

10. If $\cos\alpha$, $\cos\beta$, $\cos\gamma$ are the roots of the equation

$$\cos^3\theta+a\cos^2\theta+b\cos\theta+c=0,$$

and if $\cos\alpha(\cos\beta+\cos\gamma)=2b$, prove that $abc+2b^3+c^2=0$.

Prove the following identities:

11. $\sec\alpha+\sec\left(\frac{2\pi}{3}+\alpha\right)+\sec\left(\frac{2\pi}{3}-\alpha\right)=-3\sec 3\alpha$.

12. $\sin^2\alpha+\sin^2\left(\frac{2\pi}{3}+\alpha\right)+\sin^2\left(\frac{4\pi}{3}+\alpha\right)=\frac{3}{2}$.

13. $\operatorname{cosec}\alpha+\operatorname{cosec}\left(\frac{2\pi}{3}+\alpha\right)+\operatorname{cosec}\left(\frac{4\pi}{3}+\alpha\right)=3\operatorname{cosec}3\alpha$.

14. $\operatorname{cosec}^2\frac{\pi}{5}+\operatorname{cosec}^2\frac{2\pi}{5}=4$.

15. $\cos\frac{2\pi}{5}+\cos\frac{4\pi}{5}=-\frac{1}{2}$, and $\cos\frac{2\pi}{5}\cos\frac{4\pi}{5}=-\frac{1}{4}$.

16. Form the equation whose roots are

$$(1)\quad \cos\frac{\pi}{7},\ \cos\frac{3\pi}{7},\ \cos\frac{5\pi}{7};$$

$$(2)\quad \sin^2\frac{\pi}{14},\ \sin^2\frac{3\pi}{14},\ \sin^2\frac{5\pi}{14}.$$

17. Form the equation whose roots are

$$\sin^2\frac{\pi}{7},\ \sin^2\frac{2\pi}{7},\ \sin^2\frac{3\pi}{7};$$

and shew that $\sum_{n=1}^{n=3}\sin^4\frac{n\pi}{7}=\frac{21}{16}$ and $\sum_{n=1}^{n=3}\operatorname{cosec}^4\frac{n\pi}{7}=32.$

18. Form the equation whose roots are

$$(1)\quad \cos\frac{2\pi}{9},\ \cos\frac{4\pi}{9},\ \cos\frac{6\pi}{9},\ \cos\frac{8\pi}{9};$$

$$(2)\quad \cos\frac{\pi}{9},\ \cos\frac{3\pi}{9},\ \cos\frac{5\pi}{9},\ \cos\frac{7\pi}{9}.$$

19. Form the equation whose roots are

$$\cos^2\frac{\pi}{9},\ \cos^2\frac{2\pi}{9},\ \cos^2\frac{3\pi}{9},\ \cos^2\frac{4\pi}{9},$$

and shew that $\sum_{n=1}^{n=4}\cos^4\frac{n\pi}{9}=\frac{19}{16}$, and $\sum_{n=1}^{n=4}\sec^4\frac{n\pi}{9}=1120.$

20. Form the equation whose roots are

$$\tan^2\frac{\pi}{9},\ \tan^2\frac{2\pi}{9},\ \tan^2\frac{3\pi}{9},\ \tan^2\frac{4\pi}{9},$$

and shew that $\cot^2\frac{\pi}{9}+\cot^2\frac{2\pi}{9}+\cot^2\frac{4\pi}{9}=9.$

21. Form the equation whose roots are

$$\cos\frac{\pi}{11},\ \cos\frac{2\pi}{11},\ \cos\frac{3\pi}{11},\ \cos\frac{4\pi}{11},\ \cos\frac{5\pi}{11};$$

and shew that $\sum_{n=1}^{n=5}\sec^2\frac{n\pi}{11}=60$, and $\prod_{n=1}^{n=5}\sec\frac{n\pi}{11}=32.$

MISCELLANEOUS EXAMPLES. I.

1. If $$a \tan \alpha + b \tan \beta = (a+b) \tan \frac{\alpha+\beta}{2},$$
prove that $$a \cos \beta = b \cos \alpha.$$

2. If $$\frac{\sin^4 \alpha}{a} + \frac{\cos^4 \alpha}{b} = \frac{1}{a+b},$$
prove that $$\frac{\sin^8 \alpha}{a^3} + \frac{\cos^8 \alpha}{b^3} = \frac{1}{(a+b)^3}.$$

3. Shew that
$$2 \tan^{-1}\left\{\tan \frac{\alpha}{2} \tan\left(\frac{\pi}{4} - \frac{\beta}{2}\right)\right\} = \tan^{-1}\left(\frac{\sin \alpha \cos \beta}{\cos \alpha + \sin \beta}\right).$$

4. If the equation
$$\frac{\sin^{2n+2} \theta}{\sin^{2n} \alpha} + \frac{\cos^{2n+2} \theta}{\cos^{2n} \alpha} = 1$$
is true when $n=1$, prove that it will be true when n is any positive integer.

5. If $a \cos \theta + b \sin \theta = c$ and $a \cos^2 \theta + b \sin^2 \theta = c$,
prove that $$4a^2b^2 + (b-c)(a-c)(a-b)^2 = 0.$$

6. Prove the following identities:

(i) $\Sigma \sin(\beta-\gamma) \cos(\alpha-\beta) \cos(\alpha-\gamma) = -\Pi \sin(\beta-\gamma)$;

(ii) $\Sigma \sin \alpha \sin(\beta-\gamma) \cos(\beta+\gamma-\alpha) = 0$;

(iii) $\Sigma \sin \alpha \sin(\beta-\gamma) \sin(\beta+\gamma-\alpha) = 2\Pi \sin(\beta-\gamma)$.

7. If P be a point within a triangle ABC, such that
$$\angle PAB = \angle PBC = \angle PCA = \omega,$$
prove that (1) $\cot \omega = \cot A + \cot B + \cot C$;

(2) $\operatorname{cosec}^2 \omega = \operatorname{cosec}^2 A + \operatorname{cosec}^2 B + \operatorname{cosec}^2 C$.

8. A hill of inclination 1 in 169 faces West. Shew that a railway on it which runs S.E. has an inclination of 1 in 239.

9. Two vertical walls of equal height a are inclined to one another at an angle α. At noon the breadth of their shadows are b and c: shew that the altitude θ of the sun is given by the equation
$$a^2 \sin^2 \gamma \cot^2 \theta = b^2 + c^2 + 2bc \cos \gamma.$$

ANSWERS.

I. Page 4.

1. ·75. 2. ·125. 3. ·375. 4. $\cdot 024\dot{1}$.
5. ·089. 6. ·0204045. 7. $70^g\,91'\,66{\cdot}7''$. 8. $21^g\,12'\,50''$.
9. $56^g\,24'\,25''$. 10. $48^g\,75'\,25''$. 11. $12^g\,23'\,40{\cdot}7''$.
12. $158^g\,6'\,94{\cdot}4''$. 13. $22'\,50''$. 14. $6'\,36{\cdot}7''$.
15. $51^\circ\,11'\,15''$. 16. $35^\circ\,9'\,22{\cdot}5''$. 17. $36^\circ\,0'\,40{\cdot}6''$.
18. $55'\,5{\cdot}8''$. 19. $2^\circ\,43'\,6{\cdot}4''$. 20. $7^\circ\,17'\,26{\cdot}1''$.
21. $3'\,22{\cdot}5''$. 22. $20'\,0{\cdot}4''$. 23. $45^\circ, 27^\circ$. 24. 72°.

II. Page 11.

1. $\frac{15}{17}, \frac{17}{8}, \frac{8}{15}, \frac{17}{15}$. 2. $\frac{12}{5}, \frac{13}{12}, \frac{5}{13}, \frac{12}{13}$.
3. 25, $\frac{4}{5}, \frac{4}{5}, \frac{3}{4}, \frac{5}{3}$. 4. 7, $\frac{7}{25}, \frac{24}{7}, \frac{25}{24}$.
5. $\frac{37}{35}, \frac{37}{12}, \frac{35}{12}, \frac{35}{37}$. 6. 12 inches, $\frac{4}{5}, \frac{3}{5}, \frac{4}{3}$.
7. 25, $\frac{24}{25}, \frac{7}{25}$. 8. 40 ft., $\frac{40}{41}, \frac{9}{40}$.
9. 20 ft., sine $=\frac{20}{29}$, cosine $=\frac{21}{29}$, tangent $=\frac{20}{21}$.
10. sine $=\frac{1}{\sqrt{5}}$, cosine $=\frac{2}{\sqrt{5}}$, tangent $=\frac{1}{2}$.
11. $\frac{12}{13}, \frac{13}{12}, \frac{77}{85}, \frac{85}{77}$. 12. $\frac{3}{5}, \frac{4}{3}, \frac{20}{29}, \frac{29}{20}$.

III. c. Page 23.

1. $\frac{2}{\sqrt{3}}, \sqrt{3}$. 2. $\frac{4}{5}, \frac{3}{5}$. 3. $\frac{1}{\sqrt{15}}, \frac{\sqrt{15}}{4}$. 4. $\frac{\sqrt{5}}{2}, \sqrt{5}$.

5. $\frac{\sqrt{48}}{7}, \frac{1}{\sqrt{48}}$. **6.** $\frac{7}{24}, \frac{25}{24}$. **7.** $\sqrt{1-\cos^2 A}, \frac{\sqrt{1-\cos^2 A}}{\cos A}$.

8. $\sqrt{1+\cot^2 a}, \frac{\cot a}{\sqrt{1+\cot^2 a}}$. **9.** $\frac{\sqrt{\sec^2\theta-1}}{\sec\theta}, \frac{1}{\sqrt{\sec^2\theta-1}}$.

10. $\operatorname{cosec} A=\frac{1}{\sin A}$, $\cos A=\sqrt{1-\sin^2 A}$, $\sec A=\frac{1}{\sqrt{1-\sin^2 A}}$,

$\tan A=\frac{\sin A}{\sqrt{1-\sin^2 A}}$, $\cot A=\frac{\sqrt{1-\sin^2 A}}{\sin A}$. **11.** $\sqrt{2}$.

13. $\frac{p}{q}$. **14.** $\frac{m^2-1}{2m}, \frac{m^2-1}{m^2+1}$. **15.** $\frac{p^2-q^2}{p^2+q^2}, \frac{p^2+q^2}{2pq}$.

16. 3. **17.** $\frac{p^2-q^2}{p^2+q^2}$.

IV. a. Page 26.

1. 5. **2.** $1\frac{1}{2}$. **3.** 0. **4.** $2\frac{1}{3}$. **5.** $\frac{1}{2}$.

6. $1\frac{1}{2}$. **7.** 9. **8.** 2. **9.** $2\frac{1}{12}$. **10.** $\frac{1}{2}$.

11. $\frac{3}{4}$. **12.** 0. **13.** $1\frac{11}{12}$. **14.** $\frac{\sqrt{3}}{2}$. **15.** 6.

IV. b. Page 28.

1. 22° 30′. **2.** 64° 59′ 30″. **3.** 79° 58′ 57″. **4.** $45°+A$.

5. $45°-B$. **6.** $60°+B$. **7.** 50°. **8.** 60°.

9. 18°. **10.** 9°. **11.** 22° 30′. **12.** 45°.

13. 30°. **14.** 15°. **30.** 1. **31.** $\tan A$.

IV. c. Page 31.

1. 45°. **2.** 60°. **3.** 60°. **4.** 45°. **5.** 60°.

6. 30°. **7.** 45°. **8.** 60°. **9.** 45°. **10.** 60°.

11. 45°. **12.** 60°. **13.** 45°. **14.** 30°. **15.** 45°.

16. 60°. **17.** 30°. **18.** 60°. **19.** 45°. **20.** 60°.

21. 30°. **22.** 30°. **23.** 60°. **24.** 45°.

25. 45° or 30°. $[(2\sin\theta-1)(\tan\theta-1)=0.]$ **26.** 60°. **28.** 1 or $\frac{1}{2}$.

MISCELLANEOUS EXAMPLES. A. PAGE 32.

1. (1) ·2537064; (2) ·704. **3.** $\frac{20}{29}, \frac{29}{21}$. **4.** $\frac{15}{8}, \frac{17}{8}$.

6. (1) $15^\circ 28' 7{\cdot}5''$; (2) $1' 37{\cdot}2''$. **7.** $41, \frac{9}{40}, \frac{41}{9}, \frac{41}{40}$.

8. (1) possible; (2) impossible; (3) possible.

10. $\frac{\sqrt{1+\cot^2 a}}{\cot a}, \sqrt{1+\cot^2 a}$. **11.** 6.

12. $\frac{m}{\sqrt{m^2+n^2}}, \frac{\sqrt{m^2+n^2}}{n}$. **16.** $\frac{20}{21}, \frac{29}{20}$.

18. 10°. **20.** (1) 30°; (2) 45°. **22.** 30°.

26. $\frac{5}{14}$. **29.** (1) 30°; (2) 30°.

V. a. PAGE 37.

1. $c=2, B=60^\circ, C=30^\circ$. **2.** $a=6\sqrt{3}, A=60^\circ, C=30^\circ$.
3. $c=8\sqrt{3}, A=30^\circ, B=60^\circ$. **4.** $c=30\sqrt{3}, B=30^\circ, C=60^\circ$.
5. $b=20\sqrt{2}, A=C=45^\circ$. **6.** $c=10\sqrt{3}, A=30^\circ, B=60^\circ$.
7. $a=2\sqrt{2}, B=C=45^\circ$. **8.** $a=9, A=60^\circ, C=30^\circ$.
9. $B=60^\circ, b=27, c=18\sqrt{3}$. **10.** $C=60^\circ, b=2, c=2\sqrt{3}$.
11. $B=30^\circ, a=4\sqrt{3}, b=4$. **12.** $B=90^\circ, a=3\sqrt{3}, c=3$.
13. $A=30^\circ, a=50, c=50\sqrt{3}$. **14.** $C=90^\circ, a=20, c=40$.
15. $A=90^\circ, a=4\sqrt{2}, b=4$. **16.** $A=90^\circ, b=4, c=4\sqrt{3}$.
17. 700. **18.** 31. **19.** 86·47. **20.** 97·8.
21. $C=54^\circ, a=73, b=124$. **22.** $B=68^\circ 17', C=21^\circ 43', b=93$.
23. $C=50^\circ 36', a=34{\cdot}3875, c=30{\cdot}435$.
24. $c=353, A=39^\circ 36', B=50^\circ 24'$.

V. b. PAGE 39.

1. $10\sqrt{3}$. **2.** $a=10\sqrt{2}, c=20$.
3. $AB=10\sqrt{3}$ ft., $AC=10$ ft., $AD=5\sqrt{3}$ ft. **4.** 12, 4.
5. $24\sqrt{3}$. **6.** $20(\sqrt{3}-1)$.
7. $20(3+\sqrt{3})$. **8.** $DC=BD=100$.

VI. a. PAGE 42.

1. 173·2 ft. 2. 277·12 ft. 3. 60°. 4. 50 ft.; 100 ft.
5. 22·5 ft.; 38·97 ft. 6. 30 ft. 7. 200 yds.
8. 51 yds., 81 yds. 9. 86·6 yds. 10. 46·19 ft.
11. 273·2 ft. 12. Each = 70·98 ft. 13. 5 miles.
14. 73·2 ft. 15. 64 ft. 16. 300 ft.
17. 1193 yds. 18. 277·12 yds.

VI. b. PAGE 47.

1. 565·6 yds.; 1131·2 yds. 2. 3·464 miles; 6 miles.
3. 29 miles. 4. 10 miles per hour.
5. 10 miles; 24·14 miles. 6. 16 miles; S. 25° W.
7. 9·656 miles. 8. 5·77 miles; 11·54 miles.
9. 295·1 knots. 10. 5·196 miles per hour; 18 miles.
11. 31 minutes past midnight. 12. 38·97 miles per hour.

VII. a. PAGE 54.

1. $\frac{\pi}{4}$. 2. $\frac{\pi}{6}$. 3. $\frac{7\pi}{12}$. 4. $\frac{\pi}{8}$. 5. $\frac{\pi}{10}$.
6. $\frac{23\pi}{72}$. 7. $\frac{2\pi}{25}$. 8. $\frac{7\pi}{16}$. 9. ·4509. 10. ·6545.
11. 1·4399. 12. 1·1999. 13. 2·7489. 14. ·9163.
15. 135°. 16. 28°. 17. 33° 20′. 18. 37° 30′.
19. 22° 30′. 20. 30°. 21. 37° 30′. 22. 165°.
23. ·638. 24. 1·332. 25. 2·0262. 26. 2·9979.

VII. b. PAGE 56.

1. $\frac{3}{4}$. 2. $\frac{1}{3\sqrt{2}}$. 3. $4\frac{1}{4}$. 4. $\frac{3}{\sqrt{2}}$. 5. 9.
6. $\frac{3}{4}$. 7. 1. 13. $\frac{\pi}{4}, \frac{2\pi}{15}$. 14. $\frac{5\pi}{6}, \frac{5\pi}{7}$.

VII. c. PAGE 60.

1. $\frac{1}{5}$. 2. 300 ft. 3. A radian.
4. 5·85 yards. 5. 330. 6. $\frac{1}{44}$ of a second.
7. $58\frac{2}{3}$. 8. 40 yds. 9. 1·15192 miles.
10. 3·581. 11. 2° 6′. 12. 45 feet.

MISCELLANEOUS EXAMPLES. B. PAGE 61.

1. $9°$. 2. 95·26. 3. $54°$. 4. 3438 inches.
6. $30°$. 8. $22\frac{1}{2}°, \frac{\pi}{8}$. 9. $67\frac{1}{2}°$.
10. $a=6\sqrt{3}$, $c=12$, perp. $=3\sqrt{3}$. 12. 17·32 ft.
14. $120°$, $36°$, $24°$. 15. $-\frac{35}{8}$.
17. (1) possible; (2) impossible. 18. 8·66 miles.
19. $\frac{\pi}{5}, \frac{\pi}{3}, \frac{7\pi}{15}$. 21. 90. 24. 4 miles per hour, 1·732 miles.
25. $\frac{\pi}{8}$. 26. (1) $30°$; (2) $30°$. 27. $\frac{5}{13}$.
29. 200 yards. 30. 33 feet.

VIII. a. PAGE 69.

1. Second. 2. Third. 3. First. 4. Third.
5. Second. 6. Second. 7. Third. 8. Third.
9. Sine. 10. Cosine. 11. Tangent. 12. Sine.
13. Sine. 14. Tangent. 15. Sine. 16. All.
17. Cosine. 18. $60°, \frac{\sqrt{3}}{2}$. 19. $30°, \frac{\sqrt{3}}{2}$. 20. $45°$, 1.
21. $45°, \sqrt{2}$. 22. $30°$, 2. 23. $60°, \frac{2}{\sqrt{3}}$. 24. $45°$, 1.
25. $60°$, 2. 26. $60°, \sqrt{3}$.

VIII. b. PAGE 72.

1. $-\sqrt{3}$. 2. $\frac{1}{\sqrt{2}}$. 3. $-\frac{1}{2}$.
4. $-\frac{12}{13}, \frac{12}{5}$. 5. $-\frac{5}{4}, -\frac{3}{4}$. 6. $\frac{4}{3}, -\frac{3}{5}$.
7. $\frac{\sqrt{3}}{2}, -\frac{1}{\sqrt{3}}$. 8. $1, -\sqrt{2}$. 9. $\pm\frac{5}{13}, \pm\frac{5}{12}$.

IX. PAGE 79.

1. $\cot A$ decreases from ∞ to 0, then increases numerically from 0 to $-\infty$, then decreases from ∞ to 0, then increases numerically from 0 to $-\infty$. 2. $\operatorname{cosec}\theta$ decreases from ∞ to 1, then increases from 1 to ∞. 3. $\cos\theta$ decreases numerically from -1 to 0, then increases from 0 to 1. 4. $\tan A$ decreases from ∞ to 0, then increases numerically from 0 to $-\infty$. 5. $\sec\theta$ decreases numerically from $-\infty$ to -1, then increases numerically from -1 to $-\infty$. 6. 3. 7. 1. 8. -2. 9. 2.

MISCELLANEOUS EXAMPLES. C. Page 81.

1. $\pm\frac{4}{5}$. 3. $A=60^\circ$, $B=30^\circ$, $a=\frac{21\sqrt{3}}{2}$. 4. $\frac{7}{24}$.

5. 1313 miles, nearly. 6. 301 feet. 7. $-\frac{7}{10}$.

8. 12·003 inches. 10. 200 feet.

X. a. Page 87.

1. $-\frac{1}{\sqrt{2}}$. 2. $\frac{1}{2}$. 3. $\sqrt{3}$. 4. $-\sqrt{2}$.

5. $-\frac{\sqrt{3}}{2}$. 6. 1. 7. -1. 8. $-\frac{1}{2}$.

9. 2. 10. -1. 11. $-\frac{\sqrt{3}}{2}$. 12. -2.

13. -2. 14. $-\frac{1}{\sqrt{2}}$. 15. $\sqrt{3}$. 16. $\sin A$.

17. $\tan A$. 18. $-\cos A$. 19. $-\sec A$. 20. $-\cos A$.

21. $-\tan A$. 22. $-\cos\theta$. 23. $\tan\theta$. 24. $-\operatorname{cosec}\theta$.

25. 1. 26. $2\sin A$. 27. 1.

X. b. Page 91.

1. $-\frac{1}{2}$. 2. $-\frac{\sqrt{3}}{2}$. 3. $\frac{1}{2}$. 4. $-\frac{1}{2}$. 5. -1.

6. 1. 7. $\frac{2}{\sqrt{3}}$. 8. $-\frac{1}{\sqrt{3}}$. 9. $-\sqrt{2}$. 10. $\frac{1}{\sqrt{2}}$.

11. 0. 12. $\frac{1}{\sqrt{2}}$. 13. $-\sqrt{2}$. 14. $-\frac{1}{2}$. 15. $-\sqrt{2}$.

16. $-\frac{1}{\sqrt{2}}$. 17. -1. 18. 2. 19. $\frac{1}{\sqrt{3}}$. 20. $\frac{2}{\sqrt{3}}$.

21. $\pm 30^\circ$, $\pm 330^\circ$. 22. 210°, 330°, -30°, -150°.

23. 120°, 300°, -60°, -240°. 24. 135°, 315°, -45°, -225°.

30. 3. 31. $\cot^2 A$. 32. -1. 34. -4.

XI. a. Page 97.

4. 1; $\frac{24}{25}$. 5. $\frac{33}{65}$; $\frac{16}{65}$. 6. $-\frac{85}{36}$.

XI. b. PAGE 100.

1. 1. 2. $\frac{1}{7}$. 3. 0; $\frac{12}{35}$. 4. $-\frac{278}{29}$; $\frac{1}{2}$.

11. $\cos A\cos B\cos C-\cos A\sin B\sin C-\sin A\cos B\sin C-\sin A\sin B\cos C$;
$\sin A\cos B\cos C-\cos A\sin B\cos C+\cos A\cos B\sin C+\sin A\sin B\sin C$.

12. $\dfrac{\tan A-\tan B-\tan C-\tan A\tan B\tan C}{1-\tan B\tan C+\tan C\tan A+\tan A\tan B}$.

13. $\dfrac{\cot A\cot B\cot C-\cot A-\cot B-\cot C}{\cot B\cot C+\cot C\cot A+\cot A\cot B-1}$.

XI. d. PAGE 104.

1. $-\frac{7}{9}$. 2. $\frac{17}{25}$. 3. $\frac{24}{25}$. 4. $\frac{3}{4}$.

5. $\frac{7}{25}$; $\frac{24}{25}$. 6. $\frac{1}{3}$. 7. $\frac{1}{7}$.

XI. e. PAGE 106.

1. $-\frac{23}{27}$. 2. $\frac{117}{125}$. 3. $\frac{9}{13}$.

XII. a. PAGE 112.

1. $\sin 4\theta+\sin 2\theta$. 2. $\sin 9\theta-\sin 3\theta$. 3. $\cos 12A+\cos 2A$.

4. $\cos A-\cos 5A$. 5. $\sin 9\theta-\sin\theta$. 6. $\sin 12\theta-\sin 4\theta$.

7. $\cos 6\theta-\cos 12\theta$. 8. $\sin 16\theta-\sin 2\theta$. 9. $\cos 13a+\cos 9a$.

10. $\cos 5a-\cos 15a$. 11. $\frac{1}{2}(\sin 11a-\sin 3a)$.

12. $\frac{1}{2}(\cos 2a-\cos 4a)$. 13. $\frac{1}{2}(\sin 2A+\sin A)$.

14. $\frac{1}{2}(\sin 6A-\sin A)$. 15. $\cos\frac{7\theta}{3}+\cos\theta$.

16. $\frac{1}{2}\left(\cos\frac{\theta}{2}-\cos\theta\right)$. 17. $\cos(a+\beta)+\cos(a-3\beta)$.

18. $\cos(2a-\beta)-\cos(4a+\beta)$. 19. $\sin(3\theta-\phi)+\sin(\theta+3\phi)$.

20. $\sin(4\theta-\phi)-\sin(2\theta+3\phi)$. 21. $\frac{1}{2}\left(\frac{\sqrt{3}}{2}-\sin 2a\right)$.

XII. b. Page 114.

1. $2\sin 6\theta\cos 2\theta$. 2. $2\cos 3\theta\sin 2\theta$. 3. $2\cos 5\theta\cos 2\theta$.

4. $2\sin 10\theta\sin\theta$. 5. $2\cos 6a\sin a$. 6. $-2\cos\frac{11a}{2}\cos\frac{5a}{2}$.

7. $2\sin 8a\cos 5a$. 8. $-2\sin 3a\sin 2a$. 9. $2\cos\frac{11A}{2}\cos\frac{7A}{2}$.

10. $-2\cos 7A\sin 4A$. 11. $-\sin 20°$. 12. $\sqrt{3}\cos 10°$.

XIII. a. Page 128.

1. 60°. 2. 120°. 3. $A=30°$, $B=120°$, $C=30°$. 4. 45°.

5. 90°. 6. $A=75°$, $B=45°$, $C=60°$.

7. $A=30°$, $B=135°$, $C=15°$. 8. 28° 56′. 9. 111° 33′.

10. $\sqrt{6}$. 11. 7. 12. 8. 13. 14. 14. 9.

15. $b=2\sqrt{6}$, $A=75°$, $C=30°$. 16. $a=\sqrt{5}+1$, $B=36°$, $C=72°$.

17. $C=75°$, $a=c=2\sqrt{3}+2$. 18. $A=105°$, $a=\sqrt{3}+1$, $c=\sqrt{3}-1$.

19. $C=30°$, $a=2$, $b=\sqrt{3}+1$. 21. 2. 22. 6. 23. 60°.

24. 105°, 45°, 30°. 25. 105°, 15°, 60°. 26. $\frac{\sqrt{3}}{2}$, 105°, 15°.

XIII. b. Page 132.

1. $B=60°, 120°$; $C=90°, 30°$; $c=2, 1$.
2. $B=60°, 120°$; $A=75°, 15°$; $a=3+\sqrt{3}, 3-\sqrt{3}$.
3. $A=45°$, $B=75°$, $b=\sqrt{3}+1$; no ambiguity. 4. Impossible.
5. $C=45°, 135°$; $A=105°, 15°$; $a=3+\sqrt{3}, 3-\sqrt{3}$.
6. $C=75°, 105°$; $A=45°, 15°$; $a=2\sqrt{3}, 3-\sqrt{3}$.
7. $A=75°, 105°$; $B=90°, 60°$; $b=2\sqrt{6}, 3\sqrt{2}$.
8. $B=90°$, $C=72°$, $c=4\sqrt{5+2\sqrt{5}}$; no ambiguity. 9. Impossible.

XIII. d. Page 136.

1. 72°, 72°, 36°; each side $=\sqrt{5}+1$.
2. $A=60°$, $a=9-3\sqrt{3}$, $b=3(\sqrt{6}-\sqrt{2})$, $c=3\sqrt{2}$.
3. $A=105°$, $B=15°$, $C=60°$. 4. $B=54°, 126°$; $C=108°, 36°$.
5. $C=60°, 120°$; $A=90°, 30°$; $a=75\sqrt{3}$. No, for $C=90°$.
6. 18°, 126°. 8. $A=90°$, $B=30°$, $C=60°$; $2c=a\sqrt{3}$.

MISCELLANEOUS EXAMPLES. D. PAGE 138.

2. 43. **3.** ∞, 1. **4.** -1. **6.** $a=2$, $B=30°$, $C=105°$.
9. $A=30°$, $B=75°$, $C=75°$.

XIV. a. PAGE 145.

1. 10, 8, $-\frac{3}{2}$, $\frac{2}{3}$, $\frac{1}{2}$, -2. **2.** $\frac{4}{3}$, $\frac{5}{4}$, $-\frac{1}{2}$, $\frac{7}{4}$.

3. 2401, $\cdot 5$, $\frac{10000}{\cdot 9}$, 1, $\frac{5}{4}$, 1000, $1\dot{0}000$.

4. 5, 3, 3, 4, 0. **5.** 0, 2, $\bar{2}$, 0, $\bar{4}$, 3, $\bar{1}$.

6. $\bar{1}\cdot 8091488$, $6\cdot 8091488$, $\bar{4}\cdot 8091488$. **7.** 3·25, 325, ·000325.

8. 2·8853613. **9.** 3·3714373. **10.** 1·5475286.

11. 1·9163822. **12.** $\bar{1}\cdot 4419030$. **13.** $\bar{2}\cdot 2922560$.

14. $\bar{1}\cdot 6989700$. **15.** 1·8125918. **16.** ·0501716.

17. $\log 2 = \cdot 3010300$. **18.** $1-\log 2 = \cdot 6989700$. **19.** 1·320469.

20. ·0260315. **21.** ·2898431. **22.** $\bar{7}\cdot 2621538$.

23. 7; 4. **24.** 2058.

XIV. b. PAGE 149.

1. 9·076226. **2.** 3·01824. **3.** 2467·266.

4. 2·23. **5.** 3·54. **6.** 1·72. **7.** 96, 79.

8. 22·2398. **9.** 3·32. **10.** 5·77. **11.** 2·05.

12. $x=2\log 2=\cdot 60206$, $y=-2\log 5=-1\cdot 39794$.

13. $x=\frac{\log 3}{\log 3-\log 2}=2\cdot 71$; $y=\frac{\log 2}{\log 3-\log 2}=x-1=1\cdot 71$.

14. $3(b-a-c+2)$, $\frac{1}{2}(2a-3c+6)$.

15. $b+c-2$, $\frac{1}{6}(3a+2b+3c-5)$.

MISCELLANEOUS EXAMPLES. E. PAGE 150.

3. $b=\sqrt{3}-1$, $A=135°$, $C=30°$. **8.** $A=105°$, $B=45°$.

XV. a. PAGE 155.

1. 6·6947485. **2.** ·5404924. **3.** 6·4547860.

4. 1·7606731. **5.** 6·7840083. **6.** 55740·83.

7. 673·5469. **8.** ·0106867. **9.** ·008287771.

10. ·2531925. **11.** 2·031324. **12.** 1·389495.

13. 2·424463. **14.** 2·069138.

XV. b. Page 159.

1. ·6164825. 2. ·7928863. 3. 1·2154838. 4. 62° 42′ 31″.
5. 30° 40′ 23″. 6. 48° 45′ 44″. 7. 9·8440554.
8. 10·1317778. 9. 9·7530545. 10. 44° 17′ 8″.
11. 55° 30′ 39″. 12. 9·6653952. 13. 10·1912872.

XV. c. Page 161.

1. 2·36952. 2. 84336. 3. 33·27475.
4. ·03803143. 5. 11218·4. 6. 1225·508.
7. 27·901983. 8. ·580303. 9. 6·848293.
10. 3·288754, 1·236122. 11. 2273·54.
12. ·5095328. 13. 7·29889. 14. ·045800373.
15. ·08603526. 16. ·0001706364. 17. ·644065.
18. 9·52912. 19. ·3175272. 20. ·335859.
21. ·4221836. 22. 124272·2. 23. 250·2357.
24. (1) 36° 45′ 22″; (2) 19° 28′ 16″. 25. ·441785. 26. 68° 25′ 6″.

XVI. a. Page 166.

6. $\frac{1}{2}$. 7. $\frac{3}{2}$.

XVI. b. Page 169.

1. 113° 34′ 41″. 2. 49° 28′ 26″. 3. 55° 46′ 16″.
4. 78° 27′ 47″. 5. 64° 37′ 23″. 6. 35° 5′ 49″. 7. 93° 35′.
8. A = 67° 22′ 49″, B = 53° 7′ 48″, C = 59° 29′ 23″.
9. A = 46° 34′ 3″, B = 104° 28′ 39″, C = 28° 57′ 18″.

XVI. c. Page 173.

1. A = 79° 6′ 24″, B = 40° 53′ 36″. 2. A = 6° 1′ 54″, B = 108° 58′ 6″.
3. A = 24° 10′ 59″, B = 95° 49′ 1″. 4. B = 78° 48′ 52″, C = 56° 41′ 8″.
5. A = 27° 38′ 45″, C = 117° 38′ 45″. 6. 82° 57′ 15″, 36° 32′ 45″.
7. A = 74° 32′ 44″, C = 48° 59′ 16″.
8. B = 100° 47′ 1″, C = 14° 12′ 59″.
9. A = 136° 35′ 21·5″, B = 13° 14′ 33·5″.

XVI. d. PAGE 174.

1. 89·646162. 2. 255·3864. 3. 92·788. 4. $b=185$, $c=192$.
5. 321·0793. 6. $a=765·4321$, $c=1035·43$.
7. $b=767·72$, $c=1263·58$.

XVI. e. PAGE 176.

1. 32° 25′ 35″. 2. 41° 41′ 28″ or 138° 18′ 32″.
3. $A=100° 33′$, $B=34° 27′$. 4. 51° 18′ 21″ or 128° 41′ 39″.
5. $A=28° 20′ 50″$, $C=39° 35′ 10″$. 6. $A=81° 45′ 2″$, or 23° 2′ 58″.
7. (1) Not ambiguous, for $C=90°$;
(2) ambiguous, $b=60·3893$ ft.;
(3) not ambiguous.

XVI. f. PAGE 180.

1. $A=58° 24′ 43″$, $B=48° 11′ 23″$, $C=73° 23′ 54″$.
2. 112° 12′ 54″, 45° 53′ 33″, 21° 53′ 33″. 3. 75° 48′ 54″.
4. 4227·4718. 5. $B=108° 12′ 26″$, $C=49° 27′ 34″$.
6. $A=105° 38′ 57″$, $B=15° 38′ 57″$. 7. 17·1 or 3·68.
8. 108° 26′ 6″, 53° 7′ 48″, 18° 26′ 6″. 9. 96° 27′, or 126° 22′.
10. $B=80° 46′ 26·5″$, $C=63° 48′ 33·5″$. 11. 70° 0′ 56″, or 109° 59′ 4″.
12. 2·529823. 13. 41° 45′ 14″.
14. $A=42° 0′ 14″$, $B=55° 56′ 46″$, $C=82° 3′$.
15. 41° 24′ 34″. 16. $A=60° 5′ 34″$, $C=29° 54′ 26″$.
17. 889·2554 ft. 18. 72° 12′ 59″, 47° 47′ 1″.
19. 44·4878 ft. 20. $A=102° 56′ 38″$, $B=42° 3′ 22″$.
21. $B=99° 54′ 23″$, $C=32° 50′ 37″$, $a=18·7254$. 22. 72° 26′ 26″.
23. $A=27° 29′ 56″$, $B=98° 55′$, $C=53° 35′ 4″$.
24. $B=32° 15′ 49″$, $C=44° 31′ 17″$, $a=1180·525$.
25. $a=20·9059$, $c=33·5917$. 26. $a=2934·124$, $b=3232·846$.
27. $B=1° 1′ 23″$, $C=147° 28′ 37″$, $a=4390$.
28. $A=26° 24′ 23″$, $B=118° 18′ 25″$, $b=642·756$.
29. 53° 17′ 55″, or 126° 42′ 5″.
30. $A=1° 51′ 56″$, $C=125° 49′ 4″$, $a=67$.

31. $b = 4028 \cdot 5$, $c = 2831 \cdot 7$.

32. $B = 75° 53' 29''$, or $104° 6' 31''$; $A = 60° 54' 19''$, or $32° 41' 17''$.

33. Base $= 2 \cdot 44845$ ft., altitude $= \cdot 713322$ ft.

34. 90°. **35.** (1) impossible; (2) ambiguous; (3) 64.

36. $\theta = 72° 31' 53''$, $c = 12 \cdot 8255$. **37.** $\theta = 60° 13' 52''$, $c = 19 \cdot 523977$.

XVII. a. PAGE 185.

1. 146·4 ft. **2.** $880\sqrt{3} = 1524$ ft. **5.** $ab/(a-b)$ ft.

6. $\frac{1}{3}\sqrt{6} = \cdot 816$ miles. **7.** $10(\sqrt{10} + \sqrt{2}) = 45 \cdot 76$ ft.

9. 1 or $\frac{1}{3}$. **10.** $9\frac{3}{7}$ ft. **12.** $48\sqrt{6} = 117 \cdot 6$ ft.

14. $750\sqrt{6} = 1837$ ft. **15.** $2640(3 + \sqrt{3}) = 12492$ ft.

XVII. b. PAGE 190.

1. 30 ft. **2.** $a\sqrt{2}$ ft. **5.** 100 ft.

12. $\sqrt{500 - 200\sqrt{3}} = 12 \cdot 4$ ft.

XVII. c. PAGE 195.

1. 1060·5 ft. **2.** $\frac{500\sqrt{6}}{3} = 408$ ft.

3. $120\sqrt{6} = 294$ ft. **5.** 106 ft.

10. Height $= 40\sqrt{6} = 98$ ft.; distance $= 40(\sqrt{14} + \sqrt{2}) = 206$ ft.

11. $50\sqrt{120 + 30\sqrt{6}} = 696$ yds.

XVII. d. PAGE 197.

1. 5 miles nearly.

2. Height $= 19 \cdot 5375$ yds.; distance $= 102 \cdot 9093$ yds.

3. 200·017 ft. **4.** Height $= 418 \cdot 4045$ ft.; distance $= 430$ ft.

5. Height $= 916 \cdot 8624$ ft.; distance $= \cdot 984808$ miles.

6. Height $= 46 \cdot 140214$ ft.; distance $= 99 \cdot 92$ ft.

7. 11·550316 or 25·9733 miles per hour.

8. Height $= 159 \cdot 4321$ ft.; distance $= 215 \cdot 6762$ ft.

XVIII. a. PAGE 206.

1. 9000 sq. ft.
2. 15390.
3. $\frac{84}{85}$.
4. 24, $\frac{117}{5}$, $\frac{936}{25}$.
5. 225 sq. ft.
6. 672 sq. ft.
7. 36 yds.
8. $r=4$, $R=8\frac{1}{8}$.
9. 12, 6, 28.
10. 12, 16, 20.

XVIII. b. PAGE 210.

1. 26·46 sq. ft.
2. 9·585 yds., 7·18875 sq. yds.
4. 216·23 sq. ft.
5. 128·352 in.
6. 203·56 ft.
7. 57·232 ft.
8. 61·803 sq. ft.

XVIII. c. PAGE 218.

17. $$\frac{\pi}{3}+(-1)^{n-1}\frac{1}{2^{n-1}}\left(A-\frac{\pi}{3}\right),$$
$$\frac{\pi}{3}+(-1)^{n-1}\frac{1}{2^{n-1}}\left(B-\frac{\pi}{3}\right),$$
$$\frac{\pi}{3}+(-1)^{n-1}\frac{1}{2^{n-1}}\left(C-\frac{\pi}{3}\right).$$

XVIII. d. PAGE 223.

1. $1\frac{5}{7}$, $2\frac{1}{2}$.
4. Diagonals 65, 56; area 1764.
5. $2\sqrt{77}+6\sqrt{11}$.

XVIII. e. PAGE 225.

2. 14142 sq. yds.
5. $\sqrt{\frac{x}{y}+\frac{y}{z}+\frac{z}{x}}$.
13. 20, 21, 29.

MISCELLANEOUS EXAMPLES. F. PAGE 228.

3. Expression $=\cot A+\cot B+\cot C$.
4. $B=45°$, $135°$; $C=105°$, $15°$; $c=\sqrt{6}+\sqrt{2}$, $\sqrt{6}-\sqrt{2}$.
6. 126.
7. 68·3 yds., 35·35 yds.
11. $C=45°$, $135°$; $A=105°$, $15°$; $a=2\sqrt{3}$, $4\sqrt{3}-6$.
12. $10(2-\sqrt{3})=2{\cdot}68$ miles; $10(\sqrt{6}-\sqrt{2})=10{\cdot}35$ miles.
24. (1) $90°-\frac{A}{2}$, $90°-\frac{B}{2}$, $90°-\frac{C}{2}$;
(2) $180°-2A$, $180°-2B$, $180°-2C$.
25. Expression $=\sin^2(\alpha-\beta)$.
28. 21·3 miles per hour.
29. 1 hr. 30′; 2 hrs. 16′.

XIX. a. Page 235.

1. $n\pi+(-1)^n\frac{\pi}{6}$. 2. $n\pi+(-1)^n\frac{\pi}{4}$. 3. $2n\pi\pm\frac{\pi}{3}$.

4. $n\pi+\frac{\pi}{3}$. 5. $n\pi-\frac{\pi}{6}$. 6. $2n\pi\pm\frac{3\pi}{4}$.

7. $n\pi\pm\frac{\pi}{4}$. 8. $n\pi\pm\frac{\pi}{6}$. 9. $n\pi\pm\frac{\pi}{3}$.

10. $2n\pi\pm\alpha$. 11. $n\pi\pm\alpha$. 12. $n\pi\pm\alpha$.

13. $n\pi$. 14. $\frac{n\pi}{3}+(-1)^n\alpha$. 15. $2n\pi$, or $\frac{2n\pi}{5}$.

16. $\frac{n\pi}{3}$, or $n\pi\pm\frac{\pi}{6}$. 17. $\frac{n\pi}{4}$, or $\frac{n\pi}{3}+(-1)^n\frac{\pi}{18}$.

18. $\frac{(2n+1)\pi}{2}$, or $2n\pi$, or $\frac{(2n+1)\pi}{5}$.

19. $\frac{(2n+1)\pi}{2}$, or $\frac{(2n+1)\pi}{4}$, or $\frac{(2n+1)\pi}{8}$.

20. $n\pi$, or $\frac{(2n+1)\pi}{14}$. 21. $\frac{n\pi}{6}$, or $\frac{n\pi}{9}$.

22. $\frac{(2n+1)\pi}{6}$, or $n\pi\pm\frac{\pi}{8}$. 23. $\frac{(2n+1)\pi}{8}$, or $\frac{n\pi}{3}+(-1)^n\frac{\pi}{9}$.

24. $(2n+1)\pi$, or $2n\pi\pm\frac{\pi}{3}$. 25. $2n\pi$, or $2n\pi\pm\frac{2\pi}{3}$.

26. $n\pi+(-1)^n\frac{\pi}{6}$, or $2n\pi+\frac{3\pi}{2}$. 27. $\frac{n\pi}{2}+\frac{\pi}{8}$.

28. $2n\pi+\frac{2\pi}{3}$. 29. $2n\pi-\frac{\pi}{4}$.

XIX. b. Page 237.

1. $\frac{(2n+1)\pi}{2(p+q)}$. 2. $\frac{(4k+1)\pi}{2(n-m)}$, or $\frac{(4k-1)\pi}{2(n+m)}$. 3. $2n\pi$, or $2n\pi-\frac{2\pi}{3}$.

4. $2n\pi+\frac{\pi}{2}$, or $(2n+1)\pi+\frac{\pi}{6}$. 5. $2n\pi+\frac{\pi}{2}$, or $2n\pi+\frac{\pi}{6}$.

6. $2n\pi+\frac{5\pi}{12}$, or $2n\pi-\frac{\pi}{12}$. 7. $2n\pi+\frac{\pi}{12}$, or $2n\pi-\frac{7\pi}{12}$.

8. $2n\pi+\frac{5\pi}{4}$, or $2n\pi-\frac{3\pi}{4}$. 9. $2n\pi+\frac{\pi}{3}$, or $(2n+1)\pi$.

10. $\frac{n\pi}{2}+(-1)^n\frac{\pi}{12}$. 11. $\frac{(2n+1)\pi}{4}$, or $n\pi\pm\frac{\pi}{6}$.

12. $n\pi$, or $n\pi \pm \frac{\pi}{6}$.

13. $n\pi$, or $\frac{n\pi}{2}$.

14. $n\pi + \frac{\pi}{4}$, or $2n\pi$, or $2n\pi + \frac{\pi}{2}$.

[*In some of the following examples, the equations have to be squared, so that extraneous solutions are introduced.*]

15. $n\pi + \frac{\pi}{4}$, or $\frac{n\pi}{2} + (-1)^{n+1}\frac{\pi}{12}$.

16. $n\pi - \frac{\pi}{4}$, or $\frac{n\pi}{2} + (-1)^n \frac{\pi}{12}$.

17. $\frac{(2n+1)\pi}{10}$, or $\frac{(2n+1)\pi}{2}$.

18. $\frac{(2n+1)\pi}{2}$, or $n\pi \pm \frac{\pi}{3}$.

19. $n\pi + \frac{\pi}{4}$, or $n\pi + \frac{\pi}{6}$.

20. $\frac{n\pi}{2} + (-1)^{n+1}\frac{\pi}{12}$, or $\frac{n\pi}{2}$.

21. $\theta = n\pi \pm \frac{\pi}{4}$, $\phi = n\pi \pm \frac{\pi}{6}$.

22. $\theta = n\pi \pm \frac{\pi}{6}$, $\phi = n\pi \pm \frac{\pi}{3}$.

23. $\theta = n\pi \pm \frac{\pi}{4}$, $\phi = n\pi \pm \frac{\pi}{3}$.

XIX. d. Page 244.

1. $\pm\frac{1}{\sqrt{2}}$.

2. ± 1.

3. ± 2.

4. $\frac{-3 \pm \sqrt{17}}{4}$.

5. 1, or $\frac{1}{2}$.

6. 0, or $\pm\frac{1}{2}$.

7. $\pm\frac{1}{\sqrt{2}}$.

8. $\frac{25}{24}$.

9. $\frac{1}{2}$.

10. $\frac{a-b}{1+ab}$.

11. $\frac{b-a}{1+ab}$.

12. $\sqrt{3}$.

13. $x = ac - bd$, $y = bc + ad$.

14. ± 1, or $\pm(1 \pm \sqrt{2})$.

15. $n\pi$, or $n\pi + \frac{\pi}{4}$.

19. $x=1$, $y=2$; $x=2$, $y=7$.

MISCELLANEOUS EXAMPLES. G. Page 246.

2. (1) $\frac{(2n+1)\pi}{2}$, $2n\pi \pm \frac{\pi}{3}$; (2) $2n\pi \pm \frac{\pi}{3}$.

6. $78° 27' 4''$.

9. 6.

10. 800 yds, 146·4 yds, 546·4 yds.

XX. a. Page 255.

2. $\sin\frac{A}{2} = \frac{5}{13}$, $\cos\frac{A}{2} = -\frac{12}{13}$.

3. $\sin\frac{A}{2} = -\frac{15}{17}$, $\cos\frac{A}{2} = \frac{8}{17}$.

4. $2\sin\frac{A}{2} = -\sqrt{1+\sin A} + \sqrt{1-\sin A}$;

$2\cos\frac{A}{2} = -\sqrt{1+\sin A} - \sqrt{1-\sin A}$.

5. $2\sin\frac{A}{2} = -\sqrt{1+\sin A} - \sqrt{1-\sin A}$;

$2\cos\frac{A}{2} = -\sqrt{1+\sin A} + \sqrt{1-\sin A}$.

6. $2\sin\frac{A}{2} = +\sqrt{1+\sin A} + \sqrt{1-\sin A}$;

$2\cos\frac{A}{2} = +\sqrt{1+\sin A} - \sqrt{1-\sin A}$.

7. $\sin\frac{A}{2} = \frac{4}{5}$, $\cos\frac{A}{2} = \frac{3}{5}$.   8. $\sin\frac{A}{2} = \frac{15}{17}$, $\cos\frac{A}{2} = -\frac{8}{17}$.

9. (1) $2n\pi - \frac{\pi}{4}$ and $2n\pi + \frac{\pi}{4}$; (2) $2n\pi + \frac{5\pi}{4}$ and $2n\pi + \frac{7\pi}{4}$;

(3) $2n\pi + \frac{3\pi}{4}$ and $2n\pi + \frac{5\pi}{4}$.

10. No; $2\sin\frac{A}{2} = \sqrt{1+\sin A} + \sqrt{1-\sin A}$.

14. (1) $= \sqrt{2}\cos\left(\theta - \frac{\pi}{4}\right)$; (2) $= 2\sin\left(\theta - \frac{\pi}{3}\right)$.

15. (1) $= -\sec 2\theta$; (2) $= \tan^2\frac{\theta}{2}$.

XX. b. PAGE 260.

3. $\frac{1}{5}$.   4. $\frac{1}{3}$.

XXI. a. PAGE 267.

1. 1440 yards.   2. $342\frac{6}{7}$ yards.   3. 22 yards.
4. 6′ 34″.   5. 13′ 45″.   6. 11 ft. 11 in.
7. 210 yards.   8. 9′ 49″.   10. 50 ft.
11. (1) $\frac{\pi}{10800}$; (2) $\frac{\pi}{648000}$.   12. πr^2.
13. $\frac{1}{2}$.   14. $m - n$.   15. $\frac{1}{2} - \frac{\sqrt{3}}{200} = \cdot 491$.
16. $\frac{1}{2} + \frac{11\sqrt{3}}{7200} = \cdot 503$.   17. $\frac{21\sqrt{3}}{55} = 39\cdot 7'$.

XXI. b. PAGE 271.

1. 12 miles. 2. 150 ft. 3. 15 miles. 4. 80 ft. 8 in.

5. 204 ft. 2 in. 6. 104 ft. 2 in. 7. 54′33″.

8. 10560 ft. 9. 610 ft., $\frac{5}{2}\sqrt{110}$ minutes $= 26'13''$.

11. 8. 12. -1. 13. (1) $\cos\alpha$; (2) $-\sin\alpha$.

14. 45° 54′ 33″, 44° 5′ 27″.

MISCELLANEOUS EXAMPLES. H. PAGE 283.

3. 18° 26′ 6″. 6. 35 miles or 13 miles per hour.

XXIII. a. PAGE 291.

1. $\dfrac{\sin 2n\alpha}{2\sin\alpha}$. 2. $\sin\dfrac{n\beta}{2}\cos\left(\alpha-\dfrac{n-1}{2}\beta\right)\Big/\sin\dfrac{\beta}{2}$.

3. $-\cos\left(\alpha+\dfrac{\pi}{2n}\right)\Big/\sin\dfrac{\pi}{2n}$. 4. $\sin\dfrac{n\pi}{2k}\cos\dfrac{(n+1)\pi}{2k}\Big/\sin\dfrac{\pi}{2k}$.

5. $\dfrac{1}{2}$. 6. $-\dfrac{1}{2}$. 7. $\cot\dfrac{\pi}{2n}$. 8. $-\cos\dfrac{2\pi}{n}$.

9. $\sin n\alpha$. 10. $\sin\dfrac{n(\theta+\pi)}{2}\sin\left(\dfrac{n+1}{2}\theta+\dfrac{n-1}{2}\pi\right)\Big/\sin\dfrac{\theta+\pi}{2}$.

11. $\sin\dfrac{n(\pi-\beta)}{2}\cos\left\{\alpha+\dfrac{(n-1)(\pi-\beta)}{2}\right\}\Big/\sin\dfrac{\pi-\beta}{2}$.

12. $\sin\dfrac{n(\pi-2\beta)}{4}\cos\left\{\alpha+\dfrac{(n-1)(\pi-2\beta)}{4}\right\}\Big/\sin\dfrac{\pi-2\beta}{4}$.

13. $\dfrac{n\cos\theta}{2}-\dfrac{\sin n\theta\cos(n+2)\theta}{2\sin\theta}$.

14. $\dfrac{\sin 2n\alpha\sin 2(n+1)\alpha}{2\sin 2\alpha}-\dfrac{n\sin 2\alpha}{2}$.

15. $\operatorname{cosec}\alpha\{\tan(n+1)\alpha-\tan\alpha\}$.

16. $\operatorname{cosec}2\theta\{\cot\theta-\cot(2n+1)\theta\}$. 17. $\tan\alpha-\tan\dfrac{\alpha}{2^n}$.

18. $\dfrac{1}{2}(\operatorname{cosec}\alpha-\operatorname{cosec}3^n\alpha)$. 19. $\dfrac{1}{2}(\tan 3^n\alpha-\tan\alpha)$.

XXIII. b. PAGE 294.

1. $\frac{n}{2}+\frac{\sin 4n\theta}{4\sin 2\theta}$. 2. $\frac{n}{2}$. 3. $\frac{n}{2}$.

4. $\dfrac{3\sin\frac{n\theta}{2}\sin\frac{(n+1)\theta}{2}}{4\sin\frac{\theta}{2}}-\dfrac{\sin\frac{3n\theta}{2}\cos\frac{3(n+1)\theta}{2}}{4\sin\frac{3\theta}{2}}$.

5. 0. 6. 0. 7. $\cot\theta-2^n\cot 2^n\theta$.

8. $\frac{1}{2}\operatorname{cosec}\alpha\{\tan(n+1)\alpha-\tan\alpha\}$. 9. $\frac{\sin 2\theta}{2}-\frac{\sin 2^{n+1}\theta}{2^{n+1}}$.

10. $\sin^2\theta-2^n\sin^2\frac{\theta}{2^n}$. 11. $\tan^{-1}x-\tan^{-1}\frac{x}{n+1}$.

12. $\tan^{-1}(n+1)-\frac{\pi}{4}$. 13. $\tan^{-1}\{1+n(n+1)\}-\frac{\pi}{4}$.

14. $\tan^{-1}n(n+1)$.

XXIV. a. PAGE 301.

2. $x=a\cos\frac{\alpha+\beta}{2}\Big/\cos\frac{\alpha-\beta}{2}$, $y=b\sin\frac{\alpha+\beta}{2}\Big/\cos\frac{\alpha-\beta}{2}$.

3. $x=a(\cos\alpha+\sin\alpha)$, $y=b(\sin\alpha-\cos\alpha)$.

13. $4\sin\frac{\alpha+\beta+\gamma}{2}\sin\frac{\beta+\gamma-\alpha}{2}\sin\frac{\gamma+\alpha-\beta}{2}\sin\frac{\alpha+\beta-\gamma}{2}$.

14. $4\sin\frac{\alpha+\beta+\gamma}{2}\sin\frac{\alpha+\beta-\gamma}{2}\cos\frac{\beta+\gamma-\alpha}{2}\cos\frac{\gamma+\alpha-\beta}{2}$.

15. $-4\cos\left(\frac{\alpha+\beta+\gamma}{2}-\frac{\pi}{4}\right)\Pi\cos\left(\frac{\beta+\gamma-\alpha}{2}+\frac{\pi}{4}\right)$.

22. (1) $(a^2+b^2)x^2-2bcx+c^2-a^2=0$;

(2) $(a^2+b^2)^2x^2-2(a^2-b^2)(2c^2-a^2-b^2)x+a^4+b^4+4c^4-2a^2b^2$
$-4a^2c^2-4b^2c^2=0$.

[*Use* $\cos 2\alpha\cos 2\beta=\cos^2(\alpha-\beta)-\sin^2(\alpha+\beta)$.]

XXV. a. PAGE 318.

1. $2\sqrt{pq}$. **2.** 4. **3.** 24. **4.** 2. **7.** $\sqrt{2}$. **8.** 2.

9. $\sqrt{a^2 - 2ab \sin \alpha + b^2}$. **10.** $\sqrt{p^2 + 2pq \sin \alpha + q^2}$.

11. Maximum $= 2 \sin \frac{\sigma}{2}$. **12.** Maximum $= \sin^2 \frac{\sigma}{2}$.

13. Minimum $= 2 \tan \frac{\sigma}{2}$. **14.** Minimum $= 2 \operatorname{cosec} \frac{\sigma}{2}$.

15. Maximum $= \frac{1}{8}$. **16.** Minimum $= \sqrt{3}$.

17. Minimum $= \frac{3}{4}$. **18.** Minimum $= 6$.

19. Minimum $= 1$. **20.** Minimum $= 1$.

21. $\frac{1}{2}(a+c) \pm \frac{1}{2}\sqrt{b^2 + (a-c)^2}$. **25.** $\frac{5}{3}$.

26. $k^2/(a^2+b^2+c^2)$; $k^2/(a+b+c)$.

XXV. b. PAGE 324.

1. $\frac{x^2}{a^2} + \frac{y^2}{b^2} = 2$. **2.** $x^2 + y^2 = a^2 + b^2$. **3.** $b^2 = a^2(2 - a^2)$.

4. $y(x^2 - 1) = 2$. **5.** $(a^2 - b^2)^2 = 16ab$. **6.** $x^{\frac{4}{3}}y^{\frac{2}{3}} - x^{\frac{2}{3}}y^{\frac{4}{3}} = 1$.

7. $a^2b^2(a^2 + b^2) = 1$. **8.** $x^{\frac{2}{3}} + y^{\frac{2}{3}} = a^{\frac{2}{3}}$. **9.** $x^{\frac{4}{5}}y^{\frac{6}{5}} - x^{\frac{6}{5}}y^{\frac{4}{5}} = a^2$.

10. $\frac{x^2}{a^2} + \frac{y^2}{b^2} = 1$. **12.** $x^{\frac{1}{2}} + y^{\frac{1}{2}} = 2$. **13.** $(x+y)^{\frac{2}{3}} + (x-y)^{\frac{2}{3}} = 2$.

16. $a^2 + b^2 = 2c^2$. **20.** $(x+y)^{\frac{2}{3}} + (x-y)^{\frac{2}{3}} = 2a^{\frac{2}{3}}$. **21.** $\frac{x^2}{b^2} + \frac{y^2}{a^2} = 1$.

22. $\frac{x^2}{a} + \frac{y^2}{b} = a + b$, or $\{a(y^2 - b^2) - b(x^2 - a^2)\}^2 = -4abx^2y^2$.

24. $xy = (y - x)\tan \alpha$. **25.** $a^2 + b^2 - 2\cos \alpha = 2$. **26.** $a + b = 2ab$.

29. $(a+b)(m+n) = 2mn$. **30.** $x^2 + y^2 = 16a^2$.

31. $(a-b)\{c^2 - (a+b)^2\} = 4abcm$.

XXV. c. Page 334.

1. $2\cos 20°$, $-2\cos 40°$, $-2\cos 80°$.

2. $2\sin 10°$, $2\sin 50°$, $-2\sin 70°$.

3. $2\cos 10°$, $-2\cos 50°$, $-2\cos 70°$.

4. $\sin 15°$, $\sin 45°$, $-\sin 75°$.

5. $\frac{1}{a}\sin A$, $\frac{1}{a}\sin(60° - A)$, $-\frac{1}{a}\sin(60° + A)$.

6. $2a\cos A$, $2a\cos(120° \pm A)$.

16. (1) $8x^3 - 4x^2 - 4x + 1 = 0$; (2) $64y^3 - 80y^2 + 24y - 1 = 0$.

17. $64y^3 - 112y^2 + 56y - 7 = 0$.

18. (1) $16x^4 + 8x^3 - 12x^2 - 4x + 1 = 0$; (2) $16x^4 - 8x^3 - 12x^2 + 4x + 1 = 0$.

19. $256y^4 - 448y^3 + 240y^2 - 40y + 1 = 0$.

20. $t^8 - 36t^6 + 126t^4 - 84t^2 + 9 = 0$.

21. $32x^5 - 16x^4 - 32x^3 + 12x^2 + 6x - 1 = 0$.

CAMBRIDGE: PRINTED BY C. J. CLAY, M.A. AND SONS, AT THE UNIVERSITY PRESS.

www.ingramcontent.com/pod-product-compliance
Lightning Source LLC
Chambersburg PA
CBHW021400210326
41599CB00011B/946

* 9 7 8 3 3 3 7 2 7 9 5 1 6 *